ADVANCES IN PLASMA PHYSICS

Volume 5

AN INTERSCIENCE SERIES

Advances in

PLASMA PHYSICS

Volume 5

Edited by

ALBERT SIMON

Department of Mechanical and Aerospace Sciences
and Department of Physics
University of Rochester
Rochester, New York

and

WILLIAM B. THOMPSON

Department of Physics
University of California at San Diego
La Jolla, California

AN INTERSCIENCE® PUBLICATION

JOHN WILEY & SONS

New York . London . Sydney . Toronto

An Interscience® Publication

Library of Congress Catalogue Card Number: 67-29541

ISBN 0-471-79196-2

Printed in the United States of America.

10 9 8 7 6 5 4 3 2 1

This volume contains the selected

Proceedings of the Conference on Plasma Theory

Kiev, October 19–23, 1971

EDITED BY

A. G. SITENKO

CONFERENCE CHAIRMAN: N. N. BOGOLIUBOV

VICE-CHAIRMEN: M. S. RABINOVICH

A. G. SITENKO

Series Preface

It is with pleasure that we devote this volume in the Advances in Plasma Physics series to the selected proceedings of the Conference on Plasma Theory held in Kiev, USSR, in October 1971. In previous volumes we have presented invited review articles on various topics in the plasma realm. These conference proceedings, consisting of all the invited papers plus several of the rapporteurs' and contributed papers, are different in that most are shorter papers or talks rather than grand reviews. Nevertheless, the demonstration of a surprisingly broad scope of application of "plasma theoretical methods" to many branches of physics and the distinguished roster of speakers led us to believe that these proceedings should be of value to the plasma community. We were happy to respond to the suggestion of the conference vice-chairmen, M. S. Rabinovich and A. G. Sitenko, that we publish these proceedings as a volume in our series.

We are very grateful to Drs. Sitenko and Rabinovich for their patient cooperation through the usual delays that occur in bringing such a project to completion, as well as through the less usual difficulties of communication across great distances and language barriers. The different economic systems and changing copyright policy added to our temporary frustrations. However, much time was saved by Drs. Sitenko and Rabinovich arranging for the translation of the conference papers into English in the USSR. Although the phrasing may not always be in the most polished English, the meaning is generally clear, and we decided not to further delay publication by attempting large-scale rewriting.

In closing, we call attention to the introductory remarks of the chairman of the conference, Professor N. N. Bogolubov: "I would like to emphasize the fact that the study of plasma and its processes is beneficial not only to the neighboring branches of physics and in the development of extremely important practical applications. In addition, investigation of the processes in plasma led to an elaboration of methods and a formation of ideas which have proved to be very fruitful when applied to questions in physics that at first sight have little in common with the plasma theme."

ALBERT SIMON
WILLIAM B. THOMPSON

Rochester, New York
La Jolla, California
August 1973

Preface

The Conference on Plasma Theory was held in Kiev, October 19-23, 1971, cosponsored by the Institute for Theoretical Physics of the Academy of Sciences of the Ukrainian SSR and the Scientific Council on Plasma Physics of the Academy of Sciences of the USSR. The conference was attended by approximately 250 theorists representing various physics centers of the Soviet Union (Moscow, Leningrad, Kiev, Kharkov, Novosibirsk, Gorky, Tbilisi), as well as 50 foreign scientists from 14 countries (The United States, the German Federal Republic, German Democratic Republic, France, Britain, Belgium, Sweden, Japan, Czechoslovakia, etc.).

The conference provided a forum for an intensive examination of the most pressing problems of up-to-date plasma theory. At 9 scheduled sessions, 19 review talks, 16 rapporteurs' talks (incorporating 115 original communications submitted to the conference), and a limited number (23) of the most interesting original reports were presented. Separate sessions were devoted to highlighting the following problems: general questions of statistical theory; equilibrium and transport processes in plasma; oscillations, emission, and stability of plasma; nonlinear processes in plasma; finite-amplitude and shock waves; turbulence and stochastic processes in plasma; mathematical simulation and numerical methods in plasma theory; electromagnetic phenomena in plasmalike media; and the general problems of controlled nuclear fusion.

The present volume contains all the review talks, a number of specially revised rapporteurs' talks, and several of the original reports given at the conference. The rest of the original communications submitted to the conference will appear in a separate issue of the *Ukrainian Physical Journal.*

A. G. SITENKO

KIEV
July 1973

Contents

Introductory Talk

N. N. BOGOLUBOV

Institute for Theoretical Physics, Academy of Sciences of the Ukrainian SSR, Kiev, USSR

It is a great pleasure for me to open the first All-Union Conference on Plasma Theory in Kiev. The conference is held on the initiative of the Institute for Theoretical Physics of the Academy of Sciences of the Ukrainian SSR in cooperation with the Council on Plasma Physics of the Academy of Sciences of the USSR. The main purpose of the conference is to discuss the most pressing problems of current plasma theory, as well as some general questions concerning statistical physics. The conference has found a broad response among theoretical physicists both in our country and abroad. The leading Soviet theorists working in plasma physics and related fields of physics, as well as many noted scientists from abroad, have come to attend the conference. Such a conference on plasma theory is the first one to be convened, not only in the Soviet Union but in the world as well.

At present, research in plasma physics is carried out on a large scale by many scientific centers. The importance of the problems confronting plasma physics and the great variety of phenomena studied have brought about a rapid development of this branch of physics within the last two decades. From a small field treating a narrow range of phenomena, plasma physics has developed into a broad, independent discipline which, at present, intrudes and penetrates intensively into other fields of physics. One can cite numerous examples of such interpenetration and intertwining of plasma physics with other, related areas.

For example, most of the problems of present-day astrophysics are directly bound up with an understanding of the physical properties of plasma. The current cosmogonic hypotheses of the origin of the universe are based on the advances in plasma physics. Astrophysics, in turn, has given rise to a magnetic hydrodynamics that describes many processes in plasma. Cosmic electrodynamics is a branch of plasma physics.

The study of plasma turbulence has contributed to the development of the mechanics of continuous media. The methods of nonlinear mechanics, on the other hand, have found wide application in plasma physics. The discovery of plasma phenomena in metals and semiconductors has opened up a new direction in solid-state physics.

Much progress has been achieved in statistical physics and physical kinetics as a result of the advances in plasma physics. The development of numerical simulation in plasma physics has stimulated the elaboration of new methods in applied mathematics.

The mutual penetration and relationships between plasma physics and the related branches of physics are most pronounced in the field of theory, since they all are based on the common methods used in different fields of physics and on an intrinsic unity of fundamental physical laws. Until now, conferences on plasma physics and other branches of physics have been held separately. The present Conference on Plasma Theory will make an attempt to establish close contact between a number of directions in theoretical physics focused on plasma physics.

I would like to emphasize the fact that the study of plasma and its processes is beneficial not only to the neighboring branches of physics and in the development of extremely important practical applications. In addition, investigation of the processes in plasma led to an elaboration of methods and a formation of ideas which have proved to be very fruitful when applied to questions in physics that at first sight have little in common with the plasma theme.

Indeed, the rapid rise and intensive development of plasma physics is due to a great extent to the employment of powerful tools of statistical physics and kinetics. Plasma theory, in turn, promotes the development of kinetics and statistical physics. Contrary to gas dynamics and hydrodynamics, where the effects due to pair collisions play a decisive role, in plasma theory attention is centered mainly on the study of collisionless processes, where the collective effects are predominant. The study of collective effects, which are mainly responsible for the specific features of plasma, has become one of the most important directions in plasma theory. The methods developed in studying the collective effects in plasma have found wide application in various other branches of physics. Let us take, for example, such a fairly current notion as that of collective oscillations, which is widely used in the theory of metals, atomic nuclei, and so on. The same is true of the method of "splitting," the higher correlation, and Green functions. It is natural, therefore, that the conference has provided a section to discuss general problems concerning statistical physics; two sessions will be devoted to questions in this area. The problems of equilibrium of plasma systems and the theory of transport processes will be discussed at a separate session. Wave distribution and various types of oscillations in plasma, as well as the stability of plasma, will be covered too.

To describe many phenomena in plasma it is not sufficient to confine oneself to a linear approximation; nonlinear effects must also be taken into account. Several reports will be devoted to the nonlinear processes in plasma, finite-amplitude and shock waves. These topics are directly associated with the problem of turbulence and stochastic processes in plasma.

A separate session will focus on the problems of mathematical simulation and the numerical methods used in plasma theory. Various electromagnetic phenomena in plasmalike media are quite important. The conference will discuss some aspects of plasma phenomena in astrophysics and solid-state physics. The concluding session will be devoted to a discussion of some present-day problems of plasma physics, and especially to an examination of the current status of controlled nuclear fusion.

I hope that our conference will be interesting and fruitful. The wide participation of leading theoretical physicists, both Soviet and foreign, will make it possible to discuss the most pressing key problems of theory and undoubtedly will contribute to progress in plasma physics, as well as in fields related to this science.

Structure of Kinetic Theory

RADU BALESCU

Faculté des Sciences, Université Libre de Bruxelles, Bruxelles, Belgium, Association Euratom-Etat Belge

Introduction

The history of kinetic theory has been characterized by a succession of periods of progress and periods of inactivity. It began brilliantly (after a few preliminary explorations) in the hands of Boltzmann, who made it, right from the start, a complete and operational science. He raised all the fundamental questions in the right perspective, even if he could not answer them fully by means of the tools available in his time. It is good to remember that next year is the hundredth anniversary of Boltzmann's *H*-theorem, which is still in the center of our preoccupations. After a period of activity lasting about 50 years, kinetic theory fell into neglect.

I was fortunate enough to begin working in this field precisely at the time when new ideas and methods opened the way to a new era of activity, which is still going on. At that time, Professor N. Bogoliubov published his fundamental paper (1) on statistical dynamics, which became a cornerstone of the theory. At the same time, but independently, two other scientists, Leon van Hove in Holland, and Ilya Prigogine in Belgium, were introducing new concepts and new tools. As a result, a feverish activity began all over the world. The beautiful developments which appeared were, however, not always coordinated. Our present purpose is to try to synthesize all the existing ideas and incorporate them into a unified theory. This has become possible with the appearance of some recent reports from our group in Brussels; their work goes a long, long way toward clarifying the basic questions of nonequilibrium statistical mechanics (4, 5).

I should like to review briefly these new ideas. In doing so, in order to preserve their generality, I shall not specifically adhere to the plasma problem. The implications of these ideas for plasma physics are, however, obvious.

I. Reduced Distribution Functions

Let me begin with a somewhat technical introduction. Our system will be, for definiteness, a classical gas whose particles interact through a Hamiltonian:

$$H = \sum_j \frac{p_j^2}{2m} + \sum_{j<n} V(|\mathbf{q}_j - \mathbf{q}_n|)$$

$$\equiv \sum_j H_j^0 + \sum_{j<n} V_{jn} \tag{1}$$

The Liouville equation for the phase-space distribution function F is then

$$\partial_t F = LF \equiv \left(\sum_j L_j^0 + \sum_{j<n} L'_{jn} \right) F \tag{2}$$

A very useful representation of this equation is in terms of *reduced distribution functions*, $f_s(x_1 \ldots x_s)$ $[x_j = (q_j, p_j)]$, which are closer to observable quantities. We may group these functions together into a kind of "vector," f:

$$f = \{f_0, f_1(x_1), f_2(x_1, x_2), \ldots\} \tag{3}$$

which obeys a matrix Liouville equation:

$$\partial_t f = \mathscr{L} f \equiv (\mathscr{L}^0 + \mathscr{L}')f \tag{4}$$

The components of this equation can be written as follows:

$$\partial_t f_s(x_1 \ldots x_s) = \sum_r \langle\!\langle (s)|\, \mathscr{L} \,|(r)\rangle\!\rangle\, f_r(x_1 \ldots x_r) \tag{5}$$

These are simply the well-known equations of the Bogoliubov-Born-Green-Kirkwood-Yvon hierarchy. At this point we can easily go to the thermodynamic limit: the vector f then has an infinite number of components, which are supposed to be finite, well-behaved functions.

In many cases, however, this description is not sufficiently detailed. We are often interested in knowing how the particles in a subgroup (or subgroups) are mutually correlated. Therefore we further decompose the reduced distribution functions:

$$f_1(x_1) = p_1(x_1;\, [O_1])$$

$$f_2(x_1 x_2) = p_2(x_1; x_2) + p_2(x_1 x_2) = p_2(x_1 x_2;\, [O_2]) + p_2(x_1 x_2;\, [C_2]) \tag{6}$$

$$f_s(x_1 \ldots x_s) = \sum_{\Gamma_s} p_s(x_1 \ldots x_s;\, [\Gamma_s])$$

The functions $p_s([\Gamma_s])$ correspond to the various possible partitions of the s particles into nonoverlapping subgroups; each partition is labeled by an index Γ_s. These functions will be called *correlation patterns.* A simple realization is the following:

$$p_s(x_1 \dots x_s; [O_s]) = \prod_{j=1}^{s} f_1(x_j) \tag{7}$$

with similar factorization properties for the other patterns. Equation 6 is then simply the well-known "cluster representation." However, this is not the most general realization (6). We now consider the distribution vector as made up of the set of correlations patterns p_s for all values of s and Γ_s. It is rather easy to construct the components of the Liouville equation in this representation (6):

$$\partial_t p_s(x_1 \dots x_s; [\Gamma_s]) = \sum_r \sum_{\Gamma_r} \langle (s)[\Gamma_s] | \mathscr{L}^0 + \mathscr{L}' | (r)[\Gamma_r] \rangle \, p_r(x_1 \dots x_r; [\Gamma_r]) \tag{8}$$

The matrix elements appearing here can be determined by simple graphical methods. Equations 8 represent a true *dynamics of correlations.*

We now note that, among all possible correlation patterns of s particles, one pattern is privileged: the pattern $p_s([O_s])$. It corresponds to completely uncorrelated particles. It is not difficult to see that this pattern is the only one normalized to *one*, all the others being normalized to zero. As this pattern carries the whole normalization of f_s, *it can never vanish.* On the contrary, correlations $p_s([\Gamma_s])$ with $\Gamma_s \neq O_s$ may or may not be present in a given statistical state of the system.

We now collect all the correlation patterns $p_s([O_s])$, for $s = 0, 1, \dots,$ and arrange them into an ordered set, of the same form as the distribution vector f. This set is clearly a subset of $\{f\}$. It will be called the *vacuum component* of the distribution vector and be denoted by Vf. The remaining correlation patterns form a complementary subset, called the *correlation component* of the distribution vector and denoted by Cf. We thus write:

$$f = Vf + Cf \tag{9}$$

where

$$\begin{aligned} Vf &= \{p_s(x_1 \dots x_s; [O_s])\} \\ Cf &= \{p_s(x_1 \dots x_s; [\Gamma_s]): \qquad \Gamma_s \neq O_s\} \end{aligned} \tag{10}$$

Note that the components of Cf for $s = 0$ and $s = 1$ are identically null: there are no correlations of less than two particles.

We now note that the separation of eq. 9 can be performed by acting on the distribution vector with two *operators,* V and C. The effect of these operators is to select a subset out of the set f according to definitions 9 and 10. The sum of these operators is simply the identity operator:

$$V + C = I \tag{11}$$

Moreover, we have the following properties:

$$V^2 = V, \qquad C^2 = C$$
$$VC = 0, \qquad CV = 0 \tag{12}$$

These equations simply express the fact that the two subsets do not overlap. We may therefore call V and C (somewhat loosely) *projection operators* on the vacuum and on the correlations, respectively.

A very important property of the subspaces $\{Vf\}$ and $\{Cf\}$ is the following: *these subspaces remain invariant under the unperturbed motion.* In other words, under the effect of the Liouvillian $\mathcal{L}^0$ a vaccuum component evolves into another vacuum component, ignoring the correlations component; the unperturbed Hamiltonian does not create correlations: only the interactions do this. This property is expressed formally by the fact that the unperturbed Liouvillian *commutes* with the vacuum projector V:

$$\mathcal{L}^0 V = V\mathcal{L}^0 \tag{13}$$

It then follows that

$$\partial_t Vf(t) = \mathcal{L}^0 Vf(t) \tag{14}$$

The vacuum component obeys a closed equation (and so does Cf): it evolves in time according to its own "subdynamics."

We now consider the truly nontrivial interacting systems. We should like to have a kinetic description, by means of a kinetic equation such as the Boltzmann equation or the Landau equation. These equations have two features in common.

a'. They are closed equations for the one-particle distribution function.
b'. They lead irreversibly to thermal equilibrium.

Property a' can be reformulated in a more general form in terms of the vacuum correlation patterns:

a. They are closed equations for the vacuum component of the distribution vector.

Now, we know from the equations of the dynamics of correlations that property a cannot possibly be true for the distribution vector describing an interacting system. We know from the arguments above that the Liouville operator $\mathcal{L}'$ has matrix elements connecting the vacuum components $P_s([O_s])$ to the correlated components. This can easily be seen by projecting the Liouville equation on the vacuum:

$$\partial_t Vf = V(\mathcal{L}^0 + \mathcal{L}')f$$

or, using eq 13,

$$\partial_t Vf = \mathcal{L}^0 Vf + V\mathcal{L}' Vf + V\mathcal{L}' Cf \tag{15}$$

Hence the equation for the vacuum component Vf involves the correlations, and vice versa. Here we have localized the crux of the difficulty: as long as this apparent contradiction is not understood, the problem of irreversibility remains open. It is clear that this question is directly related to Boltzmann's *Stosszahlansatz.*

As the *Stosszahlansatz* cannot be true in exact dynamics, the next simplest guess we can make is the following: we may assume that the distribution vector can be split into two terms:

$$f(t) = \bar{f}(t) + \hat{f}(t) \tag{16}$$

The splitting would be such that the term $\bar{f}(t)$ would have typical kinetic behavior: in particular, its vacuum part would obey a closed equation of evolution, describing an irreversible approach to equilibrium. Moreover, hopefully, the remainder, $\hat{f}(t)$, should be unimportant, at least for the problems studied in kinetic theory.

Many of the older approaches can be formulated in this way. They consisted in *approximating* the exact Liouville equation by means of various devices (coarse graining in phase space, time smoothing, asymptotic approximations, truncation of the hierarchy, etc.); the result of such an approximation is a kinetic equation. Hence, by these methods, the exact $f(t)$ is replaced by an approximate distribution vector obeying a kinetic equation and playing the role of $\bar{f}(t)$. In none of these theories was much thought given to the complementary part, $\hat{f}(t)$.

II. Separation of Distribution Vector

It must be realized, however, that, without further specification, the separation (eq. 16) of $f(t)$ into a kinetic and a nonkinetic part is trivial and does not prove anything; it is always possible to write a number A as the sum of a given number B and of $A - B$. In order to make eq. 16 the basis of a true theory, we must require it to reflect an intrinsic and self-consistent structure that is forced upon us (rather than the desired results being forced by us into the theory!).

As a first property, we may ask that separation 16 have a geometrical meaning. Considering the space of all possible distribution vectors $\{f\}$, it may be possible to find two time-independent operators, $\mathbf{\Pi}$ and $\hat{\mathbf{\Pi}}$, which separate the space into two complementary subspaces, one containing all kinetic parts, the other all nonkinetic parts. These operators, applied to an arbitrary element of the space, $f(t)$, would automatically perform the separation of eq. 16:

$$\begin{aligned} &\bar{f}(t) = \mathbf{\Pi} f(t); \qquad \hat{f}(t) = \hat{\mathbf{\Pi}} f(t) \\ &f(t) = \mathbf{\Pi} f(t) + \hat{\mathbf{\Pi}} f(t) \end{aligned} \tag{17}$$

The two terms would then appear as true components of the vector $f(t)$. For self-consistency, these operators must have the properties of "projected operators":

$$\begin{aligned} \mathbf{\Pi}^2 &= \mathbf{\Pi}; & \hat{\mathbf{\Pi}}^2 &= \hat{\mathbf{\Pi}} \\ \mathbf{\Pi}\hat{\mathbf{\Pi}} &= 0; & \hat{\mathbf{\Pi}}\mathbf{\Pi} &= 0 \end{aligned} \tag{18}$$

together with the completeness relation

$$\mathbf{\Pi} + \hat{\mathbf{\Pi}} = I \tag{19}$$

These properties ensure that the two subspaces $\{\bar{f}\}$ and $\{\hat{f}\}$ are complementary and not overlapping.

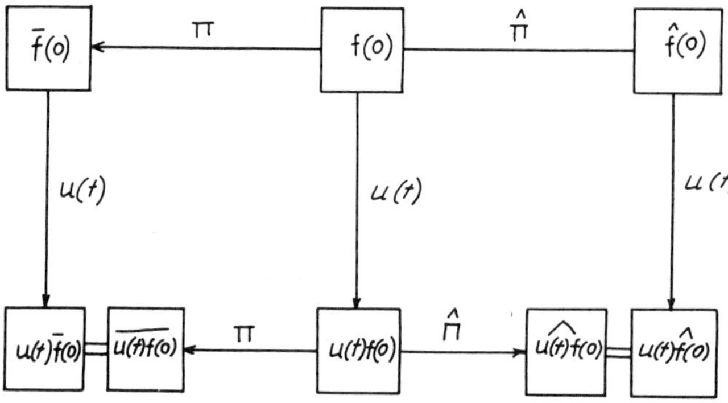

Fig. 1. Time-translation invariance of the basic decomposition (eq. 16).

But the most important property, which would really convince us of the deep nature of the theory, would be an *invariance property*. Here is what we mean (see Fig. 1). Consider a system described at time zero by a distribution vector $f(0)$. We split it into two components according to eq. 17:

$$\begin{gathered} f(0) = \bar{f}(0) + \hat{f}(0) \\ \bar{f}(0) = \mathbf{\Pi} f(0), \qquad \hat{f}(0) = \hat{\mathbf{\Pi}} f(0) \end{gathered} \tag{20}$$

We now let the system evolve. At time t its distribution vector is obtained by the action of the propagator $U(t)$ of the Liouville equation on the initial condition $f(0)$:

$$f(t) = U(t) f(0) \tag{21}$$

The vector $f(t)$ is then decomposed according to eq. 17 with the result:

$$\begin{aligned} f(t) &= \bar{f}(t + \hat{f}(t) \\ \bar{f}(t) &= \mathbf{\Pi} f(t) = \mathbf{\Pi} U(t) f(0) \\ \hat{f}(t) &= \hat{\mathbf{\Pi}} f(t) = \hat{\mathbf{\Pi}} U(t) f(0) \end{aligned} \tag{22}$$

On the other hand, we may wonder what happens to the separate components $\bar{f}(0)$ and $\hat{f}(0)$ when they are taken as an initial condition and when their evolution in time is studied. Clearly

$$\begin{aligned}\bar{f}(0) &\rightarrow U(t)\bar{f}(0) = U(t)\mathbf{\Pi}f(0)\\ \hat{f}(0) &\rightarrow U(t)\hat{f}(0) = U(t)\hat{\mathbf{\Pi}}f(0)\end{aligned} \tag{23}$$

At this point it should be clear that a theory can be self-consistent only if, at time t, the $\mathbf{\Pi}$ component of $f(t)$ coincides with the result of the evolution of the initial $\mathbf{\Pi}$ component—in other words, if the corresponding right-hand sides of eqs. 22 and 23 are identical. If this were not the case, the separation would depend on the time at which it was performed; there would be a privileged instant of time. But such an instant could not be singled out by any special physical property, and hence a theory of this type would be physically untenable.

Comparing eqs. 22 and 23, we see that the condition of *invariance of the separation under time translations* is expressed in the following simple form:

$$\mathbf{\Pi}U(t) = U(t)\mathbf{\Pi}, \qquad \hat{\mathbf{\Pi}}U(t) = U(t)\hat{\mathbf{\Pi}} \tag{24}$$

The operators $\mathbf{\Pi}$ and $\hat{\mathbf{\Pi}}$ must *commute* with the propagator $U(t)$. As a direct consequence, these operators also commute with the (complete) Liouvillian.

$$\mathbf{\Pi}\mathscr{L} = \mathscr{L}\mathbf{\Pi}, \qquad \hat{\mathbf{\Pi}}\mathscr{L} = \mathscr{L}\hat{\mathbf{\Pi}} \tag{25}$$

It then follows that the components $\bar{f}(t)$ and $\hat{f}(t)$ obey *separate* equations of evolution: there is no mixing between the $\{\bar{f}\}$ and $\{\hat{f}\}$ subspaces. Indeed,

$$\begin{aligned}\partial_t\bar{f}(t) &= \partial_t\mathbf{\Pi}U(t)f(0)\\ &= \partial_t U(t)\mathbf{\Pi}f(0) = \mathscr{L}U(t)\mathbf{\Pi}f(0)\end{aligned}$$

or

$$\partial_t\bar{f}(t) = \mathscr{L}\bar{f}(t) \tag{26}$$

and similarly

$$\partial_t\hat{f}(t) = \mathscr{L}\hat{f}(t) \tag{27}$$

Hence the components $\bar{f}$ and $\hat{f}$ evolve in time, ignoring each other. We will say again that each obeys its own *subdynamics.*

We now note that in the particular case of a noninteracting system we have found a pair of operators having precisely all the properties listed above: they are the operators V and C. Hence it is reasonable to expect that, if an operator $\mathbf{\Pi}$, having all the required properties, can be constructed for a general interacting system, it should reduce to the operator V when the interactions are switched off:

$$\mathbf{\Pi} \rightarrow V, \qquad \hat{\mathbf{\Pi}} \xrightarrow[]{\text{no interactions}} C \tag{28}$$

Having summarized this discussion, we can now start on the following program: Given a system of interacting degrees of freedom, to construct an operator $\mathbf{\Pi}$ having the following properties (as a minimum requirement):

A. $\mathbf{\Pi}$ commutes with $U(t)$ for all t.
B. $\mathbf{\Pi}$ reduces to V in the limit of no interactions.
C. $\mathbf{\Pi}$ is a projection operator.
D. The vacuum part of $\mathbf{\Pi} f(t)$ obeys a closed evolution equation.
E. The stationary solution of this evolution equation coincides with the equilibrium distribution vector.
F. The complementary component $\hat{f}(t)$ is irrelevant in certain well-defined problems of physical interest.

We cannot consider giving here any details or proofs for the realization of this program; for these, the reader is referred to Refs. 5 and 7. When we started working on this program, we thought, naturally enough, that by imposing the conditions one after another we would define classes of operators that would be more and more restricted as more conditions were added, until we would, hopefully, end with a single operator having all the desired properties. But a big surprise emerged from this investigation. Actually, *condition A combined with B turns out to be so strong as to determine a single operator* $\mathbf{\Pi}$. Hence we are given no freedom: either this unique operator possesses the remaining properties, C to F, or it does not. This unexpected feature now changes our strategy completely: we can no longer choose from among a number of items "on the market" the one that suits us best; on the contrary, the requirement of invariance offers us a single possibility, and it is now up to us to *prove* that this possibility has anything to do with kinetic theory.

We cannot emphasize strongly enough the fact that requirements A and B *have nothing to do with irreversibility.* Hence, if we do succeed (and we did!) in proving that the unique operator $\mathbf{\Pi}$, defined by conditions A and B, possesses all the properties C to F, we will be in the presence of a *truly objective* theory of irreversibility. The result can be reformulated as follows: The only invariant decomposition of the distribution function is one in which the component $\bar{f}(t)$ has a time evolution of the kinetic type.

Before leaving this subject, let us note the following. Recently, the theory has been extended to relativistic systems (8). It is then possible to show that separation 17, realized on the basis of the time-translation invariance, is automatically *invariant under all the transformations of the Lorentz group.* With this result there is no longer any doubt that eq. 17 reflects a fundamental, independent of the choice of the observer.

Let us conclude this discussion with the following remark. It would be completely unreasonable if we could construct an operator $\mathbf{\Pi}$ for *all* possible systems of interacting particles. A system of three interacting particles does not approach equilibrium. A system of particles interacting through unscreened long-range forces (e.g., gravitational forces) very probably cannot reach equilibrium either. There must be some restriction to the validity of our scheme. Sure enough, it turns out that a nontrivial operator $\mathbf{\Pi}$ (defined by conditions A and B) exists only if certain additional *auxiliary conditions* are satisfied by the system. These conditions precisely involve the thermodynamic limit, on the one hand, and the nature of the interactions, on the other.

We now come to point D of the program; here the crucial property is the following. It can be proved that the correlation part of the kinetic component, $C\bar{f}(t)$, is related to the corresponding vacuum part by an operator CCV:

$$C\bar{f}(t) = CCV\bar{f}(t) \tag{29}$$

This is an important property. The reader will recognize here the condition postulated by Bogoliubov 25 years ago as defining the "kinetic regime": the correlations change in time only through their functional dependence on the one-particle distribution function (or, more generally, on the vacuum). In the present theory, relation 29 appears as an exact property of the kinetic component of the distribution vector, which is exactly propagated in time by the equations of motion.

With property 29 we can immediately show that the vacuum part of the kinetic component obeys a closed equation of motion:

$$\begin{aligned}\partial_t V\bar{f}(t) &= V(\mathscr{L}^0 + \mathscr{L}')V\bar{f}(t) + V\mathscr{L}'C\bar{f}(t)\\ &= V(\mathscr{L}^0 + \mathscr{L}')\, V\bar{f}(t) + V\mathscr{L}'CCV\bar{f}(t)\end{aligned}$$

or

$$\partial_t V\bar{f}(t) = V\mathbf{\Gamma}V\bar{f}(t) \tag{30}$$

This is the most general *kinetic equation.* Combined with the factorization assumption (which is shown to be exactly maintained in time by eq. 30), it reduces to a closed nonlinear equation for $f_1(x_1; t)$. It can be easily shown that all the known kinetic equations are particular cases of eq. 30. We must insist that the theory provides completely explicit expressions for the operators CCV and $V\mathbf{\Gamma}V$. For instance, the ordinary kinetic equation for the one-particle distribution function is simply obtained by using the correlation pattern formalism:

$$\partial_t \bar{f}_1(x_1; t) = \sum_s \langle\langle(1)|\, V\mathbf{\Gamma}V|(s)\rangle\rangle \prod_{j=1}^{s} \bar{f}_1(x_j; t) \tag{31}$$

The operator $V\mathbf{\Gamma}V$ is a combination of operators $\mathscr{L}^0$, $\mathscr{L}'$, V, C; therefore all the matrix elements entering this expression are known from the formalism of the dynamics of correlations explained above. Their evaluation in particular cases is just a matter of substitution—and of patience.

Coming now to point E, we can prove the following property. In equilibrium, the distribution vector belongs entirely to the kinetic subspace; its nonkinetic part is rigorously zero:

$$\mathbf{\Pi} f^{\mathrm{eq}} = f^{\mathrm{eq}} \tag{32}$$

Hence all equilibrium thermodynamic properties are described in terms of the kinetic part of the distribution function alone.

But there is more in $\bar{f}(t)$ (this is part of point F): It can be shown that the *transport coefficients* are defined in terms of the kinetic part of the distribution vector alone (g). This is a much less trivial property, because in this case (i.e., in a stationary nonequilibrium state, in the presence of a time-independent external constraint) *both* $\bar{f}$ and $\hat{f}$ are nonzero.

We have now a nice picture of the evolution process. There is no contradiction whatever between irreversible and mechanical evolution. Whenever the auxiliary conditions are satisfied, the distribution vector can be decomposed in one and only one invariant way, in order to have a description valid for all possible observers. It is then proved that one of the components in this invariant decomposition obeys a kinetic equation of evolution, whereas the other component undergoes a kind of phase-mixing process, which wipes out the corresponding component as the system approaches equilibrium. Since the important physical quantities are defined in terms of the kinetic component alone, the kinetic equation provides all the relevant information for their calculation, although it does not describe the complete evolution of the system.

All the problems of irreversibility are not yet solved, of course. For instance, a general H-theorem is still lacking; also the connections between kinetic theory and ergodic theory are not yet clear (to cite just a few unanswered problems). However, it is felt that the present approach is very promising in its generality and especially in its provision of an objective theory of time evolution, independent of any arbitrary assumptions.

References

1. N. N. Bogoliubov, *Problems of a Dynamical Theory in Statistical Physics,* Gostekhizdat, Moscow, 1946.
2. L. van Hove, *Physica*, **21**, 517 (1955).
3. I. Prigogine and R. Brout, *Physica*, **22**, 621 (1956); I. Prigogine and R. Balescu, *Physica*, **25**, 281 (1959).
4. I. Prigogine, C. George, and F. Henin, *Physica*, **45**, 418 (1969).
5. R. Balescu and J. Wallenborn, *Physica* (in press).
6. R. Balescu, *Physica* (in press).
7. R. Balescu, *Statistical Mechanics* (monograph in preparation).
8. R. Balescu and L. Brenig, *Physica* (in press).
9. R. Balescu, L. Brenig, and J. Wallenborn, *Physica*, **52**, 29 (1971).

Method of Inversion of the Fluctuation–Dissipation Ratio in Plasma Theory

A. G. SITENKO AND I. P. YAKIMENKO

Institute for Theoretical Physics,
Academy of Sciences of the Ukrainian SSR, Kiev, USSR

Introduction

As it is well known, in the case of equilibrium systems the fluctuation-dissipation ratio gives the connection between fluctuations and dissipative properties. Hence the electromagnetic fluctuations in the equilibrium plasma are determined completely by the dielectric permittivity tensor. Conversely, if the fluctuations of the noninteracting particles are known, the inversion of the fluctuation-dissipation ratio may be used to find the tensor of the dielectric permittivity. Such an approach makes it possible to describe completely the electrodynamic properties of the equilibrium plasma without utilization of the kinetic equation. This is essential for the theory of bounded plasma and in some other cases as well.

For a nonequilibrium plasma (we assume, however, that the nonequilibrium state is stationary and stable) the dielectric permittivity tensor is not sufficient for describing fluctuations; we need to know also the correlation function of the noninteracting particles. In this report the method of consideration of the electromagnetic properties of equilibrium and nonequilibrium plasma systems is described. This method was developed on the basis of a generalization of the fluctuation–dissipation ratio to the case of nonequilibrium systems and utilization of the joint probability functions for the systems of the noninteracting particles. The method has been used effectively in the theory of the electromagnetic properties of bounded plasma in equilibrium as well as nonequilibrium states.

I. Inversion of Fluctuation–Dissipation Ratio

Let us consider the random fluctuations of some quantity distributed continuously in space. We can take the current density as such a quantity and denote it as $\mathbf{j}(\mathbf{r}, t)$. Let us assume that the selected quantity $\mathbf{j}(\mathbf{r}, t)$ is real, and that its mean value is zero in the absence of external effects. Correlation functions, defined as the mean values of the product of the fluctuations of selected quantities at different points of space at different times, are introduced for the fluctuation characteristics. If the stationary states are considered, the quadratic space-time correlation function will depend only on the absolute value of the time segment between the points at which the fluctuations are examined:

$$\langle j_i(\mathbf{r}, t) j_j(\mathbf{r}', t) \rangle \equiv \langle j_i(\mathbf{r}) j_j(\mathbf{r}') \rangle_{t-t'} \tag{1}$$

The brackets $\langle \ldots \rangle$ in the left-hand side of the equality denote the averaging operation. The Fourier component in time on the right side of eq. 1 we define by using the equality

$$\langle j_i(\mathbf{r}) j_j(\mathbf{r}') \rangle_\omega \equiv \int dt \, e^{i\omega t} \langle j_i(\mathbf{r}) j_j(\mathbf{r}') \rangle_t \tag{2}$$

If the system is in equilibrium state, the fluctuation–dissipation theorem establishes the connection between the correlation function of the fluctuating quantities and the dissipative properties of the system. The fluctuations of the quantity $\mathbf{j}$ are determined by a linear relation, with coefficients α_{ij}, between the quantity $\mathbf{j}$ and the appropriate potential $\mathbf{A}$. If the medium is spatially inhomogeneous, the linear connection between the Fourier components in time of the quantity $\mathbf{j}$ and $\mathbf{A}$ may be represented as the integral relationship

$$j_i(\omega, \mathbf{r}) = \int d\mathbf{r}' \, \alpha_{ij}(\omega; \mathbf{r}, \mathbf{r}') A_j(\omega, \mathbf{r}') \tag{3}$$

Having selected the potential $\mathbf{A}$ in the form that corresponds to the following expression for the change in the internal energy of the system

$$\frac{\partial U}{\partial t} = -\int d\mathbf{r} \, \mathbf{A}(\mathbf{r}, t) \mathbf{j}(\mathbf{r}, t), \tag{4}$$

we may write the fluctuation–dissipation relationship as

$$\langle j_i(\mathbf{r}) j_j(\mathbf{r}') \rangle_\omega = i \frac{T}{\omega} \left[\alpha_{ji}^*(\omega; \mathbf{r}', \mathbf{r}) - \alpha_{ij}(\omega; \mathbf{r}, \mathbf{r}') \right] \tag{5}$$

This relationship is obtained directly from the connection between the correlation function for the current density fluctuations and the average energy absorbed by the system as a result of dissipation [1, 2]. Formula 5, which is a generalization of the Nyquist fluctuation-dissipation theorem, completely defines the fluctuations of the distributed quantities in an equilibrium system.

As we are interested in the fluctuations of the current density $\mathbf{j}$, we must take the quantity $-i\omega^{-1}\mathbf{E}$ as a potential $\mathbf{A}$ ($\mathbf{E}$ is the electric-field intensity). By using the Maxwell equations and the material equations, we may express the coefficients α_{ij} in terms of the electric susceptibility tensor K_{ij}.

$$\alpha_{ij}(\omega; \mathbf{r}, \mathbf{r}') = \frac{\omega^2}{4\pi}\left[\Lambda_{ij}{}^0(\omega, \mathbf{r})\,\delta(\mathbf{r}-\mathbf{r}') - \Lambda_{ik}{}^0(\omega, \mathbf{r})\Lambda_{kl}{}^{-1}(\omega; \mathbf{r}, \mathbf{r}')\Lambda_{lj}{}^0(\omega, \mathbf{r}')\right] \quad (6)$$

$$\Lambda_{ij}{}^0(\omega, \mathbf{r}) \equiv \frac{c^2}{\omega^2}\left(\frac{\partial}{\partial r_i}\frac{\partial}{\partial r_j} - \delta_{ij}\Delta\right) + \delta_{ij} \quad (7)$$

$$\Lambda_{ij}(\omega; \mathbf{r}, \mathbf{r}') \equiv \Lambda_{ij}{}^0(\omega, \mathbf{r})\,\delta(\mathbf{r}-\mathbf{r}') + 4\pi K_{ij}(\omega; \mathbf{r}, \mathbf{r}')$$

Instead of the electric susceptibility tensor K_{ij}, we may use for describing the electrodynamic properties of plasma the dielectric permittivity tensor ϵ_{ij}:

$$\epsilon_{ij}(\omega; \mathbf{r}, \mathbf{r}') = \delta_{ij}\,\delta(\mathbf{r}-\mathbf{r}') + 4\pi K_{ij}(\omega; \mathbf{r}, \mathbf{r}') \quad (8)$$

Hence, according to eq. 5, the current fluctuations in an equilibrium plasma are determined by the anti-Hermitian part of the tensor of the linear coefficients, which is determined, in turn, by the plasma dielectric permittivity tensor ϵ_{ij}.

If the correlation function for the current fluctuations is determined with the aid of microscopic theory, the fluctuation-dissipation theorem may be utilized to find the macroscopic parameters of a substance.† In the case of plasma the fluctuation-dissipation theorem enables us to find the dielectric permittivity tensor. The special advantage of such a determination of ϵ_{ij} is the possibility of taking account of thermal effects without utilizing the kinetic equation.

By expanding the left and right sides of eq. 5 in power series in e^2 and retaining the basic terms, we have

$$\langle j_i(\mathbf{r})j_j(\mathbf{r}')\rangle_\omega{}^0 = i\omega T\{K_{ji}^*(\omega; \mathbf{r}, \mathbf{r}') - K_{ij}(\omega; \mathbf{r}, \mathbf{r}')\} \quad (9)$$

where $\langle j_i(\mathbf{r})j_j(\mathbf{r}')\rangle_\omega{}^0$ is the correlation function for the system of noninteracting particles.‡ Relationship 9 determines directly the anti-Hermitian part of the dielectric permittivity tensor. The Hermitian part of this tensor is easily found

† Such a method of determining the macroscopic parameters of a substance was proposed by Shafranov (3), Kubo (4), and Nakano (5).

‡ Relationship 9 was derived first in the paper of Leontovich and Rytov (6), where the quantity $\langle j_i j_j\rangle_\omega$ was interpreted as the correlation function for the secondary currents.

by utilizing the integral Kramers-Kronig relationship. We finally obtain the following general formula determining the electric susceptibility tensor of a substance, $K_{ij}(\omega; \mathbf{r}, \mathbf{r}')$, in terms of the known spectral distribution of the current fluctuations in the medium, $\langle j_i(\mathbf{r}) j_j(\mathbf{r}')\rangle_\omega{}^0$:

$$K_{ij}(\omega; \mathbf{r}, \mathbf{r}') = \frac{1}{2\pi\omega T} \int_{-\infty}^{\infty} d\omega' \, \frac{\langle j_i(\mathbf{r}) j_j(\mathbf{r}')\rangle_{\omega'}{}^0}{\omega' - \omega - i0} \tag{10}$$

The remarkable feature of this relationship consists in the fact that, according to eq. 10, the electric permittivity of a substance is expressed in terms of the spectral distribution of the current fluctuations in the medium without taking into account the interaction between the charged particles. This means that the plasma dielectric permittivity tensor may be determined by finding the spectral distribution of the current fluctuations in an ideal gas.

If plasma is spatially homogeneous, the correlation function for the fluctuations depends only on the relative distance between the points at which the fluctuations are examined:

$$\langle j_i(\mathbf{r}) j_j(\mathbf{r}')\rangle_\omega{}^0 = \langle j_i j_j\rangle_{\mathbf{r}-\mathbf{r}', \omega} \tag{11}$$

In this case it is convenient to utilize the space Fourier transform. For the Fourier components relationship 10 is as follows:

$$K_{ij}(\omega, \mathbf{k}) = \frac{1}{2\pi\omega T} \int_{-\infty}^{\infty} d\omega' \, \frac{\langle j_i j_j\rangle_{\mathbf{k}\omega'}{}^0}{\omega' - \omega - i0} \tag{12}$$

where

$$\langle j_i j_j\rangle_{\mathbf{k}\omega}{}^0 \equiv \int d\mathbf{r}\, e^{-i\mathbf{k}\mathbf{r}} \langle j_i j_j\rangle_{\mathbf{r}\omega} \tag{13}$$

This approach, grounded on the inversion of the fluctuation–dissipation ratio, was used in Ref. 7 for investigations of the electrodynamic properties of unbounded plasma and the different electromagnetic processes in such plasma. It is obvious, on the basis of the inversion of the fluctuation–dissipation theorem (10), that further development of the theory, taking into account the inhomogeneity of the medium and the influence of the boundary, is possible. According to Tolmachev (10), finding the permittivity of a bounded plasma is reduced to finding the correlation function for the current fluctuations in the system of noninteracting particles in the presence of the boundary conditions.

II. Correlation Function for Current Fluctuations

Let us consider the system of N identical particles with charge e and mass m, which is embedded inside the volume V. We will assume that the total electrical charge of the system is canceled by the charge of the opposite sign. The state of the particles we will characterize by the coordinate **r** and velocity **v**. Let us restrict ourselves by considering stationary states of the system as a whole, denoting the one-particle distribution function by $f_0(\mathbf{r}, \mathbf{v})$. This function we will normalize according to the condition

$$\int d\mathbf{r} \int d\mathbf{v}\, f_0(\mathbf{r}, \mathbf{v}) = V \tag{14}$$

If the system is spatially homogeneous, the one-particle function $f_0(\mathbf{v})$ does not depend on the space coordinates **r** and satisfies the next condition of the normalization:

$$\int d\mathbf{v}\, f_0(\mathbf{v}) = 1$$

Let us introduce the joint probability $W_t(\mathbf{r}, \mathbf{v}; \mathbf{r}', \mathbf{v}')$ that at time t the particle will be in state $(\mathbf{r}, \mathbf{v})$ and at time t' ($t' = 0$) in state $(\mathbf{r}', \mathbf{v}')$. It is clear that the joint probability function should satisfy the condition

$$\int d\mathbf{r} \int d\mathbf{v}\, W_t(\mathbf{r}, \mathbf{v}; \mathbf{r}', \mathbf{v}') = 1 \tag{15}$$

Using the one-particle distribution function and the joint probability $W_t(\mathbf{r}, \mathbf{v}; \mathbf{r}', \mathbf{v}')$, we can write the correlation function of microcurrents as

$$\langle j_i(\mathbf{r}, t) j_j(\mathbf{r}', t')\rangle \equiv e^2 n_0 \int d\mathbf{v} \int d\mathbf{v}'\, \mathrm{v}_i \mathrm{v}_j' f_0(\mathbf{r}', \mathbf{v}') W_{t-t'}(\mathbf{r}, \mathbf{v}; \mathbf{r}', \mathbf{v}') \tag{16}$$

where n_0 is the average particle density ($n_0 \equiv N/V$). Hence, if the one-particle distribution function is known, calculation of the correlation function for the current fluctuations is reduced to determining the joint probability, $W_t(\mathbf{r}, \mathbf{v}; \mathbf{r}', \mathbf{v}')$.

Let us notice that in addition to the quantity $W_t(\mathbf{r}, \mathbf{v}; \mathbf{r}', \mathbf{v}') \equiv W_t^{1}(\mathbf{r}, \mathbf{v}; \mathbf{r}', \mathbf{v}')$, which determines the probability of finding the particle in the state $(\mathbf{r}, \mathbf{v})$ at the moment t, if at first (at the moment $t' = 0$) the particle has been in the state $(\mathbf{r}', \mathbf{v}')$, it is possible also to introduce the quantity $W_t^{2}(\mathbf{r}_1, \mathbf{v}_1, \mathbf{r}_2, \mathbf{v}_2; \mathbf{r}', \mathbf{v}')$, which determines the probability of finding two particles in states $(\mathbf{r}_1, \mathbf{v}_1)$ and $(\mathbf{r}_2, \mathbf{v}_2)$ at the moment t, if at first (at the moment $t' = 0$) one of the particles has been in the state $(\mathbf{r}', \mathbf{v}')$, and so on. In the general case of arbitrary pairwise interaction between the particles, it is possible to obtain a hierarchy of equations for the probability $W_t^{s}(\mathbf{r}_1, \mathbf{v}_1, \ldots, \mathbf{r}_s, \mathbf{v}_s; \mathbf{r}', \mathbf{v}')$ $(s = 1, 2, \ldots)$, which is similar to Bogolubov's hierarchy (9). Such a hierarchy for the probability $W_t^{s}(\mathbf{r}_1, \mathbf{v}_1, \ldots, \mathbf{r}_s, \mathbf{v}_s; \mathbf{r}', \mathbf{v}')$ was obtained in Ref. 10. But, if we exclude from consideration the interaction between the particles (not only the electromagnetic, but also the close-range interaction), then for the joint probability $W_t(\mathbf{r}, \mathbf{v}; \mathbf{r}', \mathbf{v}')$ we obtain the single differential equation

$$\left(\frac{\partial}{\partial t} + \mathbf{v}\frac{\partial}{\partial \mathbf{r}} + \mathbf{k}\frac{\partial}{\partial \mathbf{v}}\right) W_t(\mathbf{r}, \mathbf{v}; \mathbf{r}', \mathbf{v}') = 0 \tag{17}$$

where $\mathbf{k}$ is the acceleration of some particle under the influence of the external power field. The interactions or the particle diffusion can be taken into account by adding the collision integral or the diffusion term into the right-hand side of eq. 17. It is necessary to complement eq. 17 by the initial condition

$$W_{t=0}(\mathbf{r}, \mathbf{v}; \mathbf{r}', \mathbf{v}') = \delta(\mathbf{r} - \mathbf{r}')\, \delta(\mathbf{v} - \mathbf{v}') \tag{18}$$

and, for bounded systems, by the appropriate boundary conditions.

Let us consider the simplest example, the unbounded homogeneous plasma, which is characterized by the equilibrium distribution function $f_0(v)$. In this case eq. 17, together with initial condition 18, constitutes the total set of equations for the determination of $W_t(\mathbf{r}, \mathbf{v}; \mathbf{r}', \mathbf{v}')$. It is easy to find the solution of this system at $\mathbf{k} = 0$:

$$W_t(\mathbf{r}, \mathbf{v}; \mathbf{r}', \mathbf{v}') = \delta(\mathbf{R} - \mathbf{v}t)\, \delta(\mathbf{v} - \mathbf{v}'), \qquad \mathbf{R} = \mathbf{r} - \mathbf{r}' \tag{19}$$

According to eq. 16, the spectral correlation function for the current fluctuations may be represented as

$$\langle j_i j_j \rangle_{\mathbf{k}\omega}{}^0 = 2\pi e^2 n_0 \int d\mathbf{v} v_i v_j f_0(v)\, \delta(\omega - \mathbf{k}\mathbf{v}) \tag{20}$$

Substituting this expression into eq. 12, we obtain the well-known expression for the electric susceptibility tensor for the free-collision plasma:

$$\mathrm{K}_{ij}(\omega, \mathbf{k}) = \frac{e^2 n_0}{\omega T} \int d\mathbf{v} \frac{v_i v_j f_0(v)}{\mathbf{k}\mathbf{v} - \omega} \tag{21}$$

The appropriate dielectric permittivity tensor, $\epsilon_{ij}(\omega, \mathbf{k})$, is expressed, as unual, in terms of the longitudinal and transverse dielectric permittivities, $\epsilon_l(\omega, k)$ and $\epsilon_t(\omega, k)$:

$$\epsilon_{ij}(\omega, \mathbf{k}) = \frac{k_i k_j}{k^2}\, \epsilon_l(\omega, k) + \left(\delta_{ij} - \frac{k_i k_j}{k^2}\right) \epsilon_t(\omega, k) \tag{22}$$

The dielectric permittivity tensor for a magnetoactive plasma can be obtained in the same manner by solving eq. 17 with

$$\mathbf{k} = \frac{e}{mc}\, (\mathbf{v}\mathbf{H}_0)$$

where $\mathbf{H}_0$ is the external magnetic field intensity.

III. Semibounded Plasma

The advantages of the approach discussed above are completely apparent if we consider bounded plasma systems. In fact, all the steps in the calculation of the dielectric permittivity tensor are considered in this case except in so far as it is necessary to add to eqs. 17 and 18 the relationships that express the boundary conditions for the joint probability. In the case of specular reflection the boundary condition for the joint probability is as follows:

$$W_t(\mathbf{r}, \mathbf{v}; \mathbf{r}', \mathbf{v}') = W_t[\mathbf{r}, \mathbf{v} - 2(\mathbf{vn})\mathbf{n}; \mathbf{r}', \mathbf{v}']_{\text{boundary}} \tag{23}$$

where $\mathbf{n}$ is the unit vector of the normal to the boundary surface in the reflection point.† Equation 17, together with initial condition 18 and boundary condition 23, constitutes the completed set of equations which make it possible to take the influence of the restricted dimensions on the electromagnetic properties of the system into account.

Let us suppose that plasma which is in the statistically equilibrium state is placed in the half-space $z > 0$. Hence the equation of the boundary surface is

$$z = 0$$

As a result of the homogeneity of the system in the transverse direction, the joint probability will depend only on the difference between the transverse components of vectors $\mathbf{r}$ and $\mathbf{r}'$:

$$W_t(\mathbf{r}, \mathbf{v}; \mathbf{r}', \mathbf{v}') \equiv W_{\mathbf{R}_\perp, t}(z, \mathbf{v}; z', \mathbf{v}'), \qquad \mathbf{R}_\perp = \mathbf{r}_\perp - \mathbf{r}_\perp'$$

The boundary conditions in the case of specular reflection take the form

$$W_{\mathbf{R}_\perp, t}(0, \mathbf{v}_\perp, v_z; z', \mathbf{v}') = W_{\mathbf{R}_\perp, t}(0, \mathbf{v}_\perp, -v_z; z', \mathbf{v}') \tag{24}$$

Solving eq. 17, while taking into account eqs. 18 and 24, we find

$$W_t^s(\mathbf{r}, \mathbf{v}; \mathbf{r}', \mathbf{v}') = \delta(\mathbf{R}_\perp - \mathbf{v}_\perp t)\, \delta(\mathbf{v}_\perp - \mathbf{v}_\perp')[\delta(Z - v_z t)\, \delta(v_z - v_z')t$$
$$+ \delta(Z^+ - v_z t)\, \delta(v_z + v_z')] \tag{25}$$

where $Z = z - z'$ and $Z^+ = z + z'$.

The first term in eq. 25 coincides exactly with the expression for the joint probability in the unbounded plasma eq. 19. Therefore the second term in eq. 25 may be interpreted as the additional joint probability due to the reflection of the particles from the wall. If information about the previous motion of the particles is entirely lost at the reflection (diffuse reflection), the additional probability is equal to zero. Therefore, at the diffuse reflection, the total joint probability $W_t^d(\mathbf{r}, \mathbf{v}; \mathbf{r}', \mathbf{v}')$ coincides with eq. 19:

$$W_t^d(\mathbf{r}, \mathbf{v}; \mathbf{r}', \mathbf{v}') = \delta(\mathbf{R} - \mathbf{v}t)\, \delta(\mathbf{v} - \mathbf{v}') \tag{26}$$

This makes it possible to determine the point probability taking into account the generalized boundary condition, which is used very often in the kinetic theory.

† Just the same boundary condition as eq. 23 was used in Ref. 11.

Let ρ denote the fraction of specular reflected particles, and $1 - \rho$ denote the fraction of diffuse reflected particles. Then the joint probability may be written as

$$W_t(\mathbf{r}, \mathbf{v}; \mathbf{r}', \mathbf{v}') = \rho W_t^s(\mathbf{r}, \mathbf{v}; \mathbf{r}', \mathbf{v}') + (1 - \rho) W_t^d(\mathbf{r}, \mathbf{v}; \mathbf{r}', \mathbf{v}') \tag{27}$$

By using this expression it is easy to find the correlation function for the current fluctuations in a half-space of plasma:

$$\langle j_i(z) j_j(z') \rangle_{\mathbf{k}_\perp \omega} = \mathcal{I}_{ij}(\omega, \mathbf{k}_\perp; Z) + \rho \epsilon_j \mathcal{I}_{ij}(\omega, \mathbf{k}_\perp; Z^+) \tag{28}$$

where

$$\mathcal{I}_{ij}(\omega, \mathbf{k}_\perp; \zeta) = e^2 n_0 \int d\mathbf{v} \frac{v_i v_j}{|v_z|} f_0(v) e^{i\Delta\zeta}, \qquad \Delta = \frac{\omega - \mathbf{k}_\perp \mathbf{v}_\perp}{v_z}, \qquad \epsilon_j = \begin{cases} 1, j \neq z \\ -1, j = z \end{cases}$$

Substituting eq. 28 into inversion formula 10, we obtain the following expression for the electric susceptibility of the half-space of plasma:

$$\mathrm{K}_{ij}(\omega, \mathbf{k}_\perp; z, z') = \mathrm{K}_{ij}(\omega, \mathbf{k}_\perp; Z) + \rho \epsilon_j \mathrm{K}_{ij}(\omega, \mathbf{k}_\perp; Z^+) \tag{29}$$

where

$$\mathrm{K}_{ij}(\omega, \mathbf{k}_\perp; \zeta) = \frac{1}{2\pi} \int dk_z e^{ik_z \zeta} \mathrm{K}_{ij}(\omega, \mathbf{k})$$

and the quantity $\mathrm{K}_{ij}(\omega, \mathbf{k})$ is determined by expression 21. Formula 29 agrees with the known results (in a number of special cases) of the kinetic theory (13).

In the analogous way one can investigate the electrodynamic properties of the half-space of plasma in the presence of an external magnetic field. If the direction of the magnetic field is perpendicular to the boundary surface, the electric susceptibility tensor is determined, as previously, by general formulas 28 and 29, in which, however, it is necessary to use the expression for the electric susceptibility tensor of the magnetoactive plasma (7).

IV. Information Model of the Correlation

Taking as a basis the results of Section III, we may formulate some rules for the calculation of the correlation functions in bounded systems, based on introducing the concept of the information transmission channel. According to eq. 28, the total joint probability that at time t the particle will be in state $(\mathbf{r}, \mathbf{v})$ and at time t' $(t' = 0)$ in state $(\mathbf{r}', \mathbf{v}')$ is equal to the sum of two probabilities, which correspond to two information transmission channels between points $\mathbf{r}'$ and $\mathbf{r}$. The first channel corresponds to the straight line connecting these points, and the second channel corresponds to the sum of two lengths: from point $\mathbf{r}'$ to the boundary and from the boundary to point $\mathbf{r}$; that is, the channels correspond to the natural trajectories of the particle from the initial point to the end. The contribution of each channel to the total correlation function is determined as

$$\langle j_i(\mathbf{r}, t) j_j(\mathbf{r}', t' = 0)\rangle_\alpha{}^0 = L_\alpha \mathcal{J}_{ij}(\mathbf{R}_\perp, \zeta_\alpha, t) \tag{30}$$

where ζ_α is the projection of the α-channel lengths on the z axis. The factor L_α shows the fraction of particles that can pass along the given channel without loss of information and contains in addition a certain combination of signs. In the case of one wall, $L_1 = 1$ and $L_2 = \rho\epsilon_j$. The tensor $\mathcal{J}_{ij}$ is determined as a product of delta functions, whose arguments reflect the particular space-time connections, depending on the type of channel.

We may formulate some general rules for determining the correlation functions in the presence of boundaries. These rules include the following sequence of operations:

1. Classification of all possible information channels.
2. Determination for every type of channel of the space-time relationships; these depend on the kind of equation of motion and boundary conditions under consideration.
3. Determination of the contribution of each channel to the total correlation function.
4. Summation of the result over all the channels.

The main advantage of the proposed method for taking account of the influence of boundaries consists in the fact that the result can be obtained in a mathematically simple and physically clear way. The main difficulty involves the classification of the information channels.

Let N be the total number of information channels. For an unbounded plasma $N = 1$, and the correlation function is described by formula 30 with $L_\alpha = 1$. In the presence of boundaries $N \geqslant 1$ because any boundary may lead to additional information channels.Therefore, in the case of bounded plasma, the correlation function is equal to the sum of the correlation function for unbounded plasma and an additional part, connected with $N - 1$ new information channels. The exact number of these channels and their structure are determined by the boundary geometry and the boundary conditions.

In the case of one boundary we have a finite set of information channels which consists of two channels only ($N = 2$). If we introduce a second boundary, we obtain not simply an increase in the power of the finite set but also replacement of this set by a denumerable set. Such a situation appears, specifically, in the case of the plane-parallel plasma layer, which we consider now.

Let the plasma layer be bounded by the two planes $z = 0$ and $z = l$. The boundary conditions on these planes may be, in general, different, and it is not difficult to take account of this distinction within the frames of our model; for simplicity, however, we suppose that the boundary conditions are the same and that they coincide with eq. 27. It is convenient to take the combination of the directions of the particle departure from point $\mathbf{r}'$ and its arrival at point $\mathbf{r}$ as a basis of classification of the information channels. Since the only distinguished direction is the z axis, we are interested only in the sign of the velocity component with respect to the normal to the boundary. The particle has two possibilities in departing point $\mathbf{r}'$: it can move with velocity $v_z' > 0$ or $v_z' < 0$, and the same possibilities exist for arrival at point $\mathbf{r}$. Hence there are four types of trajectories, which correspond to the following combination of signs in

Channel Type	ζ_α	ν	L_α
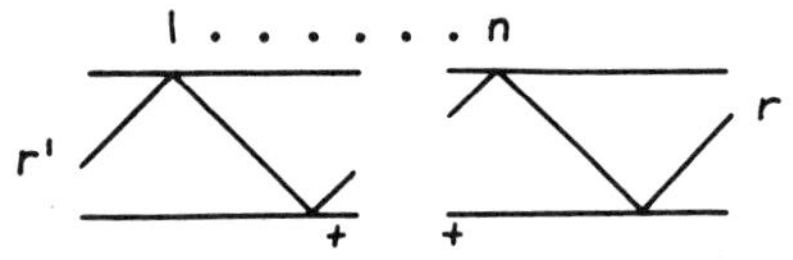	$2nl + z$	$2n$	ρ^{2n}
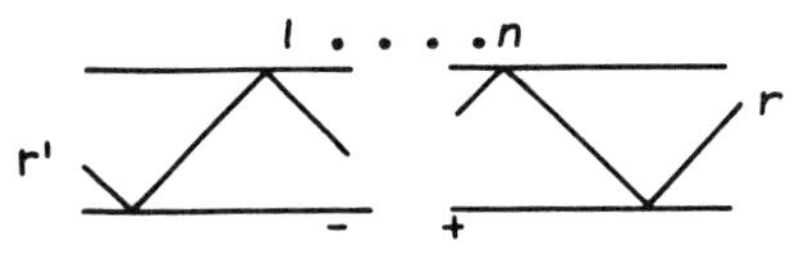	$2nl + z^+$	$2n + 1$	$\epsilon_j\rho^{2n+1}$
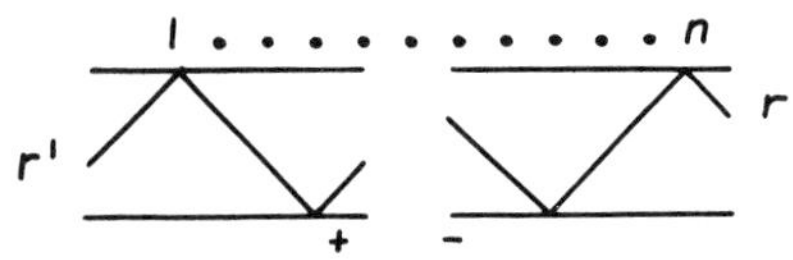	$2nl - z^+$	$2n - 1$	$\epsilon_i\rho^{2n-1}$
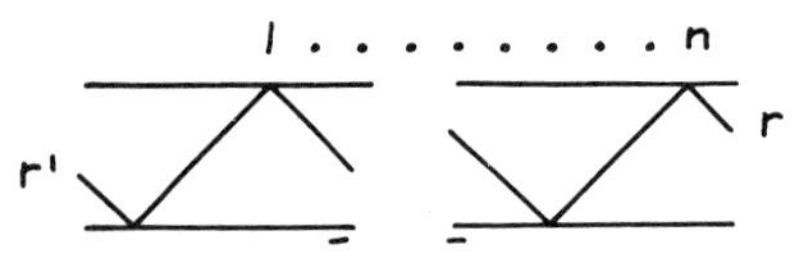	$2nl - z$	$2n$	$\epsilon_i\epsilon_j\rho^{2n}$

regard to points $\mathbf{r}'$ and $\mathbf{r}$: ++, −+, +−, and − −. These four types of information channels are shown in the table (the z axis is directed up). The set of channels inside each type is denumerable. This important circumstance makes it possible to reduce the problem to a series summation.

Every channel is characterized by a certain number ν of points on the boundary surfaces, which determine the quantity ζ_α and the factor L_α for each combination of delta functions. The quantities ν and ζ and corresponding factors L_α are indicated in the table.

With this classification of the channels as a basis, it is not difficult to write immediately the following expression for the correlation function of the current fluctuation in the plasma layer:

$$\langle j_i(\mathbf{r}, t) j_j(\mathbf{r}', t' = 0)\rangle^0 = e^2 n_0 \int d\mathbf{v} v_i v_j f_0(v)\, \delta(\mathbf{R}_\perp - \mathbf{v}_\perp t)$$

$$\Bigg[\sum_{n=0}^{\infty} \rho^{2n}\, \delta(2nl + Z - v_z t)$$

$$+ \epsilon_j \sum_{n=0}^{\infty} \rho^{2n+1}\, \delta(2nl + Z^+ - v_z t)$$

$$+ \epsilon_i \sum_{n=1}^{\infty} \rho^{2n-1}\, \delta(2nl - z^+ - v_z t)$$

$$+ \epsilon_i \epsilon_j \sum_{n=1}^{\infty} \rho^{2n}\, \rho(2nl - Z - v_z t)\Bigg] \qquad (31)$$

Selecting the contributions of the direct channel and the channels with single-multiple reflection, we easily derive the following formula for the electric susceptibility of the plasma layer:

$$\mathrm{K}_{ij}(\omega, \mathbf{k}_\perp; z, z') = \mathrm{K}_{ij}(\omega, \mathbf{k}_\perp; Z) + \rho\epsilon_j \mathrm{K}_{ij}(\omega, \mathbf{k}_\perp; Z^+)$$

$$+ \sum_{n=1}^{\infty} \rho^{2n} [\mathrm{K}_{ij}(\omega, \mathbf{k}_\perp; 2nl + Z) + \rho\epsilon_j \mathrm{K}_{ij}(\omega, \mathbf{k}_\perp; 2nl + Z^+)$$

$$+ \rho^{-1} \epsilon_i \mathrm{K}_{ij}(\omega, \mathbf{k}_\perp; 2nl - z^+) + \epsilon_i \epsilon_j \mathrm{K}_{ij}(\omega, k_\perp; 2nl - Z)] \qquad (32)$$

The first term in eq. 32 describes the electric susceptibility of the unbounded plasma. The combination of the first and second terms determines the electric susceptibility of the half-space of plasma; all the other terms in eq. 32 reflect the influence of the second boundary. An analogous formula for the electric susceptibility tensor holds also in the case of magnetoactive plasma, if the direction of the magnetic field is perpendicular to the plane plasma boundaries.

By using the explicit expression for the electric susceptibility tensor, it is easy to carry out the summation in eq. 32. As a result, we obtain the following formula:

$$\mathrm{K}_{ij}(\omega, \mathbf{k}_\perp; z, z') = \frac{e^2 n_0}{2\pi\omega T} \int dk_z \int dv\, \frac{v_i v_j f_0(v)}{\mathbf{kv} - \omega}\, [e^{ik_z(Z-l)}$$

$$+ \rho\epsilon_j\, e^{ik_z(Z^+ - l)} + \rho\epsilon_i\, e^{-ik_z(Z^+ - l)}$$

$$+ \rho^2 \epsilon_i \epsilon_j\, e^{-ik_z(Z-l)}] (e^{-ik_z l} - \rho^2 e^{ik_z l})^{-1} \qquad (33)$$

To generalize the results obtained for the case of many-component bounded plasma is not difficult; for this purpose it is necessary to repeat the analogous computation for each species of charged particles. In doing this it is necessary, of course, to suppose that inside every species of particle the equilibrium distribution exists, although the temperatures for different species may not be equal (the practically important example of such a plasma is the nonisothermal electron-ion plasma). Since the fluctuations of electrons and ions are independent in a system of noninteracting particles, it is sufficient to restrict ourselves to summation over the particle species to find the total correlation function for the current fluctuations in a many-component system of charged particles.

V. Waves and Fluctuations in a Bounded Plasma

After the dielectric permittivity tensor of the bounded equilibrium plasma has been determined, it is not difficult to investigate by the usual methods electromagnetic waves, fluctuations, and other electromagnetic phenomena in such a plasma. Determination of the fluctuation spectrum is the most general problem of this kind, the solution of which includes also the solution of the dispersion equation, which conveys the main information about the wave properties of plasma.

As an example, we determine the fluctuation field in a plasma half-space and investigate the fluctuation spectrum. Let us suppose that plasma fills the half-space $z > 0$ and that the other half-space, $z < 0$, is filled by a homogeneous medium with dielectric permittivity $\tilde{\epsilon}(\omega)$. Introducing the secondary current $\mathbf{j}^0$ into the Maxwell equations, and using the usual boundary conditions (continuity of the tangential field components on the boundary) and, at last, expression 29 for the electric susceptibility tensor, we obtain in the case of specular reflection the following expression for the fluctuation field in the entire space:

$$\mathbf{E}_\omega(\mathbf{r}) = \begin{cases} \int d\mathbf{k}\, e^{i\mathbf{k}\mathbf{r}} \mathbf{E}_{\mathbf{k}\omega}, & z > 0 \\ \int d\mathbf{k}_\perp\, e^{i(\mathbf{k}_\perp \mathbf{r}_\perp - \tilde{k}_z z)} \tilde{\mathbf{E}}_{\mathbf{k}_\perp\omega}, & z < 0 \end{cases} \tag{34}$$

where

$$E_i(\omega, \mathbf{k}) = -\frac{4\pi i}{\omega}\left\{\frac{k_i k_j}{k^2}\,\epsilon_l^{-1}(\omega, k) + \left(\delta_{ij} - \frac{k_i k_j}{k^2}\right)\left[\epsilon_t(\omega, k) - \frac{k^2 c^2}{\omega^2}\right]^{-1}\right\}$$

$$\times\, [j_j{}^0(\omega, \mathbf{k}) + \tilde{j}_j(\omega, \mathbf{k}_\perp)]$$

$$\tilde{\mathbf{E}}(\omega, \mathbf{k}_\perp) = -\frac{4\pi i}{\omega}\left\{\frac{\tilde{k}_z{}^2}{k_\perp{}^2}\,[k_x j_p(\omega, \mathbf{k}_\perp) + k_y j_s(\omega, \mathbf{k}_\perp)],\right.$$

$$\left.\frac{\tilde{k}_z{}^2}{k_\perp{}^2}\,[k_y j_p(\omega, \mathbf{k}_\perp) - k_x j_s(\omega, \mathbf{k}_\perp)],\ \tilde{k}_z j_p(\omega, k_\perp)\right\}$$

$$\tilde{\mathbf{j}}(\omega, \mathbf{k}_\perp) = -i\frac{\tilde{k}_z c^2}{\omega^2 k_\perp^2}\,[k_x \tilde{k}^2 j_p(\omega, \mathbf{k}_\perp) + k_y \tilde{k}_z{}^2 j_s(\omega, \mathbf{k}_\perp),$$

$$k_y \tilde{k}^2 j_p(\omega, k_\perp) - k_x \tilde{k}_z{}^2 j_s(\omega, k_\perp), 0] \qquad \tilde{k}^2 = \frac{\omega^2 \tilde{\epsilon}}{c^2},$$

$$\tilde{k}_z = \sqrt{\tilde{k}^2 - k_\perp{}^2}, \qquad \operatorname{Re}\tilde{k}_z > 0$$

Here $\epsilon_l(\omega, k)$ and $\epsilon_t(\omega, k)$ coincide with the longitudinal and transverse dielectric permittivities (eq. 22), and j_s and j_p are the linear functionals of the secondary fluctuation current:

$$j_s(\omega, \mathbf{k}_\perp) = \frac{1}{[1 - r_s(\omega, \tilde{k}_\perp)]\,k_z{}^2}\int dk_z \frac{1}{\epsilon_t(\omega, k) - (k^2c^2/\omega^2)} \times [k_x j_y{}^0(\omega, \mathbf{k}) - k_y j_x{}^0(\omega, \mathbf{k})]$$

$$j_p(\omega, \mathbf{k}_\perp) = \frac{1}{[1 - r_p(\omega, k_\perp)]\,\tilde{k}_z{}^2}\int dk_z \left\{\left[\frac{1}{\epsilon_l(\omega, k)} - \frac{1}{\epsilon_t(\omega, k) - (k^2c^2/\omega^2)}\right] \times \frac{k_\perp{}^2}{k^2}\,k_j{}^0(\omega, \mathbf{k}) + \frac{1}{\epsilon_t(\omega, k) - (k^2c^2/\omega^2)}\,k_\perp j^0(\omega, \mathbf{k})\right\}$$

where r_s and r_p are the ratio of plasma impedance and external medium for waves of s and p polarization (13):

$$r_s(\omega, k_\perp) = -i\frac{\tilde{k}_z c^2}{\pi\omega^2}\int dk_z \frac{1}{\epsilon_t(\omega, k) - (k^2c^2/\omega^2)}$$

$$r_p(\omega, k_\perp) = \frac{\omega^2\tilde{\epsilon}}{k_z{}^2c^2}\,r_s(\omega, k_\perp) - i\frac{\tilde{\epsilon}}{\pi\tilde{k}_z}\int dk_z \frac{k_\perp{}^2}{k^2}\left[\frac{1}{\epsilon_l(\omega, k)} - \frac{1}{\epsilon_t(\omega, k) - (k^2c^2/\omega^2)}\right] \tag{35}$$

Taking the square of the expressions for the field and using the correlation function for the fluctuations of secondary current (eq. 9), one can find any characteristics of the fluctuation field.

The average value of Poynting's vector in the normal direction with respect to the boundary characterizes the intensity of the thermal radiation from a half-space of plasma. The spectral intensity of the thermal radiation is as follows

$$P_\omega = -\frac{T}{2\pi^2}\int_0^{\tilde{k}(\omega)} dk_\perp k_\perp [1 - R(\omega, k_\perp)] \tag{36}$$

where

$$R(\omega, k_\perp) = \frac{1}{2}\left[\left|\frac{1 + r_s(\omega, k_\perp)}{1 - r_s(\omega, k_\perp)}\right|^2 + \left|\frac{1 + r_p(\omega, k_\perp)}{1 - r_p(\omega, k_\perp)}\right|^2\right]$$

Formula 36 is of the same type as the formula for the spectral intensity of thermal radiation that was derived in Ref. 8 without taking account of the space dispersion.

The correlation function for the charge density fluctuations in a half-space of plasma, taking into account the self-consistent interaction between the particles, is determined by the expression:

$$\langle \rho_{k_z} \rho^*_{k_z'} \rangle_{\mathbf{k}_\perp \omega} = \frac{k^2 T}{4\pi^2 \omega}\left\{ \frac{\operatorname{Im} \epsilon_l(\omega, k)}{|\epsilon_l(\omega, k)|^2} [\delta(k_z - k_z') + \delta(k_z + k_z')] \right.$$
$$\left. -2\pi \frac{k_\perp{}^2 \operatorname{Re} \tilde{k}_z r_p(\omega, k_\perp)}{k^2 \tilde{k}_z{}^2 |1 - r_p(\omega, k_\perp)|^2} \frac{\tilde{\epsilon}}{\epsilon_l(\omega, k)\epsilon_l^*(\omega, k')} - \phi(k, k') - \phi^*(k', k) \right\} \tag{37}$$

where

$$\phi(k, k') \equiv i \frac{k_\perp{}^2 \tilde{\epsilon} \operatorname{Im} \epsilon_l(\omega, k)}{k^2 \tilde{k}_z [1 - r_p(\omega, k_\perp)] \epsilon_l(\omega, k) |\epsilon_l(\omega, k')|^2}$$

The first term in eq. 37 describes the fluctuations in the unbounded plasma, taking account of the space dispersion; the last terms are connected with the presence of the boundary and reflect its influence on the charge density fluctuations in a plasma. The influence of the boundary is most significant in the high-frequency range, where resonances connected with the collective oscillations occur.

The separate terms in eq. 37, which describes the influence of the boundary, have different natures. It is easy to see that the second term, which differs from the first term only by the argument of the delta function, is connected with the volume oscillations and reflects the effect of the strengthening of the fluctuations in consequence of the reflection of the fluctuation waves from the boundary.

In the high-frequency range the spectral distribution of the fluctuations connected with the volume oscillations is characterized by a sharp maximum at the frequencies and wave vectors ω and $\mathbf{k}$ which satisfy the dispersion equation for the plasma oscillation, $\epsilon_l(\omega, k) = 0$:

$$\langle \rho_{k_z} \rho^*_{k_z'} \rangle_{\mathbf{k}_\perp \omega} = \frac{k^2 T}{4\pi\omega} \delta[\epsilon_l(\omega, k)] [\delta(k_z - k_z') + \delta(k_z + k_z')] \tag{38}$$

The space correlation function for the fluctuations connected with the volume oscillations in a half-space of plasma is

$$\langle \rho(\mathbf{r})\rho(\mathbf{r}')\rangle_\omega = \frac{2\pi}{3}\frac{e^2 n_0 k_0^2}{\omega}\left(\frac{\sin k_0 R}{R} + \frac{\sin k_0 R^+}{R^+}\right) \tag{39}$$

where

$$k_0 = \frac{1}{\sqrt{3}a}\sqrt{1 - (\Omega^2/\omega^2)}$$

According to eq. 39, the contribution of the fluctuational waves propagating between points $\mathbf{r}'$ and $\mathbf{r}$ with reflection from the surface of the plasma (the second term in eq. 39) turns out to be essential for distances from the boundary of the order of the Debye radius. The relative value of this term decreases with increase of the distance from the boundary of the plasma proportionally to the ratio R/R^+ and independently of the value of the Debye radius.

The third term in general formula 37 describes the fluctuations connected with the surface oscillation of plasma. In the high-frequency region the spectral intensity of the fluctuations connected with the surface oscillation is characterized by a sharp maximum near the frequencies and wave vectors ω and $\mathbf{k}$, which satisfy the dispersion relationship for the surface oscillations $r_p(\omega, k_\perp) = 1$:†

$$\langle \rho_{k_z}\rho^*_{k_z'}\rangle_{\mathbf{k}_\perp\omega} = -\frac{k_\perp T}{2\pi\omega\tilde{k}_z}\frac{\tilde{\epsilon}}{\epsilon_l(\omega, k)\epsilon_l^*(\omega, k')}\,\delta[1 - r_p(\omega, k_\perp)] \tag{40}$$

The dispersion equation for the surface oscillations $r_p(\omega, k_\perp) = 1$ expresses the equality of the impedances of plasma and vacuum, respectively. In the case of cold plasma the natural oscillation frequency is determined by the well-known relationship

$$\omega \simeq \frac{1}{\sqrt{2}}\,\Omega \tag{41}$$

[The surface oscillations are possible under the condition $(\omega^2/c^2)\,\tilde{\epsilon} - k_\perp^2 < 0$.] The space correlation function of the fluctuations connected with the surface oscillations in a half-space of plasma is essentially different from zero near the boundary of the plasma only and decreases on withdrawal from the boundary more quickly than the space correlation function of the fluctuations connected with the volume fluctuations. All these features are illustrated in Figs. 1 to 3.

The last two terms in eq. 37 take account of the interference between the space and surface fluctuational oscillations. Insofar as the natural frequencies

† The spectral distribution of the fluctuations in a half-space of plasma was derived in the hydrodynamic approximation in Ref. 15, where surface waves were first taken into account.

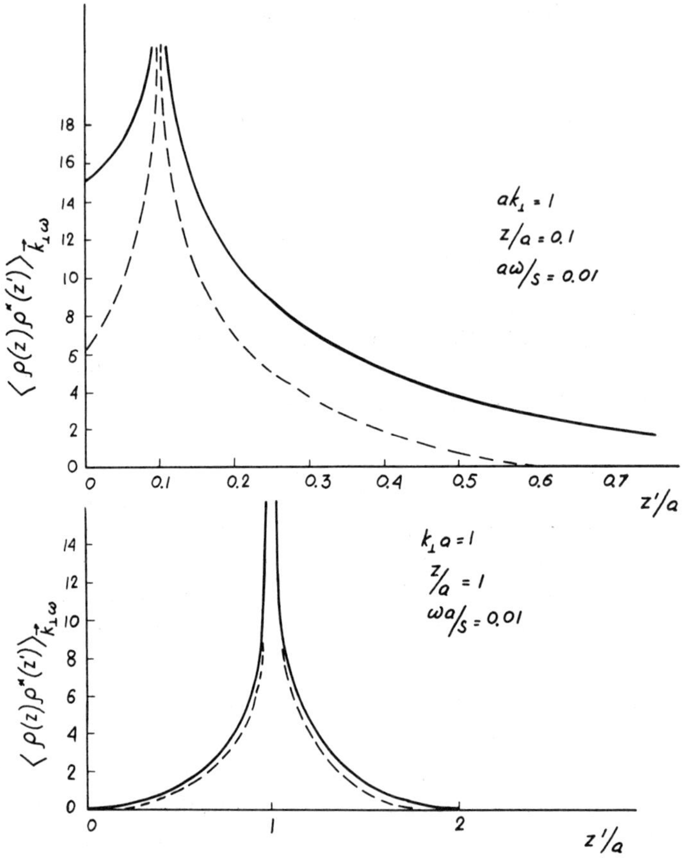

Fig. 1. The correlation function for the charge density fluctuations in the low-frequency range (in units $e^2 n_0/\hbar s$, where s is the thermal velocity). The solid curves correspond to the case of semibounded plasma (plasma fills the half-space $z > 0$), and the dotted lines correspond to the unbounded plasma. The curves are quite different at $z/a = 0.1$, but this difference is barely noticeable at $z/a = 1$. In both cases the correlation is essential only for distances of the order of a Debye-radius.

of the space and surface oscillations do not coincide, we may neglect the contribution of these terms to the total correlation function (eq. 37).

The formula obtained above also makes it possible to take into account the influence of the ion motion on the fluctuation spectrum of a bounded plasma. This influence turns out to be essential, however, only in the low-frequency band.

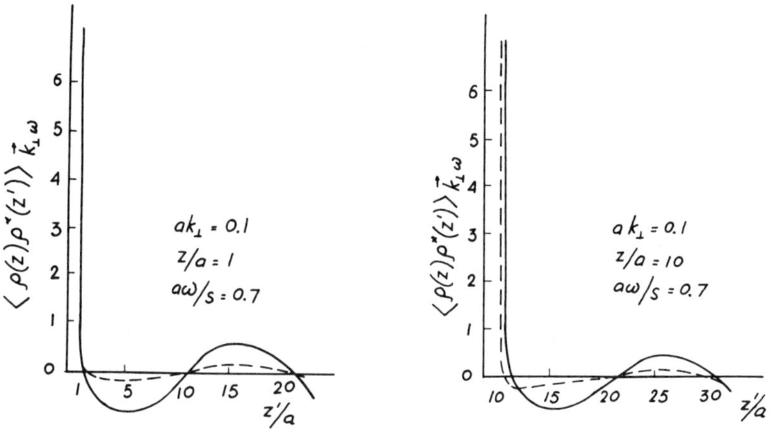

Fig. 2. The correlation function for the charge density fluctuations in the electron plasma in the range of existence of the collective oscillation. The effect of "infiltrating" the correlation on the large distances connected with the wave character of fluctuations is evident. The influence of the boundary is essential even at great distances from the boundary.

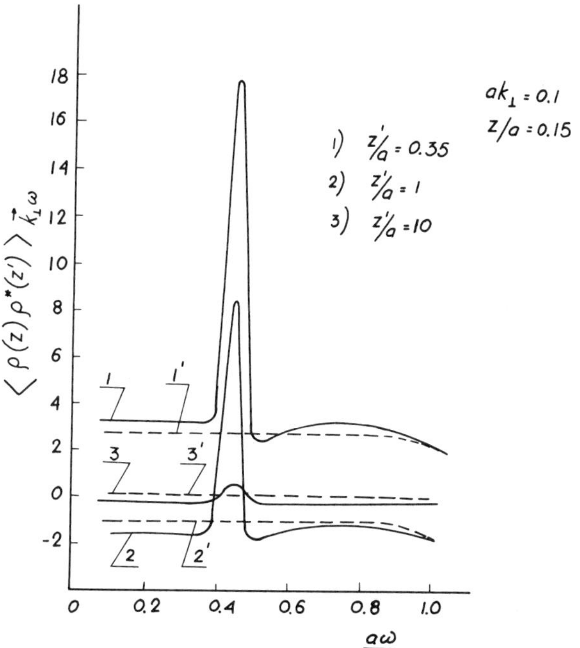

Fig. 3. The frequency spectrum of the charge density fluctuations in a plasma half-space (solid curves) and in an unbounded plasma (dotted curves). The influence of the boundary consists in changing the quantitative value of the fluctuations over the entire frequency range and in producing an additional resonance connected with the possible existence of surface fluctuational waves.

VI. Generalized Fluctuation–Dissipation Ratio

In the process of deriving the fluctuation–dissipation ratio for equilibrium systems (eq. 5), we used the relation between the correlation function for the current fluctuations and the average energy absorbed by the system as a result of dissipation. Using the analogous relationship in the nonequilibrium case, we may generalize the fluctuation–dissipation ratio to nonequilibrium systems, see Ref. 16.

Let us consider first a spatially homogeneous system, and let us suppose that the system is in a nonequilibrium but stationary stable state. The average energy absorbed per unit time is equal to

$$Q = \frac{i\omega}{4\hbar} \sum_{\mathbf{k}} \{\langle j_i j_j \rangle_{\mathbf{k}\omega}{}^{\hbar\omega} - \langle j_i j_j \rangle_{\mathbf{k}\omega}\} A_i(\omega, \mathbf{k}) A_j^*(\omega, \mathbf{k}) \tag{42}$$

where $\langle j_j j_j \rangle_{\mathbf{k}\omega}{}^{\hbar\omega}$ is the correlation function in which the statistical averaging is carried out over the distribution shifted in energy by the value $\hbar\omega$ (we use here the quantum-mechanical formulation of the problem). The correlation function in the quantum case is determined by the expression

$$\langle j_i j_j \rangle_{\mathbf{k}\omega}\, \delta(\mathbf{k} - \mathbf{k}') = \frac{1}{4\pi^2} \sum_{m,n} f_0(E_n) j_i(\mathbf{k})_{nm} j_j^*(\mathbf{k}')_{mn}\, \delta\left(\omega - \frac{E_n - E_m}{\hbar}\right) \tag{43}$$

where indices n and m characterize the state of the system. On the other hand, the absorbed energy may be expressed as in the equilibrium case by the parameters $\alpha_{ij}(\omega, \mathbf{k})$, which characterize the dissipative properties of the system. Comparing these two expressions for Q, we obtain the generalization of the fluctuation–dissipation theorem:

$$\langle j_i j_j \rangle_{\mathbf{k}\omega}{}^{\hbar\omega} - \langle j_i j_j \rangle_{\mathbf{k}\omega} = i\hbar[\alpha_{ji}^*(\omega, \mathbf{k}) - a_{ij}(\omega, \mathbf{k})] \tag{44}$$

This relationship is meaningful for system in thermodynamic equilibrium, as well as for systems in nonequilibrium but stable, stationary states.

It follows directly from eq. 44 that the anti-Hermitian part of the electric susceptibility tensor is determined by the difference of the correlation functions for the noninteracting currents, shifted in energy:

$$(\langle j_i j_j \rangle_{\mathbf{k}\omega}{}^{\hbar\omega})^0 - \langle j_i j_j \rangle_{\mathbf{k}\omega}{}^0 = i\hbar\omega^2 [\mathrm{K}_{ji}^*(\omega, \mathbf{k}) - \mathrm{K}_{ij}(\omega, \mathbf{k})] \tag{45}$$

Therefore, if we know the correlation function for noninteracting currents in a nonequilibrium plasma, we may determine, on the basis of the inversion of the fluctuation–dissipation theorem, the dielectric permittivity tensor:

$$\epsilon_{ij}(\omega, \mathbf{k}) = \left(1 - \frac{\Omega^2}{\omega^2}\right) \delta_{ij} + \frac{2}{\hbar\omega^2} \int_{-\infty}^{\infty} d\omega' \frac{(\langle j_i j_i \rangle_{\mathbf{k}\omega'}{}^{\hbar\omega'})^0 - \langle j_i j_j \rangle_{\mathbf{k}\omega'}{}^0}{\omega' - \omega - i0} \tag{46}$$

The transition to the classical formulation of the fluctuation-dissipation relationship may be achieved by expanding the distribution function $f_0(E - \hbar\omega)$ in a power series in energy $\hbar\omega$. In the limiting case $\hbar \to 0$ we have

$$\frac{\partial}{\partial E} \langle j_i j_j \rangle_{\mathbf{k}\omega}{}^0 = i\omega [\mathrm{K}_{ij}(\omega, \mathbf{k}) - \mathrm{K}_{ji}^*(\omega, \mathbf{k})] \tag{47}$$

where

$$\frac{\partial}{\partial E} \langle j_i j_j \rangle_{\mathbf{k}\omega}\, \delta(\mathbf{k} - \mathbf{k}') \equiv \frac{1}{4\pi^2} \sum_{m,n} \frac{\partial f_0(E_n)}{\partial E_n} j_i(\mathbf{k})_{nm} j_j^+(\mathbf{k}')_{mn}\, \delta\left(\omega - \frac{E_n - E_m}{\hbar}\right)$$

$$= \frac{e^2 n_0}{m\omega} \int d\mathbf{v} \int d\mathbf{v}'\, v_i v_j' \mathbf{k} \frac{\partial f_0}{\partial \mathbf{v}'} W_{\mathbf{k}\omega}(\mathbf{v}; \mathbf{v}')\, \delta(\mathbf{k} - \mathbf{k}') \tag{48}$$

(The state of the particle is denoted by the value of the velocity **v**.) In the case of an equilibrium plasma relationship 47 leads to eq. 9. The expression for the dielectric permittivity tensor in the classical case is as follows:

$$\epsilon_{ij}(\omega, \mathbf{k}) = \left(1 - \frac{\Omega^2}{\omega^2}\right) \delta_{ij} - \frac{2}{\omega^2} \int_{-\infty}^{\infty} d\omega' \frac{\omega' (\partial/\partial E) \langle j_i j_j \rangle_{\mathbf{k}\omega'}{}^0}{\omega' - \omega - i0} \tag{49}$$

It is not difficult to generalize the above result to the case of the spatially inhomogeneous plasma. Here the generalized fluctuation-dissipation relationship takes the form

$$\frac{\partial}{\partial E} \langle j_i(\mathbf{r}) j_j(\mathbf{r}') \rangle_\omega{}^0 = i\omega [\mathrm{K}_{ij}(\omega, \mathbf{r}, \mathbf{r}') - \mathrm{K}_{ji}^*(\omega, \mathbf{r}', \mathbf{r})] \tag{50}$$

where

$$\frac{\partial}{\partial E} \langle j_i(\mathbf{r}) j_j(\mathbf{r}') \rangle_\omega{}^0$$

$$\equiv \frac{e^2 n_0}{m\omega} \int d\mathbf{v} \int d\mathbf{v}'\, v_i \frac{\partial f_0}{\partial v_l'} W_\omega(\mathbf{r}, \mathbf{v}; \mathbf{r}', \mathbf{v}') [(\omega + i\mathbf{v}' \nabla')\, \delta_{lj} - i v_j' \nabla_l'] \tag{51}$$

The dielectric permittivity tensor for the spatially inhomogeneous nonequilibrium plasma is determined by the expression

$$\epsilon_{ij}(\omega, \mathbf{r}, \mathbf{r}') = \delta_{ij}\, \delta(\mathbf{r} - \mathbf{r}') - \frac{2}{\omega} \int_{-\infty}^{\infty} d\omega' \frac{(\partial/\partial E) \langle j_i(\mathbf{r}) j_j(\mathbf{r}') \rangle_{\omega'}{}^0}{\omega' - \omega - i0} \tag{52}$$

By determining the joint probability $W_\omega(\mathbf{r}, \mathbf{v}; \mathbf{r}', \mathbf{v}')$ as a result of solving eq. 17 together with the initial condition (eq. 18) and the corresponding boundary conditions, we may investigate, on the basis of the expressions derived, the electrodynamic properties of nonequilibrium bounded plasma systems.

References

1. H. Callen and T. Welton, *Phys. Rev.*, **83**, 34 (1951).
2. L. D. Landau and E. M. Lifshits, *Electrodynamics of Continuous Media*, Addison-Wesley, Massachusetts, 1960.
3. V. D. Shafranov, *Plasma Physics and the Problem of Controlled Thermonuclear Reactions*, Vol. 4, *Izd. Akad. Nauk SSSR*, Moscow, 1958, p. 416.
4. R. Kubo, *J. Phys. Soc. Japan*, **12**, 570 (1957).
5. H. Nakano, *Prog. Theor. Phys.*, **17**, 145 (1957).
6. M. A. Leontovich and S. M. Rytov, *Zh. Eksp. Teor. Fiz.*, **23**, 246 (1952).
7. A. G. Sitenko, *Electromagnetic Fluctuations in Plasma*, Academic Press, New York, 1967.
8. M. L. Levin and S. M. Rytov, *Theory of Equilibrium Thermal Fluctuations in Electrodynamics*, Nauka, Moscow, 1967.
9. N. N. Bogoliubov, *Problems of the Dynamic Theory in Statistical Physics*, Gostekhizdat, Moscow, 1946.
10. V. V. Tolmachev, *Dokl. Akad. Nauk SSSR*, **112**, 842 (1957).
11. M. A. Leontovich, *Zh. Eksp. Teor. Fiz.*, **5**, 211 (1935).
12. G. Reuter and E. Sondheimer, *Proc. Roy. Soc. (London)*, **A195**, 336 (1948).
13. V. P. Silin and E. P. Fetisov, *Zh. Eksp. Teor. Fiz.*, **41**, 159 (1961).
14. I. P. Yakimenko, *Zh. Eksp. Teor. Fiz.*, **54**, 255 (1968).
15. N. Ya. Kotsarenko and A. M. Fedorchenko, *Ukr. Fiz. Zh.*, **12**, 1531 (1967).
16. A. G. Sitenko, *Ukr. Fiz. Zh.*, **11**, 1161 (1966).

The Kinetic Equations for the Nonideal Gas and the Nonideal Plasma

Yu L. KLIMONTOVICH

Lomonosov State University, Moscow, USSR

Introduction

One of the problems involved in the statistical theory of the nonequilibrium processes is the determination of the conditions under which it is possible to obtain closed equations for the simplest one-particle distribution functions, $f_a(x, t) \equiv f_a(\mathbf{r}, \mathbf{p}, t)$ (here a is a component index). Such equations are called kinetic equations.

By this time, various methods for deriving the kinetic equations for the ideal gas and plasma have been developed. They are described in Refs. 1 to 9. Three kinetic equations are used in the classical theory of gases and plasma: the Boltzmann equation, the Landau equation, and the Balescu-Lenard equation.

These equations may be written in the following form:

$$\left(\frac{\partial}{\partial t} + \mathbf{v}\frac{\partial}{\partial \mathbf{r}} + \mathbf{F}_a \frac{\partial}{\partial \mathbf{p}}\right) f_a(x, t) = \mathcal{I}_a(x, t) \tag{1}$$

Here $\mathbf{F}_a$ is the external force, $\mathcal{I}_a$ the collision integral.

We shall write the expressions for the collision integrals in the Boltzmann, Landau, and Balescu-Lenard equations with spatial uniformity. We write the

Boltzmann collision integral in the Bogoliubov form:

$$\mathcal{I}_a(x_a, t) = \sum_b n_b \int \frac{\partial \phi_{ab}}{\partial \mathbf{r}_a} \frac{\partial}{\partial \mathbf{p}_a} f_a[\mathbf{P}_a(-\infty), t]\, f_b[\mathbf{P}_b(-\infty), t]\, dx_b \tag{2}$$

Here $\mathbf{P}_a(-\infty)$, $\mathbf{P}_b(-\infty)$ are the initial momenta of the two particles, colliding at time t; ϕ_{ab} is the potential interaction energy of the particles.

The Landau collision integral is

$$\mathcal{I}_a(x_a, t) = \sum_b 2e_a{}^2 e_b{}^2 n_b \frac{\partial}{\partial p_{ai}} \int \frac{k_i k_j}{k^4} \delta(\mathbf{kv} - \mathbf{kv}') \times \left(\frac{\partial}{\partial p_{aj}} - \frac{\partial}{\partial p_{bj}}\right) f_a(\mathbf{p}_a, t)\, f_b(\mathbf{p}_b, t)\, d\mathbf{p}_b\, d\mathbf{k} \tag{3}$$

The integration over the values of the vector $\mathbf{k}$ is performed in the limits $k_{\max} \geqslant k \geqslant k_{\min}$. The value $k_{\min}$ is determined by the Debye radius. In transforming eq. 3 to the usual form of Landau collision integral, one must recognize that

$$\int \frac{k_i k_j}{k^4} \delta(\mathbf{kv} - \mathbf{kv}')\, d\mathbf{k} = \frac{\pi}{|\mathbf{v} - \mathbf{v}'|^3} [(\mathbf{v} - \mathbf{v}')^2 \delta_{ij} - (\mathbf{v} - \mathbf{v}')_i (\mathbf{v} - \mathbf{v}')_j] \ln\left(\frac{r_d}{r_{\min}}\right)$$

The Balescu-Lenard collision integral is determined by the expression

$$\mathcal{I}_a = \sum_b 2e_a{}^2 e_b{}^2 n_b \frac{\partial}{\partial p_{ai}} \int \frac{k_i k_j \delta(\mathbf{kv} - \mathbf{kv}')}{k^4 |\epsilon(\mathbf{kv}, \mathbf{k})|^2} \left(\frac{\partial}{\partial p_{aj}} - \frac{\partial}{\partial p_{bj}}\right) f_a(\mathbf{p}_a, t) f_b(\mathbf{p}_b, t)\, d\mathbf{k}\, d\mathbf{p}_b \tag{4}$$

This expression differs from eq. 3 by taking into consideration the polarization of the medium. The polarization of the plasma in formula 4 is taken into account by means of the dielectric constant, $\epsilon(\omega, \mathbf{k})$. For further discussion it is convenient to use the following expression for $\epsilon(\omega, \mathbf{k})$:

$$\epsilon(\omega, \mathbf{k}) = 1 - i \sum_a \frac{4\pi e_a{}^2 n a}{k^2} \int_0^\infty \int e^{-\Delta\tau + i(\omega - \mathbf{kv})\tau}\, \mathbf{k} \frac{\partial f_a}{\partial \mathbf{p}} (\mathbf{p}, t)\, d\tau\, d\mathbf{p} \tag{5}$$

All three collision integrals (eqs. 2, 3, and 4) have the following properties:

1. The collision integrals cancel when the Maxwell distribution is inserted into them.
2. The integral

$$I(t) = \sum_a n_a \int \varphi_a(\mathbf{p}_a) \mathcal{I}_a\, d\mathbf{p}_a = 0$$

when

$$\varphi = 1, \mathbf{p}, p^2/2m_i \tag{6}$$

This property provides for the fulfilment of the conservation laws for the number of particles, total momentum, and total kinetic energy of the particles of all the gas or plasma components.

3. The gas or plasma entropy, expressed by

$$S(t) = -k \sum_a n_a \int \ln(f_a) \cdot f_a \, d\mathbf{p}_a \tag{7}$$

does not decrease with time, that is, $dS/dt \geqslant 0$.

In the Boltzmann, Landau, and Balescu-Lenard equations the interaction between the particles is not taken into account completely. This interaction reveals itself in these equations only in the dissipative processes, for example, in the process of establishing the equilibrium state. The contribution of interaction to the thermodynamic functions is not taken into account. This is seen, in particular, from expression 7 for the entropy, which does not take the correlations into account and is valid, therefore, only for the ideal gas. From the properties of eq. 6 it follows that the full integral energy is not conserved in the process of establishing the equilibrium state, but only the kinetic energy of the particles. The equations of state obtained from the Boltzmann, Landau, and Balescu-Lenard kinetic equations are those of the ideal gas.

A number of authors (see Refs. 1 to 9) have studied the generalized Boltzmann, Landau, and Balescu-Lenard equations. The generalization consists in taking into consideration the spatial nonuniformity in the collision integral.

One such generalization is that of Enskog. He took into account the correlation between the particles (the balls) due to their finite sizes. The generalized Boltzmann equation for the spatially nonuniform gas was obtained by Bogoliubov. For the rigid-ball model the Enskog equation follows from it.

Consideration of the spatial nonuniformity is facilitated when the nonuniformity is weak. In the case of a plasma it means that the distribution function changes little for distances of the order of a Debye-radius. Consideration of the spatial nonuniformity makes it possible to take partially into account the contributions of the interaction to the thermodynamic functions, in particular, the equations of state, energy, and momentum fluxes. However, this still does not take into account the contribution of the correlations due to the interaction to the internal energy and entropy. This is connected with the fact that in the above expressions for the collision integrals the retardation effects are completely neglected. It is assumed, in other words, that the collision integrals at time t are determined by the values of distribution functions f_a, taken at the same moment in time.

Taking into account the retardation effects, along with the nonuniformity in collision integrals, makes it possible to obtain the kinetic equations for the nonideal gas and plasma. To reveal the role of the retardation effect more explicitly, we shall confine our discussion to the case of spatially uniform distribution.

The generalization of the Boltzmann equation (taking into account triple and higher-order collisions) was done by Uhlenbeck and Ford (2) and by Cohen (see, e.g., Cohen's work Ref. 9). However, the contribution of triple collisions, as in the Boltzmann equation itself, was not taken into account completely; the contribution of these collisions to the internal energy and other thermodynamic functions was ignored.

I. Boltzmann Equation for Nonideal Gas in the Binary Collision Approximation

In Ref. 10 the expression for the binary distribution function $f_2(x_1, x_2, t)$ was obtained for the case of a single-component spatially uniform gas. In the first approximation in the small quantity $\tau_{\mathrm{col}}(\partial/\partial t)$.

$$f_2(x_1, x_2, t) = f_1[\mathbf{P}_1(-\infty), t] f_1[\mathbf{P}_2(-\infty), t) - \frac{\partial}{\partial t}\int_0^\infty \tau \frac{d}{d\tau} f_1[\mathbf{P}_1(-\tau), t] f_1[\mathbf{P}_2(-\tau), t]\, d\tau \tag{8}$$

The second term in this expression arises because of taking the time retardation into account.

In accordance with expression 8, the collision integral consists of the two parts:

$$\mathscr{I}(x_1, t) = \mathscr{I}_1 + \mathscr{I}_2 \tag{9}$$

The integral $\mathscr{I}_1$ for the single-component gas coincides with the collision integral of eq. 2 and determines the dissipative processes.

The integral $\mathscr{I}_2$ is expressed by

$$\mathscr{I}_2(x_1, t) = -n\frac{\partial}{\partial t}\int_0^\infty\int \tau \frac{\partial\phi(1,2)}{\partial \mathbf{r}_1}\frac{\partial}{\partial \mathbf{p}_1}\frac{d}{d\tau} f_1[\mathbf{P}_1(-\tau), t] f_1[\mathbf{P}_2(-\tau), t]\, d\tau\, dx \tag{10}$$

This additive contribution to the Boltzmann collision integral changes the properties of this equation.

Thus, instead of eq. 6, we have now

$$I(t) = n\int \varphi(\mathbf{p}_1)\mathscr{I}\, d\mathbf{p}_1 = 0 \tag{11}$$

only when $\varphi = 1, \mathbf{p}$. When $\varphi = p^2/2m$,

$$n\int \frac{p_1{}^2}{2m}\mathscr{I}\, d\mathbf{p}_1 = -\frac{n^2}{2}\frac{\partial}{\partial t}\int \phi(1,2) f_1[\mathbf{P}_1(-\infty), t] f_1[\mathbf{P}_2(-\infty), t]\, d\mathbf{r}_2\, d\mathbf{p}_1\, d\mathbf{p}_2 \tag{12}$$

This property leads to the full energy conservation law. In the binary collision approximation this law has the form

$$\frac{\partial}{\partial t}\left\{n\int \frac{p_1{}^2}{2m} f_1 \, \mathbf{dp}_1 + \frac{n^2}{2}\int \phi(1,2) f_1[\mathbf{P}_1(-\infty),t]\, f_1[\mathbf{P}_2(-\infty),t]\, d\mathbf{r}_2\, d\mathbf{p}_1\, dp_2\right\} = 0 \tag{13}$$

In Ref. 10 the kinetic equation in the binary collision approximation, taking into account the time retardation as a weak nonuniformity, is also given. The corresponding equations of gas dynamics now contain in the thermodynamic functions (the internal energy and pressure) the corrections due to the interaction between the gas particles.

II. H-Theorem for Nonideal Gas in the Binary Collision Approximation

Let us write expression 7 for the density of entropy S_{id} for the ideal single-component gas in the symmetrized form:

$$S_{\text{id}}(t) = -\frac{kn}{2}\int \ln (f_1 f_1)\cdot f_1 f_1 \, \mathbf{dp}_1 \, \mathbf{dp}_2 \tag{14}$$

Here we used the normalization condition $(1/V)\int f_1 \, d\mathbf{r}_1 \, d\mathbf{p}_1 = 1$.

Let S_{int} be the part of the entropy due to the particle-particle interaction. The particle-particle correlation enters into S_{int} under the logorithm [through $\ln (f_2/f_1 f_1)$, directly through the function f_2].

We multiply the equation for $(\partial(f_1 f_1)/\partial t)$ by

$$\left[-\frac{kn}{2}\ln (f_1 f_1) - \frac{kn^2}{2} V \ln\left(\frac{f_2}{f_1 f_1}\right)\right]\frac{dx_1\, dx_2}{V^2}$$

and integrate it over x_1, x_2. Neglecting the term proportional to n^3, in the right-hand side, we obtain

$$-kn\int \ln (f_1)\, f_1 \, \mathbf{dp}_1 - \frac{kn^2}{2}\int \ln\left(\frac{f_2}{f_1 f_1}\right)\frac{\partial (f_1 f_1)}{\partial t}\frac{dx_1\, dx_2}{V} \tag{15}$$

$$= -kn\int \ln (f_1)\, \mathscr{I}_1 \frac{dx_1\, dx_2}{V} - kn\int \ln (f_1)\, \mathscr{I}_2 \frac{dx_1\, dx_2}{V}$$

The integral $\mathscr{I}_1$ coincides with the Boltzmann collision integral, so that

$$-kn\int \ln f_1 \mathscr{I}_1 \frac{dx_1\, dx_2}{V} \geqslant 0 \tag{16}$$

Using for $\mathcal{I}_2$ expression 10, we obtain

$$- nk \int \ln f_1 \mathcal{I}_2 \frac{dx_1\, dx_2}{V} = - \frac{kn^2}{2} \int \ln (f_1 f_1) \frac{\partial g}{\partial t} \frac{dx_1\, dx_2}{V} \tag{17}$$

Here $g = f_2 - f_1 f_1$ is the correlation function.

This expression may be written in another form if one takes into account the fact that the equation for function f_2 in the binary collisions approximation has the form (see eq. 12 in Ref. 10).

$$\frac{\partial f_2}{\partial t} + H(1, 2) f_2 = \frac{\partial}{\partial t} (f_1 f_1) \tag{18}$$

From eq. 18 it follows that

$$\int \ln (f_2) \frac{\partial}{\partial t} g \frac{dx_1\, dx_2}{V} = 0$$

Therefore, instead of eq. 17, we may write

$$- nk \int \ln (f_1) \mathcal{I}_2 \frac{dx_1\, dx_2}{V} = k \frac{n^2}{2} \int \ln \left(\frac{f_2}{f_1 f_1} \right) \frac{\partial g}{\partial t} \frac{dx_1\, dx_2}{V} \tag{19}$$

Taking eq. 19 into account, we write eq. 15 in the form

$$\begin{aligned} &- kn \int \ln (f_1) f_1\, d\mathbf{p}_1 - \frac{kn^2}{2} \int \ln \left(\frac{f_2}{f_1 f_1} \right) \frac{\partial}{\partial t} f_2\, d\mathbf{r}_2\, d\mathbf{p}_1\, d\mathbf{p}_2 \\ &= - kn \int \ln f_1\, \mathcal{I}_1 \frac{dx_1\, dx_2}{V} \end{aligned} \tag{20}$$

Because of

$$\int f_1 \frac{\partial}{\partial t} \ln f_1\, d\mathbf{p}_1 = 0; \qquad \int \frac{\partial}{\partial t} \ln \left(\frac{f_2}{f_1 f_1} \right) f_2 \frac{dx_1\, dx_2}{V} = - \int \frac{\partial g}{\partial t} \frac{dx_1\, dx_2}{V}$$

and taking eq. 16 into account, we obtain (11)

$$\frac{\partial}{\partial t} (S_{\text{id}} + S_{\text{int}}) \geqslant 0 \tag{21}$$

where

$$S_{\text{int}} = - \frac{kn^2}{2} \int \left[\ln \left(\frac{f_2}{f_1 f_1} \right) f_2 - g \right] \frac{dx_1\, dx_2}{V} \tag{22}$$

In the perturbation theory approximation, from eq. 22 we obtain

$$S_{\text{int}} = - \frac{kn^2}{2} \int \frac{g^2}{2 f_1 f_1} \frac{dx_1\, dx_2}{V}$$

In the local thermodynamic equilibrium state the following expression follows from eq. 22:

$$S_{\text{int}} = 2\pi k n^2 \int_0^\infty \left[\frac{\phi}{kT} e^{-(\phi/kT)} + e^{-(\phi/kT)} - 1\right] r^2\, dr$$

From this, for the model of the balls with weak attraction, in the quadratic approximation in $|\phi|/kT \ll 1$, we obtain

$$S_{\text{int}} = -kn^2 b - \pi k n^2 \int_0^\infty \frac{\phi^2(r)}{(kT)^2} r^2\, dr$$

where $b = (2\pi/3) r_0{}^3$ is the van der Waals constant.

III. Kinetic Equation for Nonideal Gas in the Triple Collision Approximation

In the triple collision approximation taking into account the retardation effects, we obtain for the second distribution function

$$\begin{aligned} f_2(x_1, x_2, t) &= S_{-\infty}{}^{(2)}(1,2) f_1 f_1 - \frac{\partial}{\partial t}\int_0^\infty \tau' \frac{d}{d\tau'} S_{-\tau'}{}^{(2)} f_1 f_1\, d\tau' \\ &+ n \int_0^\infty \int S_{-\tau'}{}^{(2)}(1,2)\{\Theta_{13}[S_{-\infty}{}^{(3)}(1,2,3) - S_{-\infty}{}^{(2)}(1,3)] \\ &+ \Theta_{23}[S_{-\infty}{}^{(3)}(1,2,3) - S_{-\infty}{}^{(2)}(2,3)]\} f_1 f_1 f_1\, d\tau'\, dx_3 \\ &- n\frac{\partial}{\partial t}\int_0^\infty d\tau\, S_{-\tau}{}^{(2)}(1,2) \int dx_3 (\Theta_{13} + \Theta_{23}) \\ &\times [\tau S_{-\infty}{}^{(3)}(1,2,3) - \int_0^\infty d\tau'\, S_{-\tau'}{}^{(3)}(1,2,3)] f_1 f_1 f_1 \end{aligned} \tag{23}$$

The insertion of the first and the second terms into the collision integral

$$\mathcal{I} = n \int \frac{\partial \phi_{1,2}}{\partial \mathbf{r}_1} \frac{\partial}{\partial \mathbf{p}_2} f_2(x_1, x_2, t)\, dx_2 \tag{24}$$

leads to the Uhlenbeck-Ford kinetic equation.

The third and the fourth terms take into account the retardation effects due to the binary and triple collisions, respectively.

The collision integral of eq. 24 has the same properties as in the case of binary collisions, with the only difference that, instead of eq. 13, we now obtain the energy conservation law, taking into account the triple collisions, as

$$\frac{\partial}{\partial t}\left[n \int \frac{p^2}{2m} f_1\, d\mathbf{p}_1 + \frac{n^2}{2}\int \phi(1,2) f_2(x_1, x_2, t)\, d\mathbf{r}_2\, d\mathbf{p}_1\, d\mathbf{p}_2\right] = 0 \tag{25}$$

For f_2 here one must use the first two terms of expression 23. In the local equilibrium approximation the expression for f_2 in eq. 25 has the following form:

$$f_2(x_1, x_2, t) = \frac{1}{(2\pi mkT)^3} e^{-[H(1,2)/kT]}$$

$$\times \left(1 + n \int [e^{-(\phi_{13} + \phi_{23})/kT} - e^{-(\phi_{13}/kT)} - e^{-(\phi_{23}/kT)} + 1] \, d\mathbf{r}_3\right) \tag{26}$$

and coincides with the corresponding terms of the density series expansion of the equilibrium function f_2.

IV. Landau Kinetic Equation for Nonideal Plasma

In the pertubation by interaction theory approximation and in the binary collisions approximation, the expression for the function f_{ab} has the form (10)

$$f_{ab}(x_a, x_b, t) = f_a(\mathbf{p}_a, t) f_b(\mathbf{p}_b, t) + \int_0^\infty \frac{\partial \phi_{ab}\,[|\mathbf{r}_a - \mathbf{r}_b - (\mathbf{V}_a - \mathbf{V}_b)\tau|]}{\partial \mathbf{p}_a}$$

$$\times \left(\frac{\partial}{\partial \mathbf{p}_a} - \frac{\partial}{\partial \mathbf{p}_b}\right) f_a(\mathbf{p}_a, t - \tau) f_b, t - \tau) \, d\tau \tag{27}$$

In the first approximation in $\tau_{\text{col}}(\partial/\partial t)$ this expression may be obtained from eq. 8. We insert this expression into the collision integral

$$\mathscr{I}_a = \sum_b n_b \int \frac{\partial \phi_{ab}}{\partial \mathbf{r}_b} \frac{\partial}{\partial \mathbf{p}_b} f_b \, dx_b \tag{28}$$

and then perform the Fourier transformation over $\mathbf{r}_a - \mathbf{r}_b$.

Taking into account the fact that $\phi_{ab}(\mathbf{k}) = 4\pi e_a e_b / k^2$, we obtain

$$\mathscr{I}_a = \sum_b \frac{2}{\pi} e_a{}^2 e_b{}^2 n_b \frac{\partial}{\partial p_{ai}} \operatorname{Re} \int_0^\infty \int \frac{k_i k_j}{k^4} e^{-\Delta\tau - i(\mathbf{k}\mathbf{v}_a - \mathbf{k}\mathbf{v}_b)\tau}$$

$$\times \left(\frac{\partial}{\partial p_{aj}} - \frac{\partial}{\partial p_{bj}}\right) f_a(\mathbf{p}_a, t - \tau) f_b(\mathbf{p}_b, t - \tau) \, d\tau \, d\mathbf{k} \, d\mathbf{p}_b \tag{29}$$

When the retardation is neglected, expression 29 coincides with eq. 3.

The equation with the collision integral (eq. 29) was used by Silin (8) for the description of fast processes. However, the slow time retardation was not taken into account; therefore the nonideality of plasma was not considered.

The kinetic equation 29 with the collision integral 29 has the properties of eqs. 11 and 12, (provided that the summation over plasma components is performed).

The energy conservation law (eq. 12) for plasma may be written in the following form:

$$\frac{\partial}{\partial t}\left(\sum_a n_a \int \frac{p^2}{2m_a} f_a \, d\mathbf{p} + \frac{\overline{\delta\mathbf{E}\,\delta\mathbf{E}}}{8\pi}\right) = 0 \tag{30}$$

The electric-field energy in the Landau approximation is determined by the expression

$$\frac{\overline{\delta\mathbf{E}\,\delta\mathbf{E}}}{8\pi} = \frac{1}{(2\pi)^3} \sum_{ab} \frac{e_a{}^2 e_b{}^2 n_a n_b}{\pi} \int_0^\infty \int e^{-\Delta\tau} \sin(\mathbf{kv} - \mathbf{kv}')\tau$$

$$\times\, \mathbf{k}\left(\frac{\partial}{\partial \mathbf{p}} - \frac{\partial}{\partial \mathbf{p}'}\right) f_a(\mathbf{p}, t-\tau) f_b(\mathbf{p}', t-\tau)\, d\tau \, \mathbf{dp}\, d\mathbf{p}' \tag{31}$$

If we limit ourselves to the first approximation in $\tau_{\text{col}}(\partial/\partial t)$, the retardation in eq. 31 may be neglected when it is inserted in eq. 30.

V. Balescu-Lenard Equation for Nonideal Plasma

The Balescu-Lenard equation differs from the Landau equation by giving fuller consideration to the plasma polarization. In the Landau equation only the static plasma polarization is taken into account by cutting off the limits of integration over $|\mathbf{k}|$ at small values. In the Balescu-Lenard collision integral the dielectric permeability (eq. 5), which itself depends on the distribution function is involved.

In the work of Resibois and Balescu (see Appendix 8 in Ref. 5) the derivation of the plasma kinetic equation with retardation is given. The properties of this equation, however, were not investigated.

Let us consider the kinetic equation for the nonideal plasma, taking into account the dynamical polarization, which is evaluated under two assumptions (13).

1. The second correlation function approximation is considered, that is, g_3, g_4, and so on are equal to zero.

2. The collision integral $\mathcal{J}_a$ is determined by the spectral function of the fast fluctuations of the phase density δN_a and of the field $\delta\mathbf{E}$. This means that only the contribution of fluctuations with correlation times shorter than the relaxation times of the functions f_a is taken into account.

In deriving the kinetic equation we shall use the method described in Refs. 6, 12, and 13. The collision integral for the spatially uniform plasma we write in the following form:

$$\mathcal{I}_a = -\frac{e_a}{n_a}\frac{\partial}{\partial \mathbf{p}}\overline{\delta N_a(\mathbf{r}, \mathbf{p}, t)\,\delta \mathbf{E}(\mathbf{r}, t)} \tag{32}$$

The equations for the fluctuations δN_a, $\delta \mathbf{E}$ in the second correlation approximation (the first assumption) may be written in the form (12)

$$\left(\frac{\partial}{\partial t} + \mathbf{v}\frac{\partial}{\partial \mathbf{r}}\right)(\delta N_a - \delta N_a^{\text{source}}) + e_a n_a \delta \mathbf{E}\frac{\partial f_a}{\partial \mathbf{p}} = 0 \tag{33}$$

$$\operatorname{div}\delta \mathbf{E} = 4\pi \sum_a e_a \int \delta N_a(\mathbf{r}, \mathbf{p}, t)\, d\mathbf{p}$$

The phase density "source" fluctuations $\delta N_a^{\text{source}}(\mathbf{r}, \mathbf{p}, t)$ satisfy the equation

$$\left(\frac{\partial}{\partial t} + \mathbf{v}\frac{\partial}{\partial \mathbf{r}}\right)\delta N_a^{\text{source}}(\mathbf{r}, \mathbf{p}, t) = 0 \tag{34}$$

To determine the correlation $(\delta N_a\, \delta N_b)^{\text{source}}_{\mathbf{r}\mathbf{p}t, \mathbf{r}'\mathbf{p}', t'}$ this equation is solved with the following initial condition ($t = t'$):

$$(\delta N_a\, \delta N_b)^{\text{source}}_{\mathbf{r}\mathbf{r}'\mathbf{p}\mathbf{p}'t't'} = \delta_{ab}\, \delta(\mathbf{r} - \mathbf{r}')\, \delta(\mathbf{p} - \mathbf{p}')\, n_a f_a(\mathbf{r}, \mathbf{p}, t') \tag{35}$$

The fluctuations δN_a, $\delta \mathbf{E}$ are functions of the "fast" and the "slow" time, that is,

$$\delta N_a = \delta N_a(\mathbf{r}, \mathbf{p}, t, \mu t), \qquad \delta \mathbf{E} = \delta \mathbf{E}(\mathbf{r}, \mathbf{p}, t, \mu t)$$

The functions δN_a and $\delta \mathbf{E}$ and their derivatives depend on the slow time only through the slowly varying functions, $f_a(\mathbf{p}, \mu t)$ (the second assumption).

From eqs. 33 for the Fourier transforms over the coordinates and the fast time we obtain

$$\delta N_a(\omega, \mathbf{k}, \mathbf{p}, \mu t) = \delta N_a^{\text{source}}(\omega, \mathbf{k}, \mathbf{p}, \mu t)$$
$$- \frac{e_a n_a}{k^2}\int_0^\infty e^{-\Delta\tau + i(\omega - \mathbf{k}\mathbf{v})\tau}\, \mathbf{k}\frac{\partial f_a(\mathbf{p}, \mu(t-\tau))}{\partial \mathbf{p}}\, \mathbf{k}\, \delta \mathbf{E}(\omega, \mathbf{k}, \mu t)\, d\tau$$

$$\delta \mathbf{E}(\omega, \mathbf{k}, \mu t) = \delta \mathbf{E}^{\text{source}}(\omega, \mathbf{k}, \mu t)$$
$$+ i\sum_a \frac{4\pi e_a^2 n_a}{k^2}\int_0^\infty\!\!\int e^{-\Delta\tau + i(\omega - \mathbf{k}\mathbf{v})\tau}\, \mathbf{k}\frac{\partial f_a(\mathbf{p}, \mu(t-\tau)}{\partial \mathbf{p}}$$
$$\times\, \delta \mathbf{E}[\omega, \mathbf{k}, \mu(t-\tau)]\, d\tau\, d\mathbf{p} \tag{36}$$

Here Δ is the time parameter, separating the fast and slow processes:

$$\omega_{\min}, 1/\tau_{\rm cor} \gg \Delta \gg 1/\tau_{\rm col}$$

In the first approximation in $\tau_{\rm col}(\partial/\partial t)$ from eq. 36 we obtain

$$\delta N_a = \delta N_a{}^{\rm source} - \frac{e_a n_a}{k^2} \int_0^\infty e^{-\Delta\tau + i(\omega - \mathbf{kv})\tau}\, \mathbf{k}\frac{\partial f_a}{\partial \mathbf{p}}[\mathbf{p}, \mu(t-\tau)] \times \left(1 - \tau \frac{\partial}{\partial t}\right) \mathbf{k}\, \delta \mathbf{E}(\omega, \mathbf{k}, \mu t)\, d\tau \tag{37}$$

$$\epsilon(\omega, \mathbf{k}, \mu t)\, \delta\mathbf{E}(\omega, \mathbf{k}, \mu t) = \delta \mathbf{E}^{\rm source} - i \frac{\partial \epsilon}{\partial \omega} \frac{\partial \delta \mathbf{E}}{\partial \mu t} \tag{38}$$

Here

$$\epsilon(\omega, \mathbf{k}, \mu t) = 1 - i \sum_a \frac{4\pi e_a{}^2 n_a}{k^2} \int_0^\infty \int e^{-\Delta\tau + i(\omega - \mathbf{kv})\tau}\, \mathbf{k}\frac{\partial f_a [\mathbf{p}, \mu(t-\tau)]}{\partial \mathbf{p}}\, d\tau\, d\mathbf{p} \tag{39}$$

This expression for the dielectric permeability differs from eq. 5 in that it does not take into account the retardation of the function f_a.

In the same approximation from eq. 38, for the expression for the derivative $\partial\delta\mathbf{E}/\partial\mu t$, we find

$$\epsilon(\omega, \mathbf{k}, \mu t) \frac{\partial \delta \mathbf{E}}{\partial \mu t} = \frac{\partial \delta \mathbf{E}^{\rm source}}{\partial \mu t} - \frac{\partial \epsilon}{\partial \mu t} \delta \mathbf{E} \tag{40}$$

From here

$$\frac{\partial}{\partial \mu t} (\delta\mathbf{E}\, \delta\mathbf{E})_{\omega, \mathbf{k}, \mu t} = \frac{\partial}{\partial \mu t} \frac{(\delta\mathbf{E}\, \delta\mathbf{E})^{\rm source}}{|\epsilon(\omega, \mathbf{k}, \mu t)|^2} \tag{41}$$

and, consequently, in this approximation the energy density of the electric field is expressed by

$$\frac{1}{8\pi} \overline{\delta\mathbf{E}(\mathbf{r}, t)\, \delta\mathbf{E}(\mathbf{r}, t)} = \frac{1}{(2\pi)^4} \int \frac{1}{8\pi} \frac{(\delta\mathbf{E}\, \delta\mathbf{E})^{\rm source}_{\omega, \mathbf{k}, \mu t}}{|\epsilon(\omega, \mathbf{k}, \mu t)|^2}\, d\omega\, d\mathbf{k} \tag{42}$$

Let us now write the expressions for the spectral functions of t, the source fluctuations. They follow from eqs. 34 and 35 and have the form

$$(\delta N_a\, \delta N_b)^{\rm source}_{\omega, \mathbf{k}, \mathbf{p}, \mathbf{p}', \mu t} = \delta_{ab}\, \delta(\mathbf{p} - \mathbf{p}') n_a\, \mathrm{Re} \int_0^\infty e^{-\Delta\tau + i(\omega - \mathbf{kv})\tau} f_a[\mathbf{p}, \mu(t-\tau)]\, d\tau \tag{43}$$

$$(\delta N_a\, \delta\mathbf{E})^{\rm source}_{\omega, \mathbf{k}, \mathbf{p}, \mu t} = \frac{i\mathbf{k}}{k^2} 4\pi e_a n_a 2\, \mathrm{Re} \int_0^\infty e^{-\Delta\tau + i(\omega - \mathbf{kv})\tau} f_a[\mathbf{p}, \mu(t-\tau)]\, d\tau \tag{44}$$

$$(\delta\mathbf{E}\,\delta\mathbf{E})^{\text{source}}_{\omega,\mathbf{k},\mu t} = \sum_a \frac{(4\pi)^2 e_a^2 n_a}{k^2}\, 2\,\mathrm{Re}\int_0^\infty \int e^{-\Delta\tau + i(\omega - \mathbf{kv})\tau} f_a[\mathbf{p}, \mu(t-\tau)]\, d\tau\, d\mathbf{p} \tag{45}$$

We write the collision integral (eq. 32) in the following form:

$$\mathcal{I}_a = -\frac{e_a}{n_a}\frac{1}{(2\pi)^4}\frac{\partial}{\partial \mathbf{p}}\int \mathrm{Re}\,(\delta N_a\,\delta\mathbf{E})_{\omega,\mathbf{k},\mu t}\, d\omega\, d\mathbf{k} \tag{46}$$

The structure of eq. 37 for δN_a shows that the collision integral $\mathcal{I}_a$ may be represented as the sum of the two parts:

$$\mathcal{I}_a = \mathcal{I}_a^{\text{ind}} + \mathcal{I}_a^{\text{source}} \tag{47}$$

The first part is proportional to the spectral density, $(\delta\mathbf{E}\,\delta\mathbf{E})_{\omega,\mathbf{k},\mu t}$ and therefore may be called the induced part. We write the expression for $\mathcal{I}_a^{\text{ind}}$ in the general form, without picking out the first derivative, as

$$\mathcal{I}_a^{\text{ind}} = \frac{e_a^2}{(2\pi)^4}\frac{\partial}{\partial \mathbf{p}}\,\mathrm{Re}\int_0^\infty \int \frac{\mathbf{k}}{k^4}\, e^{-\Delta\tau + i(\omega - \mathbf{kv})\tau}$$
$$\times (\delta\mathbf{E}\,\delta\mathbf{E})_{\omega,\mathbf{k},\mu(t-\tau),\mu t}\, \mathbf{k}\,\frac{\partial f_a[\mathbf{p}, \mu(t-\tau)]}{\partial \mathbf{p}}\, d\tau\, d\omega\, d\mathbf{k} \tag{48}$$

The argument $\mu(t-\tau)$ in the field spectral function corresponds to the first factor, and μt to the second one.

The second term in eq. 47,

$$\mathcal{I}_a^{\text{source}} = -\frac{e_a}{n_a}\frac{1}{(2\pi)^4}\frac{\partial}{\partial \mathbf{p}}\int \mathrm{Re}\,(\delta N_a^{\text{source}}\,\delta\mathbf{E})_{\omega,\mathbf{k},\mu t}\, d\omega\, d\mathbf{k} \tag{49}$$

is determined by the spectral function of phase density fluctuations of the source, $\delta N_a^{\text{source}}$ and the field fluctuation $\delta\mathbf{E}$. With the help of eq. 40 we express $\delta\mathbf{E}$ through $\delta\mathbf{E}^{\text{source}}$ and the derivative $\partial\delta\mathbf{E}^{\text{source}}/\partial\mu t$:

$$\delta\mathbf{E} = \left(1 - i\frac{\partial\epsilon}{\partial\omega}\frac{\partial}{\partial\mu t}\frac{1}{\epsilon}\right)\frac{\delta\mathbf{E}^{\text{source}}}{\epsilon} - i\frac{\partial\epsilon}{\partial\omega}\frac{1}{\epsilon^2}\frac{\partial\delta\mathbf{E}^{\text{source}}}{\partial\mu t} \tag{50}$$

We shall insert eq. 50 into eq. 49 and use formulas 43 throughout 45 for the spectral functions of the source fluctuations. The result is:

$$\mathcal{I}_a^{\text{source}} = \left\{-\frac{4\pi e_a^2}{(2\pi)^4}\frac{\partial}{\partial\mathbf{p}}\int_0^\infty\int \frac{\mathbf{k}}{k^2}\left[\frac{i}{\epsilon^*}\left(1 + i\frac{\partial\epsilon^*}{\partial\omega}\frac{\partial}{\partial\mu t}\frac{1}{\epsilon^*}\right)\right]\right.$$
$$\times 2\,\mathrm{Re}\, e^{-\Delta\tau + i(\omega - \mathbf{kv})\tau} f_a[\mathbf{p}, \mu(t-\tau)]$$
$$\left. + i\frac{\partial\epsilon^*}{\partial\omega}\frac{1}{\epsilon^{*2}}\, 2\,\mathrm{Re}\,(-i)\, e^{-\Delta\tau + i(\omega - \mathbf{kv})\tau}\frac{\partial}{\partial\mu t} f_a[\mathbf{p}, \mu(t-\tau)]\right\} d\tau\, d\omega\, d\mathbf{k} \tag{51}$$

Note that the division of the collision integral $\mathcal{I}_a$ into two parts, one of which is proportional to $(\delta\mathbf{E}\,\delta\mathbf{E})_{\omega,\mathbf{k},\mu t}$, is, of course, the conditional procedure. Actually, with the help of formulas 50 and 40 for $\delta\mathbf{E}$ and $\partial\delta\mathbf{E}/\partial\mu t$, the collision integral may be completely expressed through the first distribution functions f_a, but with the retardation taken into account.

If one neglects the retardation in the collision integral of eqs. 48 and 51, the Balescu-Lenard collision integral follows from these expressions. Without the polarization, but with the retardation taken into account, expression 29 follows from eqs. 48 and 51.

The kinetic equation with the collision integral of eqs. 48 and 51 leads to the energy conservation law in the form given by eq. 30. Now, however, in stead of eq. 31 we have

$$\frac{\overline{\delta\mathbf{E}\,\delta\mathbf{E}}}{8\pi}=\frac{1}{(2\pi)^4}\int\frac{(\delta\mathbf{E}\,\delta\mathbf{E})_{\omega,\mathbf{k},\mu t}}{8\pi}\,d\omega\,d\mathbf{k} \tag{52}$$

where

$$(\delta\mathbf{E}\,\delta\mathbf{E})_{\omega,\mathbf{k},\mu t}=\frac{(\delta\mathbf{E}\,\delta\mathbf{E})^{\text{source}}_{\omega,\mathbf{k},\mu t}}{|\epsilon(\omega,\mathbf{k},\mu t)|^2} \tag{53}$$

The spectral function $(\delta\mathbf{E}\,\delta\mathbf{E})^{\text{source}}_{\omega,\mathbf{k},\mu t}$ is determined by expression 45.

In the local statistical equilibrium state the expression for the spatial spectral function of the field follows from eq. 53:

$$(\delta\mathbf{E}\,\delta\mathbf{E})_{\mathbf{k},\mu t}=4\pi\,\frac{kT}{1+r_d{}^2k^2}$$

The following spatial correlation function corresponds to this expression:

$$g_{ab}(\mathbf{r})=-\,\frac{e_a e_b}{kT}\,\frac{e^{-(r/r_d)}}{r}$$

which describes the Debye correlation of the plasma particles.

Thus the kinetic equation with the collision integral of eqs. 48 and 51 is valid for the nonideal plasma in the first approximation in the plasma parameter.

VI. Kinetic Equations for Nonideal Plasma in the Plasmon Approximation

Expressions 48 and 51 through 53 take into account the contribution of all the fast fluctuations with correlation times shorter than the relaxation times of the functions f_a, that is, shorter than the collision times. Among these fluctuations, one may distinguish those the frequencies of which are connected with $\mathbf{k}$ by the dispersion equation:

$$\text{Re}\,\epsilon(\omega,\mathbf{k})=0$$

and with the decay rate:

$$\omega_k \gg \gamma_k \gg 1/\tau_{\text{col}}$$

that is, pick out the plasmon contribution.

The kinetic equations for the nonideal plasma in the plasmon approximation were considered for the first time in Ref. 6. In this approximation only the plasmon energy contribution is taken into account in eq. 52. The corresponding kinetic equations are discussed in detail in Chapter 5 of Ref. 6.

In describing plasma processes the so-called quasilinear approximation plays a special role. This is considered in detail in the works of Vedenov (14), Kadomtsev (15), and Zavoisky and Rudakov (16), as in Section 58 of the book by Silin (8) and in other publications.

VII. Conclusion

All that has been said above, of course, does not exhaustively solve the problem of deriving the kinetic equations of the nonideal gas and nonideal plasma. For example, it is interesting to derive the kinetic equations for the nonideal plasma in strong fields. The equations for turbulent plasma are also the equations for the nonideal plasma.

In developing the theory of the dense plasma, there arises also the necessity to investigate the nonideal plasma, taking into account bound states.

References

1. N. N. Bogoliubov, *Problems of a Dynamical Theory in Statistical Physics*, Gostekhizdat, Moscow, 1946; *Selected Papers*, Kiev, 1970.
2. G. E. Uhlenbeck and G. W. Ford, *Lectures in Statistical Mechanics*, 1963.
3. K. P. Gurov, *The Foundations of Statistical Theory*, Nauka, Moscow, 1966.
4. I. Prigogine, *Nonequilibrium Statistical Mechanics*, Interscience, New York, 1963.
5. R. Balescu, *Statistical Mechanics of Charged Particles*, Interscience, New York, 1963.
6. Y. L. Klimontovich, *The Statistical Theory of the Nonequilibrium Processes in Plasma*, Pergamon Press, New York, 1967.
7. D. N. Zubarev, *The Nonequilibrium Statistical Thermodynamic*, Nauka, Moscow, 1971.
8. V. P. Silin, *Introduction to the Kinetic Theory of Gases*, Nauca, Moscow, 1971.
9. W. Brittin, Ed., *Lectures in Theoretical Physics*, Vol. IX: *Kinetic Theory*, Gordon and Breach, New York, 1967.
10. Y. L. Klimontovich, The Boltzmann Kinetic Equation for the Nonideal Gas, *Zh. Eksp. Teor. Fiz.*, **60**, 1352 (1971).
11. Y. L. Klimontovich, The Boltzmann H-Theorem in Binary Collisions Approximation (in press).

12. Y. L. Klimontovich, The Statistical Theory of Atoms-radiation Interaction, *Sov. Phys. Usp.*, **101**, 577 (1970).
13. Y. L. Klimontovich, The Kinetic Equations for the Nonideal Plasma (in press).
14. A. A. Vedenov, *The Theory of Turbulent Plasma*, Moscow, 1965.
15. B. B. Kadomtsev, The Plasma Turbulence, *The Problems of Plasma Theory*, Vol. 4, Atomizdat, Moscow, 1964.
16. E. K. Zavoisky and L. I. Rudakov, The Collective Processes in Plasma and Turbulent Heating, Znanie, Moscow, 1967.

Statistical Theory of Equilibrium Systems of Charged Particles

I. R. YOUKHNOVSKY

Institute for Theoretical Physics, Academy of Sciences of the Ukrainian SSR, Lvov, USSR

Introduction

The general statistical theory of the system of charged particles in equilibrium is in fact a universal problem, which encompasses a wide range of phenomena, including first the theory of Fermi systems of degenerate electrons in metals and charged Bose gases, then the theory of electrolytes, where classical notions may be used, and finally the statistical theory of the high-temperature plasma in equilibrium.

Each of these theories has its own special difficulties. In the theory of low-temperature plasma, it is necessary to take quantum effects into account. In the theory of electrolytes we have to deal with ions of more or less complicated electrostatic structure, and ones that exist only because of the presence of a great number of dipole molecules of solvent. In high-temperature plasma we have bound states of nuclei and electrons and interactions between excited ions. From a quantum point of view the electron subsystem needs particular attention because of the small mass of each particle.

The equilibrium statistical theory may be developed when the free energy of the system or the hierarchy of the distribution functions is known. Distribution functions depend on the thermodynamical variables (temperature, volume, number of particles of types a, b, etc.), on the specific parameters of particles such as charges $z_a e, z_b e, \ldots$ and masses $m_a, m_b, \ldots$, and on the Planck constant $\hbar$. Different combinations having dimensions of length and energy can be constructed from these quantities. The ratios of these lengths are the parameters of the system. If some of the parameters are small, one can construct a particular model of the system. Such parameters are $\lambda/\langle r\rangle$, λ/r_d, and λ/d, where $\lambda = \sqrt{\hbar^2/m\theta}$ is the de Broglie wavelength of a particle with mass m, $\langle r\rangle$ is average distance, $r_d \simeq \sqrt{\theta/N4\pi e^2}$ is the screening radius, θ is average energy per particle, and ze is the charge of the particle $d = e^2/\theta$. The conditions under which these parameters are equal to unity divide the N/V, θ plane into regions where quantum or classical statistics is valid.

In addition there are plasma regions for the systems of charged particles, where expansion of the thermodynamical functions in the $\langle r\rangle/r_d$ parameter is valid. The region where $\langle r\rangle/r_d \ll 1$ is that of the Debye approximation. In that region the free energy consists of an ideal part and a Debye term. In the case of absolute zero temperature, the Debye term was calculated by K. Brueckner; for high temperatures, by Debye. This approximation, termed the self-consistent field approximation, now is widely adopted in the theory of charged particles. It is correct for very high temperatures and very low densities or for very low temperatures and very high densities of charged particles.

In the classical region, where $\langle r\rangle/r_d < 1$, Bogolubov's plasma parameter expansion is valid, for equilibrium state as well as nonequilibrium state theories (1). But in this method the influence of short-range forces is not taken into account (1, 2).

Success has been achieved by means of the collective variables method, in which the Coulomb interaction is considered in a space of collective variables, and the short-range interaction in a space of individual coordinates with the help of functional expansions. That method also was originated by N. Bogolubov.

As a result a virial series for the free energy and for the binary distribution function has been obtained (3, 4, 5):

$$F_{ab} = \exp\left(g_{ab} - \frac{\varphi_{ab}}{\theta}\right)\left\{\quad - \quad + \cdots\right\}$$

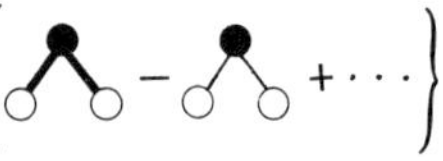

where the function $\exp(g_{ab} - \varphi_{ab}/\theta) - 1$ is denoted by thick line, φ_{ab} is a potential of short-range forces, $-\theta g_{ab}$ is the screening potential, the function g_{ab} is denoted by a thin line, and every field vertex is related to the sum over species $\Sigma_a N_a/V$ and the integral over coordinates $\int d\mathbf{r}$.

Generalization of this method in the quantum case led to the method of displacements and collective variables (SKP method), (6).

The main point of this method consists in splitting-off of the symmetry-independent operator which is connected with the interaction between screened packets from the statistical operator:

$$\exp -\beta H = (\exp U)\cdot B$$

where U and B are determined by the equations

$$\frac{\partial U}{\partial \beta} = -V + \sum_{i=1}^{N} \frac{\hbar^2}{2m}(\Delta_i U + \nabla_i U \nabla_i U)$$

$$B = T \exp\left(\int_0^\beta \sum_{i=1}^{N} \frac{\hbar^2}{2m}(\Delta_i + 2\nabla_i U \nabla_i)\, d\beta'\right)$$

These equations were named the transformations of displacements. Here U is a symmetric function of the coordinates of particles and therefore may be expressed in the collective variables, . . ., ρ_k, . . .:

$$\rho_k = \frac{1}{\sqrt{N}} \sum_{i=1}^{N} \exp -i\mathbf{k}\mathbf{r}_i$$

the ρ_k are the Fourier components of the density.

In these variables U may be written in the form of a power series:

$$U = a_0 + \tfrac{1}{2}\sum_k a_2 \rho_k \rho_{-k} + \frac{1}{3!} \sum_{k_1+k_2+k_3=0} a_3 \rho_{k_1}\rho_{k_2}\rho_{k_3} + \cdots$$

This series is put into the equations for U. Then we come to a chain of coupled equations for $a_0, a_2, a_3, \ldots$, etc.; a_2 describes the two-particle interaction. The equation for a_2 includes an essential nonlinear term.

In the papers of M. Vavrooh the SKP (Smestchenij-Koljektivnych-Peremennych) method was generalized to the case of the two-species system of electrons and of ions (7). In the SKP method a new characteristic length, an amplitude of collective oscillations, $l = (\hbar/m\omega_0)^{1/2}$, appears in a natural way, where $\omega_0 = [k^2\phi(k)Nm^{-1}V^{-1}]^{1/2}$ is the Fourier transform of the potential of interaction forces. If l is taken as a unity of length, all integrals in U become dimensionless. The ratio $\langle r\rangle/l$ is the expansion parameter in the SKP method.

We described the methods by means of which the thermodynamical functions and distribution functions of the systems of charged particles were computed.

I. Classical System of Charged Particles

First let us consider a statistical theory of classical systems of charged particles. In works prepared by the author in association with M. Golovko,

A. Necrot, and V. Vysotchansky, the many-species systems of particles, each of a complex electrostatic structure, have been considered (5, 8–11).

To each particle species a a complex multipole charge is attributed. This charge is proportional to the particle structure factor:

$$Q_a(k) = \int \psi_0^x(\ldots r_i \ldots) \exp\left[\sum_{\mu=1}^{y_a} z_\mu e^{i\mathbf{kr}_\mu} - \sum_{\nu=1}^{x_a} e^{i\mathbf{kr}_\nu}\right] \psi_0(\ldots r \ldots) \prod_i d\mathbf{r}_i$$

where y_a is the number of nuclei, x_a is the number of electrons, and ψ_0 is the ground-state function of the electrons. The charge $Q_a(k)$ may be presented in the form of a multipole expansion.

The system under consideration is described by a statistical sum:

$$Z = S_p \exp - \beta(H_0 + V) \text{ where } H_0 = -\frac{\hbar^2}{2m} \Sigma_i \Delta_i$$

is the energy of noninteracting particles. A transition from the cartesian coordinates to an infinite number of collective variables is accomplished by means of a rigorous Jacobian (12).

When the integration in new variables is performed, screening appears automatically.

For $\ln Z$ we get a typical virial expansion of the system of charged particle. The free energy consists of an ideal part, the Debye free energy for the charged particles, and a virial series. The virial series is a sum of group integrals. Sub-integral functions consist of the products of Mayer's functions and screened electrostatic potentials. Each of Mayer's functions contains a short-range force potential and a screened electrostatic potential.

The form of the screening potential of the system of charged particles, $g_{ab}(r)$, is very important:

$$g_{ab} = Q_a(\nabla)Q_b(\nabla) \frac{1}{\epsilon r} (\exp - C)(1 + A(r) \exp - Dr)$$

Here C, D, and $A(r)$ are known functions of temperature, concentration, and the parameters of particles. This form of potential g_{ab} is a result of the screening caused by electrostatic interactions. In the case of small concentrations this influence can be considered differentially for each species electrostatic interaction.

The Coulomb force screening consists in substitution of the potential $1/r$ by the potential $(1/r) \exp -Kr$, where

$$\mathrm{K} = r_d^{-1} = \left[\frac{V}{N} kT\left(4\pi \sum_c z_c e^2 n_c\right)^{-1}\right]^{-1/2}$$

The screening caused by dipole interactions introduces a dielectric constant:

$$Q_a Q_b \frac{1}{r} \to Q_a Q_b \epsilon_{\text{scr}}^{-1} \frac{1}{r}$$

Here ϵ_{scr} is a part of the dielectric constant of the system (screening part):

$$\epsilon_{\text{scr}} = 1 + 4\pi \sum_c \frac{N_c}{V} \left(\frac{p_c{}^2}{3kT} + \ldots \right)$$

Quadripolar force screening generates a quadripolar packet

$$\frac{1}{r} \to \frac{1 - \exp -r/q}{r}$$

where q is the radius of the quadripolar packet:

$$q^2 \simeq \frac{4\pi}{3} \frac{1}{k^4} \sum_{a,i} \{(3\Theta V)^{-1} \langle 0 \mid (\Theta_a : kk)^2 \mid 0 \rangle + \ldots\}$$

where Θ_a is quadripolar momentum of a particle.

Thus in the theory of electrostatic screening we have a succession of radii: Debye's radius r_d, quadripolar radius q, octopolar radius, and so on. The first is a radius of screening; the others are radii of packets. There is an important difference between them. The r_d is inversely proportional to the square root of the concentration, and the radius of n-polar screening, $n \geqslant 2$, is proportional to the root of $2n$ degree of a sum of two expressions: one of them depends on $N/V\theta$ and the second is connected with the corresponding quantum transition and does not depend on temperature.

When concentrations of particles are high and ions have quadripolar momentum, the radius of screening is different from Debye's radius. It does not depend on the concentration of ions and is proportional to the mean value of the quadripolar momentum.

Often in physics papers the following form of potential is discussed:

$$\frac{1}{\epsilon r} [1 - A(r) \exp -\alpha r]$$

Indeed, if multipolar screening is taken into account, such a form may be obtained; ϵ, $A(r)$ and α are definite functions of the temperature and of the concentration. In addition, we can say that $A(r)$ cannot be a polynomial in powers of r.

As mentioned, a virial series for the free energy and binary functions of systems of particles with complex electrostatic interactions has been obtained. There are some methods of summation of the group and their classification. But first we have to investigate the necessity of considering the multipolar structure of particles. This question was regarded from the point of view of the stability conditions of the system and from the standpoint of the behavior of the binary distribution function.

As was shown (8), calculation of the multipolar screening strengthens the stability of the system. The quadrupolar screening has an influence on the form of the activity coefficient curves, bringing them into agreement with the experimental ones.

It is worth mentioning that, when we use a potential of type $1/r \times (1 - \exp - \alpha r)$, where α is a constant, the stability conditions become worse in comparison with the pure Coulomb potential. The situation changes diametrically when we take $\alpha = q^{-1}$, where q is the quadrupolar screening radius determined previously. This is so because the derivative $\partial/\partial V$ in the expressions $\partial p/\partial V$ also varies with the function q^{-1}.

II. Mixed Ion-Dipole Systems

We shall now consider the behavior of the curves of the binary distribution function for mixed ion-dipole systems. We need to make a detailed investigation of the virial expansion of the binary distribution function.

The contributions from the second and third virial coefficients have been computed. The calculations were made by A. Necrot for a system of dipoles and by V. Vysotchansky, M. Golovko, and this author for a mixed system of ions and dipoles. Short-range forces were taken into account.

In the system of particles considered there are three kinds of interactions: short-range, Coulomb, and dipolar interactions. The investigated binary distribution functions contain screening potentials. The importance of screening is appreciable in the region of small concentrations of ions and diminishes with an increase in ion concentration.

The influence of concentration is apparent in two ways: through the factors N_c/V, which appear before all the group integrals, and through the radius of screening, r_d. For investigating the influence of the ion concentration on the properties of the considered system, all other parameters were taken constant. We shall take, for the radius of short-range forces, $b = 2$ Å, and, for dipole concentration, $(N_s/V)(4\pi/3)\sigma^3 = 0.5$.

Let us consider the graphs in the case in which the concentration of ions is 5 moles/liter, and the dipole subsystem is present. The drawn curves describe the following approximations for $F_{+-}(r)$: (*a*) the inclusion of the second virial coefficient, and (*b*) the inclusion of the third virial coefficient.

In Fig. 1 there are two curves for approximation *a*. Curve 1 describes the function

$$F_{+-}^{(2)} = \exp\left(\frac{d}{r} \exp - \mathrm{K}r\right)$$

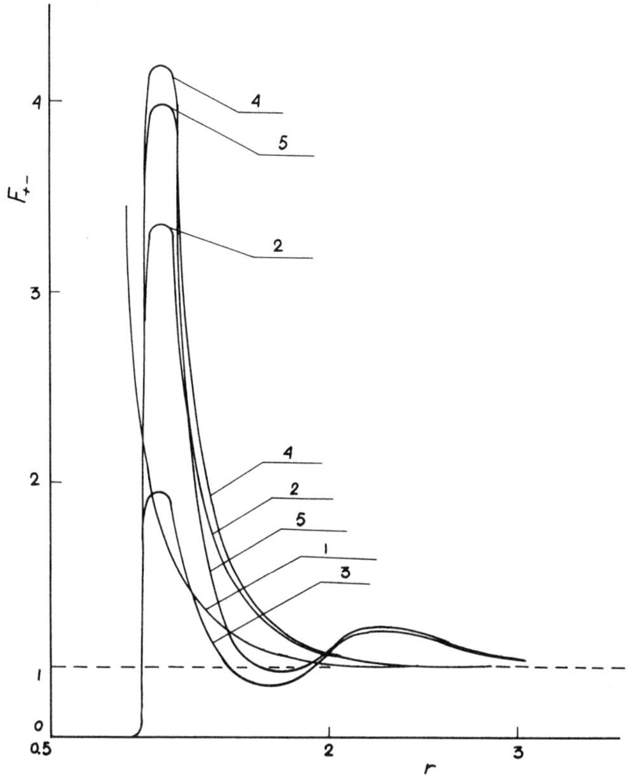

Fig. 1. Binary distribution function variation with distance. Curves 1 and 2 describe results without the third virial coefficient being included. Curve 1 has a simple screened coulomb law and curve 2 includes short-range forces. Curve 3 shows the Lenard-Jones interaction. Curve 4 is for the pure ion system with a dissolvent included only by dielectric constant. Curve 5 is for an ion-dipole system with all interactions included.

Curve 2 describes the same function, but with short-range forces taken into account

$$F_{+-}^{(2)} = \exp\left\{-\frac{4}{T^*}\left[\left(\frac{\sigma}{r}\right)^{12} - \left(\frac{\sigma}{r}\right)^{6}\right] + \frac{d}{r}\exp -Kr\right\}$$

Only one maximum is located at point $r/\sigma = \sqrt[6]{2}$ (at that distance the potential energy reaches its minimum).

There are three curves of type b. Curve 3 describes F_{+-} for the system with the Lenard-Jones interactions.

The binary distribution function for the pure ion system is shown by curve 4. Dissolvent is taken into account only by dielectric constant ϵ_{scr}. A second maximum is absent.

The differences arise when the dipole subsystem is taken into account in detail. Curve 5 describes F_{+-} for the ion-dipole system, when all interactions are taken into account. The presence of the second maximum shows that there is a near order in the disposition of two ions and that a dipole can be located

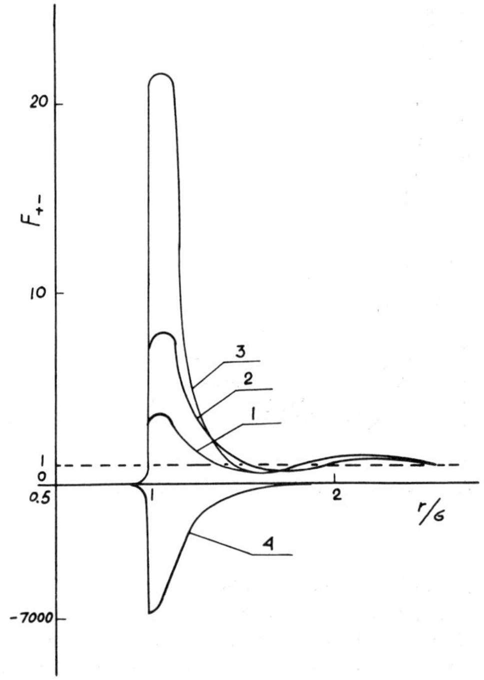

Fig. 2. Ion-Dipole binary distribution function for various ion concentrations; (1) 5 moles/liter; (2) 2.5 moles/liter; (3) 1 moles/liter; (4) 0.01 moles/liter.

between two ions. Hence the dipoles help in ordering the ionic subsystem. Also the altitude of the first maximum is decreased.

A situation analogous to that considered in regard to the dipole-dipole binary distribution function arises during the analysis of the ion-dipole distribution function (11).

The altitudes of the first and the second maxima change, depending on the orientation of the moments of two dipoles or of a dipole and an ion.

A paradoxal situation occurs when the concentration of ions decreases and the concentration of dipoles remains constant: The influence of virial coefficients becomes stronger, and the altitude of the first maximum increases.

This is shown in Fig. 2. Curves 1, 2, 3, and 4 describe the binary distribution function for the following concentrations of ions: 5, 2.5, 1, and 0.01 moles/liter. The concentration of dipoles is $(4\pi\sigma^3/3)(N/V) = 0.5$. Therefore, if we let the ionic concentration tend to zero, two physical phenomena arise: the contribution to the binary distribution function from the third virial coefficient becomes large, negative, and in absolute value much greater than the contribution from the second virial coefficient.The altitude of the first maximum increases.

The virial series becomes unsuitable for small concentrations of ions. Both parts of the group integrals (ion-dipole group integrals and pure ion-ion group integrals) lead to unphysical results. The first unphysical result can be avoided by means of the replacement of the virial series by the plasma parameter series; the second one is eliminated by correcting the potential of the ion-ion and ion-dipole interactions. Some new functions describing indirect interaction between two ions and dipoles are added to this potential.

Therefore, when the radius of screening is large, the plasma parameter expansion and the osmotic theory for an ionic system are suitable. We shall consider the main features of the two theories. They were developed by the author in association with M. Golovko and V. Vysotschansky.

III. Plasma Parameter Expansion

This expansion was proposed by N. Bogolubov for systems with Coulomb interaction. In Bogolubov's work the first approximation to the binary distribution function was computed directly. The higher approximations have been found with the help of Bogolubov's method as presented in Ref. 2. It was shown that the binary distribution function in the nth approximation in the plasma parameter expansion is a sum of terms, which start with $g_{ab}{}^n/n!$ and finish with some irreducible cluster integrals, including $2n$ field vertices and two fixed vertices, a and b. Such integrals appear as a result of the functional differentiation of the $2n + 2^{nd}$ virial coefficient of the free energy. Hence the transition from the virial expansion to the plasma parameter expansion requires essential reconstruction of the virial series.

In the proposed method, the usual form of plasma parameter expansion is retained, but the "complex" field vertices and fixed vertices are introduced. The expression for the binary distribution function to the first order in the plasma parameter is as follows:

$$F_{ab} = F_{ab}^{\mathrm{sh}} + v^x \left\{ G_{ab}^{(0)} F_{ab}^{\mathrm{sh}} + \sum_c \frac{N_c}{V} \int G_{ac}^{(0)} (F_{abc}^{\mathrm{sh}} - F_{ab}^{\mathrm{sh}} F_{ac}^{\mathrm{sh}})\, d\mathbf{r}_c \right.$$
$$\left. + \sum_{cd} \frac{N_c}{V} \frac{N_d}{V} \int G_{cd}^{(0)} (F_{abcd}^{\mathrm{sh}} - \cdots)\, d\mathbf{r}_c\, d\mathbf{r}_d \right\} + v^{x\,2} \{\ldots\} + \cdots$$

where $v^x G_{ab}^{(0)} = g_{ab}$, and $F_{ab}{}^{\mathrm{sh}}$ and so forth denote the distribution functions only for short-range forces. Functions $F_{ab}{}^{\mathrm{sh}} \ldots$ may be expanded in the usual way in the degree of concentration. In particular, if we put

$$F_{ab}^{\mathrm{sh}} = F_{ab}^{\mathrm{sh}\,(0)} = e^{-\varphi_{ab}/\theta} = \begin{cases} 0, & r < \sigma \\ 1, & r \geqslant \sigma \end{cases}$$

then

$$F_{ab}^{(r)} = \exp \frac{-\varphi_{ab}}{\theta} \Biggl\{ 1 + v^x G_{ab}^{(0)} \Biggl[1 - \sum_c \frac{N_c}{V} \frac{e_c}{e_a} e^{-\varphi_{ac}(r)/\theta} \frac{4\pi\sigma^3}{3}$$

$$+ \sum_{c,d} \frac{N_c N_d}{V^2} \frac{e_c e_d}{e_a e_b} e^{-\varphi_{cd}(r)/\theta} \left(\frac{4\pi\sigma^3}{3} \right)^2 + \cdots \Biggr] + v^{x2} [\cdots] + \cdots \Biggr\}$$

where σ is the particle's radius, $v^x = V/Nr_d{}^3$.

The decrease of the altitude of the first maximum is due to the change in the form of the potential of interaction and to the influence of the third particle, belonging to the subsystem of dipoles, on the interactions between two ions. Hence we shall put the binary distribution function in the form:

$$F_{ab}(r) = \exp \left\{ g_{ab} - \frac{\varphi_{ab}}{\theta} + \right.$$  $$\left. + \cdots \right\} \{1 + \cdots\}$$

where the diagram  marks the contribution of the third virial coefficient of the ion-dipole correlation to the interaction between two ions, a and b. This contribution acts essentially on the altitude of the first maximum of the curve of the function F_{+-}. The latter circumstance is connected with the osmotic theory of electrolytes.

IV. An Osmotic Theory of Systems of Charged Particles

The problem is to find an effective potential of interaction between the particles of the first component (dissolved substance) when the distribution of the particles belonging to the second component (dissolvent substance) is arbitrary. This problem was solved in studies by M. Golovko and the present author.

We shall integrate the Gibbs function in phase space of particles of the dissolvent:

$$\int \exp -\beta [H_{ii} + H_{id} + H_{dd}] \, d\Gamma_d = \int \exp -\beta H_{ii} \left\{ \prod_i f_i \right\} d\Gamma_d$$

Implementing an exact integration and introducing the distribution functions for the dipolar subsystem, we obtain a general expression for an effective interaction potential between ions in a polar medium. This potential consists of the sum of many-particle interactions:

$$U = \sum_{a,i} U_a(r_i) + \sum_{\substack{ab \\ i<j}} U_{ab}(r_{ij}) + \sum_{\substack{a,b,c \\ i,j,l}} U_{abc}(r_i, r_j, r_l) + \cdots$$

In schematic designation $U_a(r_i)$ is the sum of diagrams

$$\frac{U_a(r_i)}{N} =$$

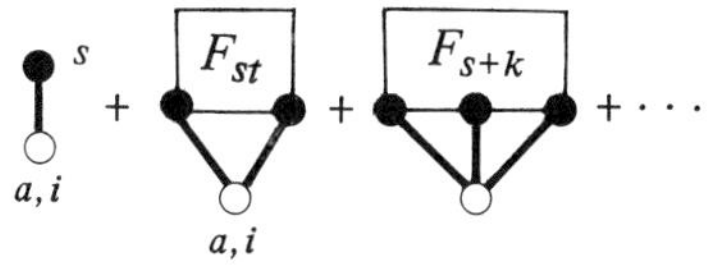

and characterizes the energy of solvation of one ion when the ionic concentration goes to zero. Each line describes the function

$$f_{as}(r) = \exp\left[-\frac{\varphi_{as}(r)}{\theta} - \frac{\phi_{as}(r)}{\theta}\right] - 1$$

Here a is the species of ions, s is the index of dipolar particles, φ_{as} is the short-range force potential, ϕ_{as} is the potential of interaction between ion and dipole, and the diagram  denotes two field vertices and the binary distribution function of the dipole subsystem.

The term $U_{ab}(r_{ij})$ describes a potential of two-particle interaction. We can give a schematic form for $U_{ab}(r_{ij})$:

$$U_{ab}(r_{ij}) = \frac{Z_a Z_b e^2}{r_{ij}} - \Theta$$

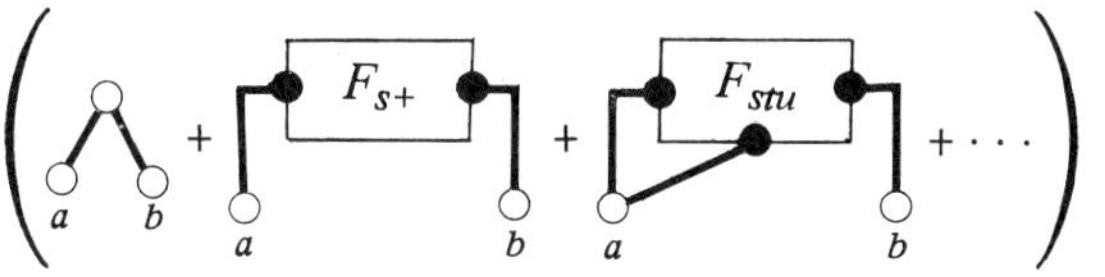

or an expression in the form of a Fourier series:

$$U_{ab}(r) = \sum_{\mathbf{k}} \frac{4\pi e^2}{k^2} \frac{1}{\epsilon(k)} e^{i\mathbf{kr}}$$

Because of the fact that the integration over phase space of the dipoles has been carried out, the potential $U_{ab}(r_{ij})$ is a spherically symmetric function. Therefore the dielectric function $\epsilon(k)$ is an even function of k. Hence we can expand $\epsilon(k)$ or $\eta(k) = \epsilon^{-1}(k)$ in a power series in k^2.

For small k such expansion leads to an expression for the dielectric constant of the dissolvent and for the values of the multipolar packets. The expressions for the dielectric constant

$$\eta(0) = \frac{1}{\epsilon(0)} = 1 + 4\pi \frac{N p^2}{V 3\Theta} \left[1 + \frac{N}{4\pi V} \int \cos \vartheta_{12} (F_{12} - 1)\, d\Gamma_1\, d\Gamma_2 \right]$$

and for the quadrupolar moment of ions

$$eZ_a \eta^1(0) = \frac{1}{45} \pi \frac{N}{V} p \left(\frac{eZ_a p}{\Theta} \right)^2$$

$$\times \left\{ 1 + 4\pi \frac{N}{V} \int \cos \vartheta_{12} (\cos \vartheta_1 + \cos \vartheta_2)(F_{12} - 1)\, d\Gamma_1\, d\Gamma_2 \right\}$$

have been found. It is known that such forms lead to stability of the system. The dependence of the dielectrical constant and of the multipolar packets on the concentration of the ions is connected with the many-particle interactions. An expression for such interactions is obtained.

V. Pair Correlation Function for an Electron Gas (13)

The pair correlation function is a very sensitive physical characteristic of the state of systems of interacting particles.

Many authors, including T. Gaskell, F. Brouers, K. S. Singui, and M. P. Tosi, were concerned with the investigation of the binary distribution function. There is a difference between classical and quantum-mechanical studies of this function. In the quantum case, a system is described by the Hamiltonian, in which the potential energy is a sum of Coulomb interactions only. The short-range interactions are not explicitly evident. They arise in the process of integration, thanks to the symmetrization of the wave function and the formation of wave packets. In the classical description, however, the phenomenological short-range forces are included in the potential energy, together with the Coulomb forces.

In order to calculate the pair correlation function it is necessary to take a trace, utilizing the antisymmetrical states. The latter is equivalent to including some additional repulsion. The problem consists in the necessity for a correct combination of quantum effects that are connected with the forming of the packets (the effects of the uncertainty principle) and with the exchange effects. This problem produces difficulties during the calculation of the pair correlation function for small distances between particles.

It is necessary to point out that the results of the various authors are very close for the correlation energy and the compressibility, but for the pair-correlation function the results are different for small interparticle separations. In many papers, however, the Debye approximation for the distribution function is used as an initial term; this corresponds to the random phase approximation.

The classical analog of the approximation is

$$f_2^{\mathrm{cl}}(r) = 1 + \frac{1}{(2\pi)^3}\frac{V}{N}\int d\mathbf{k}\left(\frac{1}{1 + (N/V\theta)\nu(k)} - 1\right)\exp i\mathbf{kr}$$

where $\nu(k) = 4\pi e^2/k^2$.

It is clear that in the classical case, for $r \to 0, f_2^{\mathrm{cl}}(r) \to -\infty$.

What happens here in the quantum-mechanical case? We shall use the SKP method to describe the difficulties that arise (13).

When one takes into account quantum displacement of the particles, the quantity $(N/V\theta)\nu(k)$ transforms into the function $a_2(k)$, which describes the interaction between the particles, where

$$a_2(k) = -\tfrac{1}{2}(\alpha - 1)\frac{1 - \exp - 2E_k\alpha\beta}{1 + \eta\exp - 2E_k\alpha\beta}; \qquad \alpha = \sqrt{(\hbar\omega_0)^2 E_k^{-2} + 1}$$

$$\omega_0 = \left[\frac{k^2\phi(k)N}{mV}\right]^{1/2}, \qquad \eta = \frac{\alpha - 1}{\alpha + 1}, \qquad E_k = \frac{\hbar^2 k^2}{2m}, \qquad \beta = \Theta^{-1} = (kT)^{-1}$$

The term $f_2(r)$ becomes equal to

$$f_2(r) = 1 + \frac{1}{(2\pi)^3}\frac{V}{N}\int d\mathbf{k}\left(\frac{1}{1 + |a_2|} - 1\right)e^{i\mathbf{kr}}$$

The comparison of two functions shows that $f_2^{\mathrm{cl}} \to -\infty$ when $r \to 0$ and $f_2(r) \to$ const when $r \to 0$. The negative value should not be astonishing, because, after including the virial integrals in the treatment, one can obtain a positive exponent form for the pair correlation function. The difficulties appear when one deals with the exchange. In this case the dispersion of the collective oscillations differs from unity. It is described by the structure factor of exchange, $1 + s_2(k)$, where

$$s_2(k) = (S_p B)^{-1} S_p\left(-\sum_{\mathbf{p}} n_{\mathbf{p}+\mathbf{k}} n_{\mathbf{p}} B\right)$$

In comparison to the previous form, the pair correlation function has one addition essential term, negative for the $r \to 0$ term. The difficulties of the modern theories of the quantum electron gas arise from the presence of this negative additional term:

$$\Delta(r) = \int_0^\infty \left\{ \frac{1 + s_2(k)}{1 + [1 + s_2(k)]\,|a_2(k)|} - \frac{1}{1 + |a_2(k)|} \right\} \frac{\sin kr}{r} k\, dk$$

R. N. Petrashko and this author have tried to avoid these difficulties by using the virial integrals. For the investigation of the regions of small interparticle distances it is enough to include the second virial coefficient in the treatment. The second quantum virial coefficient was used with the quantum screening potential, which included the exchange:

$$g(r) = \frac{1}{N} \sum_k e^{i\mathbf{kr}} g(\mathbf{k}) = \frac{1}{N} \sum_k e^{i\mathbf{kr}} \frac{|a_2(k)|}{1 + [1 + s_2(k)]\,|a_2(k)|}$$

In addition, by construction of the second coefficient, the exchange has been calculated in the transition from the individual degrees of freedom to the collective variables. If the latter is not done, we obtain the highest possible value of the pair correlation function for $r \to 0$:

$$F_2^{\max}(r) = [\exp - g(r)] - f(r)$$

$$f(r) = \frac{1}{N} \sum_{\mathbf{k}} \frac{s_2(k)[1 - a_2(k)]}{1 + |a_2(k)|\,[1 + s_r(k)]} e^{i\mathbf{kr}}$$

If we calculate the exchange in the transition, admitting the possibility of the crossing of two adjacent Fermi spheres $\left[\text{to take } \sum_{\mathbf{p}} n_{\mathbf{p}}(1 - n_{\mathbf{p+q}}) \text{ instead of the classical } \sum_{\mathbf{p}} n_{\mathbf{p}} = N\right]$, we obtain the minimum possible value of the pair correlation function for $r \to 0$:

$$F_2^{\min}(r) = [\exp -g(r)] f_1(r) + f_2'(r)$$

where

$$f_1(r) = 1 + \frac{g''(r)}{g(r)} + \frac{4}{g^2(r)} [g(r) - g_1(r)]$$

$$f_2'(r) = -f(r) + \frac{g''(r)}{g(r)} [-1 + g(r)] + \frac{4}{g^2(r)} [g(r) - g_1(r)]$$

$$g''(r) = \frac{1}{N} \sum_{\bar{k}} e^{i\mathbf{kr}} g(k) s_2^2(k)$$

$$g_1(r) = \frac{1}{N} \sum_{\bar{k}} e^{i\mathbf{kr}} g(k)[1 + s_2(k)]$$

The numerical results for various values of r_s are given in the table. For comparison other authors' results are also listed.

Source	$r_s = 2$	$r_s = 4$	$r_s = 6$
RPA	−0.53	−1.33	−2.04
F. Brouers, *Phys. Stat. Sol.*, **19**, 867 (1967)	−0.16	−0.68	−1.14
Singwi and Tosi, *Phys. Rev.*, **176**, 589 (1968)	0.11	0.006	−0.03
Minimum result	0.20	0.06	0.01
Maximum result	0.26	0.20	0.20

VI. Electron-Ion System

In several papers M. Vavrukh investigated the system of ions and electrons by the SKP method. Ions and electrons are considered equally in their own right. There are two types of collective variables: one for electrons and a second for ions. Then for the partial diagonalization of the Hamiltonian of excited states new collective variables are considered:

$$\rho_k{}^e, \rho_k{}^i \rightarrow \rho_k{}^{(1)}, \rho_k{}^{(2)}$$

The operator of free collective movements is represented in the new collective variables in the form of a sum with the separable variables:

$$\sum_{\mathbf{k}} \hbar\omega_k{}^{(1,2)} \rho_{\mathbf{k}}{}^{(1,2)} \frac{\partial}{\partial \rho_{\mathbf{k}}{}^{(1,2)}}$$

where the function $\hbar\omega_k{}^{(1,2)}$ represents the collective oscillations branch of the spectrum.

The detection of correlation between $\rho_{\mathbf{k}}{}^{(1)}, \rho_{\mathbf{k}}{}^{(2)}$ and $\rho_{\mathbf{k}}{}^e, \rho_{\mathbf{k}}{}^i$ affords an opportunity to conclude that $\rho_k{}^{(1)}$ describes the oscillation of the total charge when ions and electrons are moving in an opposite phase, $\hbar\omega_k{}^{(1)}$, defines the energy of quantum plasmons. In the limit $k \rightarrow 0$

$$\hbar\omega_k{}^{(1)}{}_{k\rightarrow 0} \rightarrow \hbar\omega_0 \left(1 + \frac{m}{M} z\right)^{1/2}$$

where ω_0 = the Langmuir frequency of electrons, and z = the valence of ions. The term $\hbar\omega_k{}^{(1)}$ transforms into the energy of free electrons when k is large.

The variable $\rho_k{}^{(2)}$ describes the oscillations of the system density, and at small values of k the frequency $\omega_k{}^{(2)}$ is a linear function of k. In Fig. 3 we see the dependence $\omega_k{}^{(2)}(s_0 k_F \sqrt[4]{r_s \gamma_s})^{-1}$ on the variable $x = k(k_F \sqrt[4]{\gamma_s r_s})^{-1}$ for several values of Brueckner's parameter and ion valence $z = 1$ ($\gamma_s = 0.884 \ldots$,

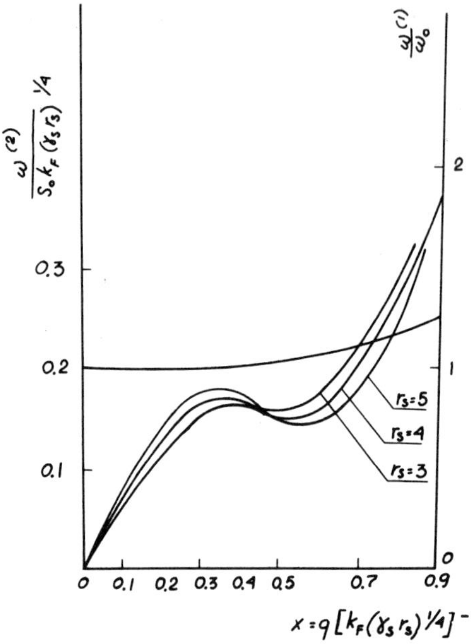

Fig. 3. Density oscillation frequency versus wavenumber. Shown for several values of the Brueckner parameter.

k_F = Fermi's wave number, $s_0 = [(m/M)(z/3)]^{1/2}(\hbar k_F/m)$, velocity of sound in the jellium model). The most remarkable feature of the figure is the characteristic twist of curve, which indicates the existence of a short-range order in the system. When the wavenumber increases, $\hbar\omega_k^{(2)}$ turns into the energy of Langmuir ion oscillations; when the wavenumber further increases, $\hbar\omega_k^{(2)}$ turns into the energy of free ions.

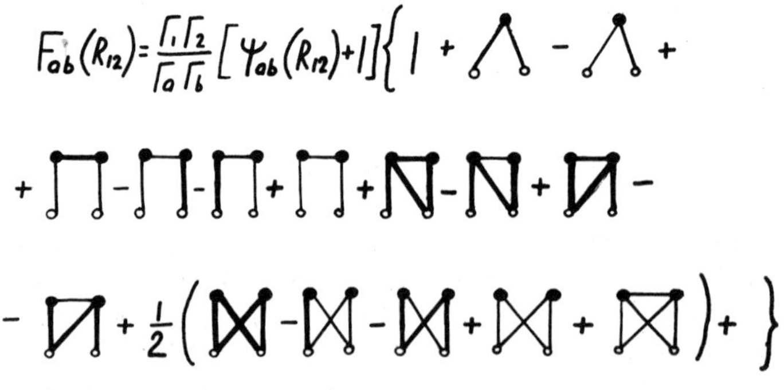

Fig. 4. Correlation function.

The low-frequency branch of collective oscillations includes all properties of the experimental spectrum of liquid metals.

The statistical operator in the SPK representation is used for the calculation of the free energy of the system at low temperature for metallic densities at which a crystal lattice is formed (Fig. 4).

References

1. N. N. Bogoliubov, *Problems of the Dynamical Theory in Statistical Physics*, Gostekhizdat, Moscow, 1946.
2. I. R. Youkhnovsky, *JETP*, **27**, 690 (1954).
3. I. R. Youkhnovsky, *JETP*, **32**, 379 (1958).
4. I. R. Youkhnovsky, *Ukr. Phys. J.*, **4**, 167 (1959).
5. I. R. Youkhnovsky and A. A. Necrot, *Ukr. Phys. J.*, **9**, 365 (1969).
6. I. R. Youkhnovsky, Preprint ITP-26P, *Ukr. Phys. J.*, **9**, Nos. 7, 8 (1964).
7. M. V. Vavrukh, *Ukr. Phys. J.*, **13**, No. 5, 733 (1968); I. R. Youkhnovsky and M. V. Vavrukh, Preprint ITP 71-32-P, Kiev, 1971.
8. I. R. Youkhnovsky and M. F. Golovko, *Ukr. Phys. J.*, **14**, 1119 (1969).
9. M. F. Golovko, *Ukr. Phys. J.*, **15**, No. 12 (1970).
10. I. R. Youkhnovsky, V. S. Vysotchansky, and M. F. Golovko, Preprint Institute for Theoretical Physics, Lvov, USSR.
11. A. A. Necrot and I. R. Youkhnovsky, *Ukr. Phys. J.*, **13**, 1636 (1968).
12. D. N. Zubaryev, *Dokl. Akad. Nauk USSR*, **95**, 757 (1954).
13. I. R. Youkhnovsky and R. N. Petrashko, Preprint Institute for Theoretical Physics, Lvov, USSR.
14. T. Gaskell, *Proc. Phys. Soc. (London)*, **77**, 1182 (1961); **80**, 1091 (1962).
15. J. Hubard, *Phys. Lett.*, **25A**, 709 (1967).
16. K. Singvi and M. Fosi, *Phys. Rev.*, **176**, 589 (1968).

The Bogoliubov Method in the Theory of Crystal State

I. P. BAZAROV

Lomonosov State University, Moscow, USSR

I. Application of the Vlasov Equation to Crystals

It is known that the behavior and the properties of a plasma that is not very dense are well described by Vlasov's kinetic equation for the self-congruent field (1):

$$\frac{\partial f(q, p, t)}{\partial t} + \frac{1}{m}(p, \nabla_q f) = (\nabla_p f, \nabla_q) \int \phi(|q - q'|) f(q', p', t)\, dq'\, dp' \quad (1)$$

where $f(q, p, t)$ is the average density for the number of particles with mass m at the point q, p at time t, and $\phi(|q - q|)$ is the potential of pair interaction.

It is also known that eq. 1 evolves directly from the first equation of Bogoliubov's chain of equations for unequilibrium distribution functions (2):

$$\frac{\partial f(q, p, t)}{\partial t} + \frac{1}{m}(p, \nabla_q f) = \int [\nabla_q \phi(|q - q'|), \nabla_p f_2(q, p, q', p', t)]\, dq'\, dp' \quad (2)$$

in the approximation of separability of the binary distribution function, $f_2(q, p, q', p', t)$:

$$f_2(q, p, q', p', t) = f(t, q, p) \cdot f(t, q', p') \quad (3)$$

Approximation 3 for plasma is based on the long-distance Coulomb interaction between particles, as a result of which the motion of each particle is determined not by its interaction with any other one particle but by the cumulative action of all the particles in the plasma.

In common gases and liquids the forces of interaction between the particles diminish quite rapidly with distance. Therefore the hypothesis of separability of the binary distribution function (eq. 3) becomes an incorrect approximation, and the kinetic equation of self-congruent field (eq. 1) is not applicable to such systems (3).

The stress on the fact that the kinetic equation of the self-congruent field is applicable only to systems with long-distance interaction between particles (plasma) and is not applicable to gases and liquids, and also the results contradictory to the experimental ones, which were obtained in Ref. 4 with the use of such an equation in the theory of crystallization—both of these factors delayed for a long time the use of this equation in other fields than plasmas, namely in application to the crystal, which represents a system with short-range interaction.

At first sight it seems that the application of the kinetic equation to the crystal is unfounded because the particles in crystal are spaced very close to each other and strongly interact between themselves. Therefore it would seem that separability of the binary distribution function (eq. 3) could not be valid. But this is not so, since this view does not take into account the specific character of motion of the particles in a crystal.

Two proofs of justification for the representation of the binary distribution function in the crystal state as separable could be mentioned. One of these proofs is qualitative (physical), and the other quantitative—an exact mathematical one, which is the major proof. Now we shall consider these proofs.

In Ref. 5, correspondingly to the experimental results, it was shown that the dimensions b of regions for the particle motion in crystal junctions are small in comparison with the distances between rhe nearest neighbors a ($b \approx 0.1\ a$). This is true up to the temperature of fusion. Therefore the motion of each particle at any given moment is determined by interaction with the nearest and also more distant neighbors, and not only with any particular one. This is the physical foundation for the separability of the binary distribution function (eq. 3) in crystal.

Let us prove this statement in another way. Using eq. 1, we obtain by the self-congruent field approximation the next equation for the variation of the distribution function:

$$\delta f = f(q, p, t) - f_0(q, p) \equiv \varphi(q, p, t)$$

where $f_0(q, p)$ is the equilibrium distribution function. This equation takes the form

$$\frac{\partial \varphi}{\partial t} + \frac{1}{m}(p, \nabla_q \varphi) = (\nabla_q u, \nabla_p f_0) + (\nabla_q u_0, \nabla_p \varphi) \qquad (4)$$

where

$$u(q, t) = \int \phi(|q - q'|)\varphi(q', p', t)\, dq'\, dp'$$

$$u_0(q) = \int \phi(|q - q'|) f_0(q', p')\, dq'\, dp'$$

In Refs. 6 to 8 it was shown that the use of the kinetic equation for the variation of the distribution function of eq. 4 in the case of a crystal makes it possible to determine the collective oscillation spectrum in the form of acoustical and optical branches (9). In the particular case of a one-dimensional lattice, taking into account the interaction only between the nearest neigbors, we obtain from eq. 4 the result of Born:

$$\omega^2 = \frac{4}{m} \phi''(a) \sin^2 \frac{ka}{2} \tag{5}$$

where k is the wave vector of possible oscillations in crystal.

This result establishes the applicability of the self-congruent field equation to the crystal. The same is true in regard to the quantum kinetic equation of the self-congruent field (10). In this case the collective oscillations in a crystal are determined by the equation of the self-congruent field for variation of the quantum distribution function:

$$i\hbar \frac{\partial \varphi}{\partial t} (q, q', t) = (H_q - H_{q'})\varphi(q, q', t) + \mathcal{D}_1{}^0(q, q') \int [\phi(|q - q_1|) - \phi(|q' - q_1|)]\varphi(q_1, q_1, t)\, dq_1$$

where $\mathcal{D}_1{}^0(q, q')$ is the one-particle equilibrium quantum distribution function,

$$H = H_q = -\frac{\hbar^2}{2m} \Delta_q + u_0(q)$$

$$u_0(q) = \int \phi(|q - q_1|) \mathcal{D}_1{}^0(q_1, q_1)\, dq_1$$

Therefore analysis of decoupling of the chain of Bogoliubov's equations (2) in regard to the crystal leads to the conclusion that the kinetic equation of the self-congruent field has, in addition to the plasma, another wide field of use, namely, the crystal state.

II. Crystal Free Energy

The determination of the oscillation spectrum for the collective vibration of crystal particles with the use of eq. 4 for the self-congruent field, obtained from eq. 1, indicates that eq. 1 takes into account to the required degree the correlation in particle motion. Therefore the published statements that the self-congruent equation does not take into account the correlation in particle motion are not correct.

It is necessary to note that the equilibrium self-congruent field equation actually describes the crystal without due regard to the correlation in particle motion. This corresponds to the Einstein model of a crystal. Nevertheless in the classical region the thermodynamic parameters of a crystal coincide both with and without regard to the collective vibration of crystal particles. This permits application of the equilibrium equation of the self-congruent field to the crystal in this region.

The distribution function $\rho(q - a_i)$, which determines the motion of a given particle in the region of some function i (with a vector a_i) satisfies the equilibrium equation of the self-congruent field:

$$\theta\,[\ln \lambda]\,\rho(q - q_i) + \sum_{j \neq i} \int \phi(|q - q'|)\rho(q' - a_j)\,dq' = 0 \tag{7}$$

In Refs. 3 and 5 are given the solutions of this equation by the iteration method and variation method, and expressions for the free energy F of crystal are obtained.

In the quasiharmonic approximation F is given by

$$F = -\frac{3}{2} N\theta \ln (2\bar{n}\theta) + \frac{N\theta}{2} \sum_{\alpha=1}^{3} \ln (\lambda_\alpha) + U_0 \tag{8}$$

where U_0 is the static energy of the crystal; λ_α are the diagonal elements of the matrix

$$\left\| \left(\frac{\partial^2 u(q)}{\partial q^\alpha\, \partial q^\beta} \right)_{q=0} \right\|$$

related to the main axis ($\alpha, \beta = 1, 2, 3$); $u(q)$ is the potential of the self-congruent field in which this particle is located; and θ is the absolute temperature, taken in units of energy.

III. Other Applications to Crystals

In conclusion it should be noted that Bogoliubov's method in the theory of the crystal state makes it possible to take a considerably larger step towards the study of crystal properties than is possible by the ordinary theory.

Particularly in regard to crystal structure, the use of this method has enabled various workers to accomplish the following:

1. A theory of unharmonic effects in the crystals, which coincides well with the experimental results (11), has been developed.

2. A curve was obtained for the limit of crystal stability over a large range of temperatures and pressures. This curve coincides very well with the curve of phase equilibrium for crystal-liquid (these results were originally obtained in Ref. 12), which in the absence of the statistical theory for the liquid state has important practical value. In addition, this result quite convincingly supports the idea that under usual conditions the superheating of solid bodies is absent.

3. For the first time the consequent statistical theory of polymorphous transformations in crystals (13) has been developed.

More detailed and subsequent discussions of all topics considered in this paper can be found in the monograph by the present author, published by Moscow University Press in 1972.

References

1. A. A. Vlasov, *JETP*, **8**, 291 (1938).
2. N. N. Bogoliubov, *Problems in the Dynamical Theory of Statistical Physics*, Gostekhizdat, Moscow, 1946
3. I. P. Basharov, Publications of Moscow University, *Phys. Astron.*, **5**, 85 (1968).
4. A. A. Vlasov, *Many Body Theory*, Gostekhizdat, Moscow, 1950.
5. I. P. Basharov, *Izv. Vuzov, Phys.*, **2**, 92 (1967).
6. I. P. Basharov, Publications of Moscow University, *Phys. Astron.*, **1**, 56 (1964).
7. I. P. Basharov, *Doklady*, **148**, 1283 (1963).
8. I. P. Basharov, Publications of Moscow University, *Phys. Astron.*, **2**, 7 (1965).
9. M. Born and M. Goeppert-Mayer, *Theory of Solid Matter*, 1938.
10. I. P. Bashorov, Publications of Moscow University, *Phys. Astron.*, **4**, 3 (1965).
11. I. P. Basharov, V. B. Glasco, and N. I. Kupik, Publications of Moscow University, *Phys. Astron.*, **1**, 21 (1969).
12. I. P. Basharov, *PTT*, **11**, 840 (1969).
13. I. P. Basharov and V. V. Kotenok, Theory of Polymorphic Transitions in Crystals with Three Particle Interactions, *Material from the Working Conference on Statistical Physics*, Institute for Theoretical Physics, Ukraine, Academy of Sciences, October 25–28, 1971.

Present Status of Controlled Thermonuclear Theory

MARSHALL N. ROSENBLUTH

Institute for Advanced Study, Princeton, New Jersey, USA

I. Free Energy Considerations

The basic problem in magnetic confinement is of course that we are attempting to create a nonequilibrium state. The distribution function f cannot be the Gibbs distribution if it is to be confined.

In thermodynamic equilibrium $f \equiv e^{-H/T}$, where the Hamiltonian H for a static magnetic field is just $H = m\mathrm{V}^2/2$. Hence the density must be uniform in thermodynamic equilibrium. As we know, true thermodynamic equilibrium is reached only through the relatively slow collisional processes; nonetheless, sources of free energy are available to drive instabilities, and we must wonder whether the various constraints on the system (e.g., Liouville's theorem and adiabatic invariants), really allow such free energy to be tapped.

One source of free energy is magnetic—the field is distorted by the plasma away from its minimum energy state. In general this gives rise to critical conditions on β and q, such as the Kruskal-Shafranov limit.

Another possible source of free energy is that the distribution may be anisotropic, especially in magnetic mirrors, and of course its energy may be lowered while conserving phase space by isotropization. Here the adiabatic invariance of the magnetic moment tells us that only high-frequency modes, $\omega \approx \Omega_i$, are effective.

Finally there is the most basic source of free energy, arising from confinement itself—expansion free energy. A simple argument based on free energy shows us that only low frequencies can tap expansion free energy.

Suppose that we have a confined plasma with a density gradient as shown in Fig. 1. We know that expansion free energy may be liberated by making the density more uniform. Suppose that some instability does this, creating the dotted profile shown. Using the fact, based on Liouville's theorem, that internal energy $E \sim \rho^{\gamma-1}$, we may easily estimate the internal energy released:

$$\frac{\Delta E}{E} \approx \mathcal{O}\left(\frac{\Delta r}{\rho}\frac{\partial \rho}{\partial r}\right)^2$$

However, the energy released must be enough to drive the instability, that is, $\Delta E > \rho V^2$, where $V \sim \omega \, \Delta r$ is the velocity in organized motion in the mode.

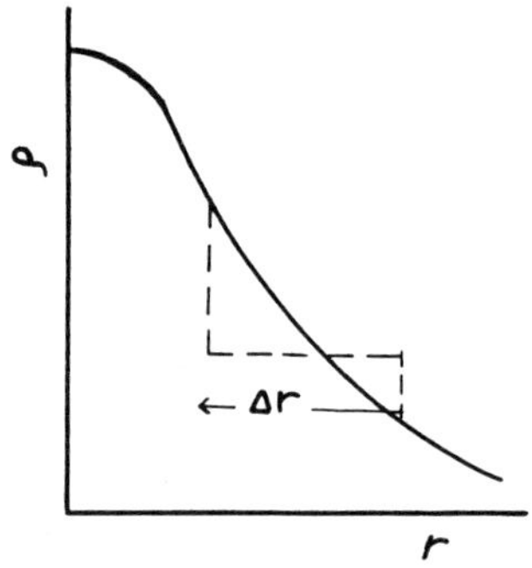

Fig. 1. Interchange mode.

Using this inequality, we see that for expansion free energy to be available we must have

$$\omega < V_{\text{th}} \frac{1}{\rho} \frac{\partial \rho}{\rho r} < \Omega_i$$

where the last inequality follows in the usual case that the system is many gyroradii in size. It follows that only low-frequency modes can tap expansion-free-energy modes—in the regime where magnetic moments are conserved, and plasma is usually frozen to flux lines.

Putting together these arguments, we see that, on the one hand, we may think of magnetic mirrors of the minimum-B type, where no expansion free energy is available but where we may fear high-frequency short-wavelength disturbances. On the other hand, for closed systems with basically isotropic distributions we need consider only low-frequency modes, such as magnetohydrodynamic, drift, and trapped particle.

II. Mirror Machines

We review briefly the situation with mirror machines. It is well known that in such devices a loss cone must exist in velocity space when $V_{\parallel}/V_{\perp}$ is too large. If we project the resulting velocity distribution on one of the axes perpendicular to B, the loss cone ensures that the projected distribution will be of the two-humped "two-stream unstable" type.

When we look at the actual dispersion relations for such an anisotropic plasma, we find easily that for an infinitely long plasma rapidly growing modes occur with short wavelength compared to an ion gyroradius, and high frequencies compared to ion gyrofrequency. These so-called loss-cone modes are, however, convective. They run rapidly down the field lines. We may estimate a critical length from $L_{cr} = (1/\gamma)\,(d\omega/dk)$. Typically critical lengths based on these estimates are some hundreds of gyroradii, the exact coefficient depending on the degree of anisotropy. Systems longer than the critical length will be unstable.

More dangerous are the so-called absolute modes that are nearly resonant with the gyrofrequency and for infinite plasma analysis grow in amplitude at a fixed point in space. Even for these modes there is a question of fitting within a finite system. Here, guided by the short wavelength of the modes, we may use the quasiclassical method to derive rather complex integral equations for the perturbing potential. Considerable numerical work has been done at Livermore and Oak Ridge, indicating critical lengths in the range of a few tens of gyroradii. These are hard problems, and it remains a somewhat open question whether the critical length for instability is greater than the minimum length needed to guarantee adiabaticity and single-particle confinement.

Let me now turn to some simple nonlinear considerations which indicate that perhaps low levels of turbulence may be tolerable in a mirror trap. At first sight this is surprising, since any small fluctuation at a multiple of the gyrofrequency changes the magnetic moment and leads to scattering into the loss cone. However, I would like to show that there is a phenomenon of super-adiabaticity which keeps such motions bounded.

Consider a flute mode of electrostatic perturbations (Fig. 2), such as the drift-cone mode, resonant with the mid-plane gyrofrequency, Ω_0, of a trapped ion. As we see, each time the ion passes through this resonant zone its magnetic moment is changed. However, the change is proportional to $\sin\psi$, where ψ is the phase angle of magnetic gyration relative to the wave.

For a long system such phases on subsequent transits would tend to be random, and a stochastic diffusion of μ would result. The phase angle, however, is determined by the transit time between reflections. This in turn depends on μ. The result is a set of equations defining a mapping of subsequent transits in the μ, ψ plane.

We may use the well-known results of mapping theory to study the nature of the mapping—a problem similar to the flux surface problem. In particular, we may analyze the stability of fixed points, those where a transit does not change the μ, ψ point. Looking in the neighborhood of such points, we find stability if $e\phi/E < (a_i/L)^{5/3}$ where a_i is the ion gyroradius. We may expect that such low-level fluctuations will lead only to small bounded excursions of the magnetic moment and no particle loss—the so-called superadiabatic behavior. Here only a single wave has been considered, and evidently the theory needs to be extended.

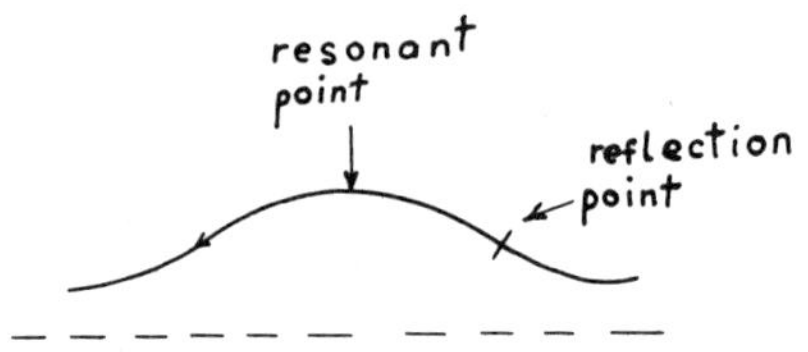

Fig. 2. Resonant electrostatic mode in mirror device. Here $\Omega = \Omega_0[1 + (x/L)^2]$; $\phi = \hat{\phi}e^{i\Omega_0 t}$.

In regard to the more difficult question of the nonlinear limit to be expected for such fluctuations, not much has been done. Among possible limiting mechanisms are the expulsion of low-energy particles. For example, in the 2-X machine $T_e \ll T_i$, and a potential of a few tenths T_e would be just about at the superadiabatic limit discussed above.

A second possible limit is given by cross-field diffusion. A damping proportional to $DK_\perp^2$ is to be expected, and in view of the short wavelengths involved the damping may be considerable. Finally, numerical calculations at Berkeley indicate rather strong saturation due to wave-wave and wave-particle scattering. I want to stress, however, that the whole question of nonlinear effects in mirrors remains largely unexplored—perhaps because of an exaggerated tendency to believe that no excess scattering into the loss cones can be tolerated.

III. Tokamaks

Let me turn now to closed systems, where most thinking of late has been guided by Tokamak results—perhaps because it is the only closed device in which we have a plasma approaching the conditions needed either for a reactor or for the validity of most present theories. Whether this indicates that Tokamaks are innately superior to stellarators or merely that a Tokamak of given aperture and aspect ratio is much cheaper and easier to heat remains an open question.

The first surprising theoretical development of the past few years was the so-called neoclassical theory described in Galeev's lecture, in which it was shown that a consideration of collisional diffusion with real particle orbits gave a considerable increase over classical values. The familiar Galeev–Sagdeev diagram was obtained for the diffusion coefficient versus collision frequency in which three idealized regions were calculated separately, that is, banana, plateau, and classical.

Hinton, Hazeltine, and I have recently succeeded in deriving the complete curve, including the transition regions, which show rather large deviations from the simple picture. The basic equations are rather simple, a balance between

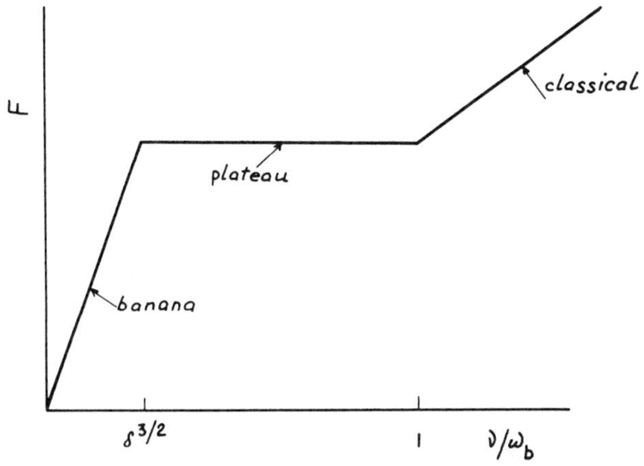

Fig. 3. Galeev-Sagdeev diagram for neoclassical particle flux, F, versus collision frequency, ν. Here $\omega_b{}^{-1}$ is the transit time for an average particle around the torus, δ is the toroidal aspect ratio.

convection along the field lines and pitch angle scattering. The basic complication arises because of different drift orbits and different boundary conditions for trapped and untrapped particles. The untrapped must be single-valued in θ, while the trapped must have right- and left-going particles equal at the reflection point, $V_{\parallel} = 0$. This gives a boundary layer of thickness proportional to the square root of collision frequency. Such a problem with discontinuous boundary conditions may be solved by the Weiner-Hopf technique, and the first correction to the Galeev-Sagdeev results obtained. The results may be extended, using the entropy variational principle, and the whole curves obtained as shown in Figs. 3 and 4. When the proper energy averages are obtained, it may be found, for example, that ion thermal transport at the banana-plateau transition is only about one-third the previous value.

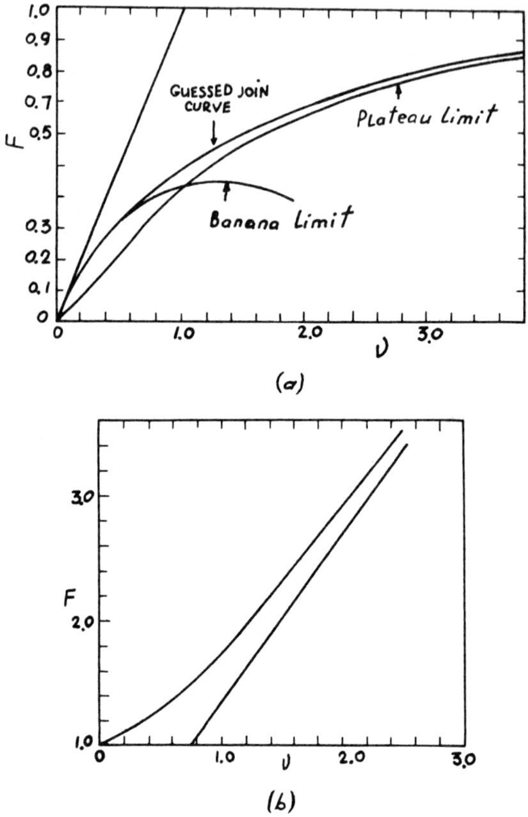

Fig. 4. Plateau-classical regime for diffusion. (*a*) Details of the banana-plateau transition. (*b*) Details of the plateau-classical transition.

Neoclassical theory may be elegant, as the name implies, but unfortunately it does not yet explain all the facts. In particular, while ion thermal transport is, within experimental error, neoclassical, electron transport is better described by the so-called pseudoclassical formula. Neoclassical diffusion is given by

$$D \approx \nu {\rho_{e\theta}}^2 \delta^{1/2}$$

where $\rho_{e\theta}$ is the electron gyroradius in the poloidal field, and δ the aspect ratio. On the other hand, the pseudoclassical formula as proposed by Artsimovich and Yoshikawa is given by

$$D \approx 5\nu {\rho_{e\theta}}^2$$

which is about an order of magnitude greater. It may perhaps be noted that, if the condition $T_e = T_i$ is obtained, neoclassical ion loss will become larger than pseudoclassical electron transport, so that perhaps its relevance will decrease.

Nonetheless the explanation of this pervasive phenomenon is of key theoretical interest, the first of four Tokamak mysteries I wish to discuss. An interesting observation of Furth is shown in the curve of Fig. 5—that almost all Tokamak and stellarator lifetime points may be fitted on the assumption that the smaller diffusion, pseudoclassical or Bohm, applies in all cases. Clearly it would be of great interest to determine whether a stellarator heated past the crossover point was pseudoclassical.

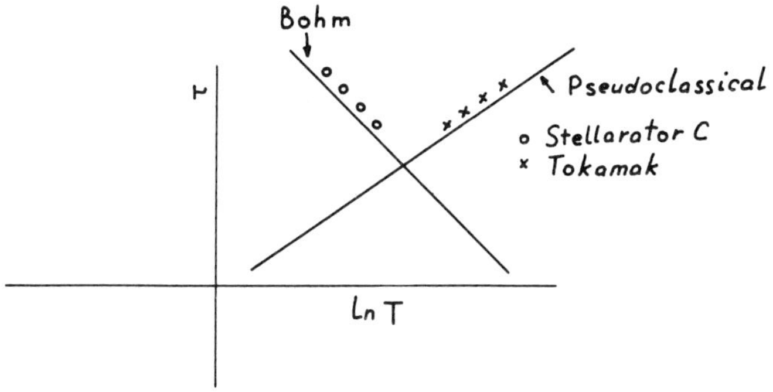

Fig. 5. Furth's universal law.

A basic theory of pseudoclassical diffusion remains a challenge. Tentative models such as those of Yoshikawa show that, if fluctuations achieve a sort of stationary plateau, disturbed by collisions, one obtains the pseudoclassical formula.

However, such analyses are only guideposts for future efforts. A better candidate may be the Kadomtsev trapped electron drift temperature gradient mode, which should be present in Tokamak banana regimes. Here the interesting fact is that, while we cannot determine the absolute level of diffusion without knowing the fluctuation level, there is a specific prediction of a 5:1 ratio for energy to particle diffusion—in agreement with experiments at Moscow, and disagreement with those at Princeton.

Another question of interest with regard to pseudoclassical diffusion is whether the striking features of neoclassical theory might be carried over in a model-independent way—entropy production, anomalous pinching, bootstrap currents, and so on. What is the relation to anomalous resistance?

The second Tokamak problem to be explained is the anomalously rapid skin penetration of the toroidal current. In order to understand the lack of a thin current skin in Tokamak experiments, it is necessary to postulate transport several orders of magnitude above pseudoclassical. Possible explanations are not lacking; three possibilities are as follows:

1. Magnetohydrodynamic kinks in the shear-free region.
2. A postulated electron viscosity with momentum carried inwards radially by unstable current-driven drift waves.
3. Temperature gradient drift modes.

As of the moment we cannot say which, if any, of these effects is operative, whether they lead to extreme heat loss in the skin phase, and whether they are adequate for skin penetration in a large system.

Another mystery was discussed in Rutherford's paper—the nature of the violent instabilities often observed to disrupt Tokamak discharges. Although the onset of these modes correlates reasonably well with Furth's suggestion of tearing modes, our nonlinear analysis indicated that the resulting transition to a stable helical equilibrium should produce a *positive* voltage spike, in contradiction to the most striking experimental fact. The experimentally observed high-density limit is an important, perhaps related question. Hopefully, in the near future measurements of the current profile, in particular the value of q on axis, and the nature of the fluctuations will allow a more meaningful study of these matters.

Finally, in addition to these problems, there remains the prediction that between present regimes and a reactor the trapped particle mode of Kadomtsev and Pogutse will occur and may be very dangerous. Fairly careful calculations of critical conditions for onset in the localized approximation have been made, but the quantitative inclusion of radial effects in a sheared magnetic field remains a problem of key importance. It appears to me that the most dangerous effects of these modes will occur when an ion temperature gradient is present. In this case one can show that ion Landau damping is destabilizing over a large range of parameters for $d \ln T_i / d \ln n > \frac{2}{3}$.

IV. High-β Devices

When we turn to high-β devices, even the problem of obtaining MHD toroidal equilibrium and stability is a difficult one because of the strong curvature forces. It has been shown by Grad and others that in the model of a sharp surface current an $l = 1$ stellarator with $l = 0$ or $l = 2$ corrector windings provides a toroidal equilibrium having only a weak $m = 1$ instability, which should be possible to feedback-stabilize. However, a distributed current profile has not yet been analyzed, and it remains somewhat of a hope that finite Larmor radius stabilization can cure higher m values. The field of microinstabilities is largely unexplored. However, the diamagnetic well dug by plasma pressure offers some hope for stabilization. It is well known that in MHD modes where plasma is frozen to flux lines the diamagnetic well is carried along during interchanges and does not provide minimum B stabilization. However, when particles and field move incoherently, as in the trapped particle mode, such stabilization may occur, offering a hope for high-β systems. This area too needs much theoretical work.

In this survey of various problems I hope I have made clear that, although a qualitative understanding of most types of instability may now exist, often even the linear theory remains to be developed with the proper inclusion of all realistic effects. Nonlinear saturation is not really understood in any practical case. At first it would seem that numerical particle simulation offers the key here, but I believe that it will prove of only limited help. The reason is that fortunately such systems as Tokamak are really highly stable. Only residual instabilities remain, with growth rates far smaller than other characteristic times of the system (e.g., stable Alfvén waves). It is very difficult to see such weak effects through numerical and physical noise. Such strong modes as a one-dimensional two-stream instability are easily studied numerically, but the generically similar mode of interest, the current-driven drift mode, is very feeble and really requires three-dimensional consideration. Drift waves are basically two-dimensional, and effects of radial convection and reflection in sheared magnetic fields introduce the third dimension. Hence I would expect numerical simulation to be most useful in the study of specific well-defined physical phenomena in simplified situations, rather than for the overall complex problem. For example, recent nonlinear theory has shown the importance of particle trapping effects, and resultant "clumps," which are quite difficult to treat analytically.

For all these reasons I believe that future theory, on the whole, should be analytic and phenomenological, based very closely on careful experimental observations of not only macroscopic transport but also microscopic fluctuations, with some guidance from numerical modeling. This has been the history of the science of aerodynamics, and I strongly believe that within a few years we will also fly.

Neoclassical Theory of Transport Processes

A. A. GALEEV

Institute of High Temperature, Academy of Sciences of USSR, Moscow, USSR

In the theory of transport processes based on the kinetic equation solution, there is a class of problems in which the terms representing some of the quasi-periodic motions appear to be the main terms of the kinetic equation. In this case our problem can be reduced to the well-known Chapman-Enskog problem by use of the averaging method (1). With this procedure, instead of considering the particles, we operate with the "quasiparticles." In the case of the straight magnetic field lines we average over the fast Larmor rotation and consider "Larmor rings" as the quasiparticles. In the toroidal magnetic field we need double, or sometimes even triple, averaging. We restrict ourselves to the axially symmetric model of the magnetic field, so that double averaging is sufficient. The need for the second averaging arises because of the presence of the trapped particles oscillating between the local magnetic mirrors. We introduce here a new quasiparticle, the "banana," the drift trajectory of the "Larmor ring." The trajectories traced by the particle in the first and second types of fast motion are fixed in space because of the conservation of the magnetic moment and the longitudinal adiabatic invariant. Violation of the invariant conservation because of particle collisions leads to a random walk of the trajectory with steps of the order of the width of the trajectory across the magnetic surface Δr. A simple estimate based on the random walk approach

$$D = \frac{\Delta r^2}{\tau} \tag{1}$$

shows that this effect should lead to an increase in the transport coefficients. Nevertheless, the quantitative theory taking into consideration the complicated nature of the particle motion in a toroidal trap, later termed the neoclassical theory, was published only in 1967 (2). In spite of the simplifications we achieved by the averaging, the kinetic equation solution had a boundary-layer-type singularity, and to solve it we used the asymptotic expansion. Our results (2, 3) are verified now by using another method (4).

I. Neoclassical Diffusion

In this report we restrict ourself to the following model of the magnetic field:

$$\mathbf{B} = B_0 [\mathbf{e}_z + \Theta(r)\mathbf{e}_\vartheta](1 - \epsilon \cos \vartheta)$$

$$\epsilon = \frac{r}{R} \ll 1, \qquad \Theta(r) = \frac{B_\vartheta}{B_z} \ll 1 \tag{2}$$

In a rarefied plasma one should distinguish the "trapped" and "transit" particles. The longitudinal velocity of the trapped particles at the outer portion of the toroidal tube is so small, $v_\| < \epsilon^{1/2} v_\perp$, that such particles cannot reach the inner portion of the tube, where the toroidal field is larger by the quantity

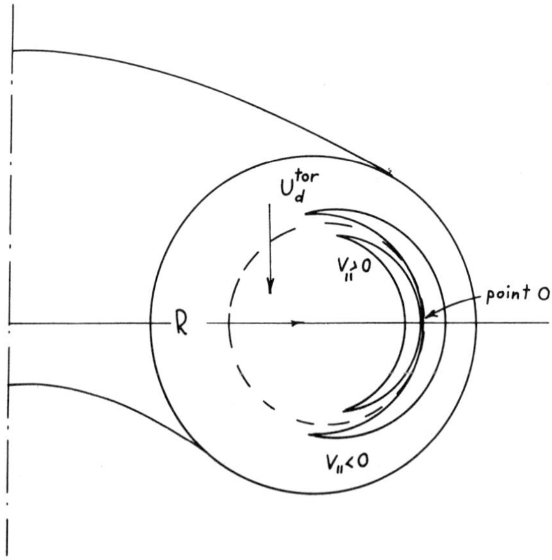

Fig. 1. Projection of a trapped particle trajectory on a perpendicular cross section of a toroidal tube.

$2\epsilon B_0$ than in the outer part. With toroidal drift taken into account, the trapped particles trace a trajectory, which we shall name "banana," according to the analogy with the "Larmor ring."

Let us consider a particle with positive velocity. It moves around the magnetic surface upward from the point O and deviates to the region inside the magnetic surface (see Fig. 1). Then it reflects from the magnetic mirror formed by the toroidal magnetic field at the inner portion of the toroidal tube and moves back. The particle still moves away from the magnetic surface until it crosses the mid-plane. Below the mid-plane it moves toward the magnetic surface, and after the second reflection it returns to the initial point. The trajectory of a trapped particle with negative velocity is situated outside the magnetic surface.

The transit particles have larger velocity, $v_\parallel > \epsilon^{1/2} v_\perp$, and move along magnetic field lines without reflections, tracing a trajectory very close to the magnetic surface. During the bounce period

$$\tau_b = \frac{r}{\Theta(r) v_{\text{th}} \epsilon^{1/2}} \tag{3}$$

the toroidal drift, $U_d{}^{\text{tor}}$, causes the trapped particle to deviate away from the surface to a distance

$$\Delta r_t = U_d{}^{\text{tor}} \tau_b = \frac{r_c}{\Theta} \epsilon^{1/2} \tag{4}$$

Taking into account the fact that a change in the longitudinal velocity by a small value of the order $v_\parallel \approx \epsilon^{1/2} v_{\text{th}}$ is sufficient to scatter the trapped particle, we find the scattering time to be of the order

$$\tau_M = \frac{\epsilon}{\nu_j}, \qquad \nu_j = \frac{16\pi^{1/2} n e^4 \ln \Lambda}{3 m_j{}^2 v_{\text{th}_j}{}^3}$$

It is easy now to estimate the diffusion coefficient. For this purpose we insert into eq. 1 the above values of Δr and τ and take into account that only a small fraction of particles, $\delta n_t = \epsilon^{1/2} n$, have such a large random walk. Then we obtain

$$D \approx \epsilon^{1/2} \nu \frac{r_c{}^2}{\Theta^2}, \qquad \frac{\nu}{\epsilon} \tau_b \ll 1 \tag{5}$$

This approach gives a rough estimate of the diffusion coefficient, but on the basis of it we are unable to discuss such fine effects as the role of the electron-electron collisions, and the presence of a plateau in the variation of $D(\nu)$. That is why we shall use a more general approach, based on the analogy of our problem concerning the equilibrium particle distribution in a toroidal magnetic field with the Landau damping problem in a rarefied plasma. As in the Landau problem, the particles in a torus move in a sinusoidal-like potential of the diamagnetic force (see Fig. 2), since the magnitude of the magnetic field varies along the line of force (the field is larger at the inner portion of the torus and smaller at the outer portion). The distribution relaxation problem for the particles moving in such a potential is in some sense additional to the Landau problem, where one considers the relaxation of the potential under the influence of the particles obeying the Maxwellian distribution. It is known that

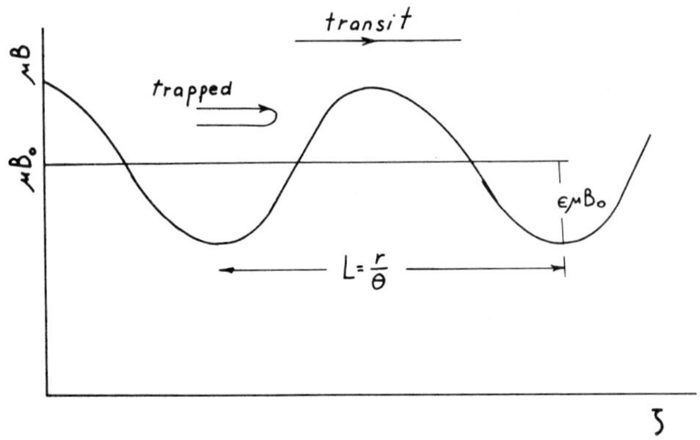

Fig. 2. Effective particle potential in a toroidal magnetic field.

the friction between wave and resonant particles, causing the wave damping, is described by the expression:

$$F_{\text{fr}}^{(0)} = -\int d^3v\ \mu\nabla_{\|}Bf_{\text{res}}$$

$$\approx \int d^3v\mu\ \nabla_{\|}B\left(\frac{\mu\nabla_{\vartheta}B}{m\omega_c}\frac{\partial f_{\text{M}}}{\partial r} + \mu\nabla_{\|}B\frac{\partial f_{\text{M}}}{m\partial v_{\|}}\right)\frac{\pi}{|\Theta|r}\Bigg|_{v_{\|}=-\frac{v_0}{\Theta}}$$

where

$$v_0 = \frac{c}{B_0}\frac{d\phi_0}{dr}$$

is the electric drift velocity. This equation is valid only within the restricted interval of the collision frequencies, where the particle distribution relaxes to the Maxwellian one ($\tau_{\text{M}} < \tau_b$).

In a quasistationary state the friction force between the wave and resonant particles is balanced by the electron-ion friction due to the presence of an additional current, $j_{\|}$:

$$F_{\text{fr},i}^{(0)} - \frac{j_{\|}}{\sigma_{\|}e} = 0, \qquad F_{\text{fr},e}^{(0)} + \frac{j_{\|}}{\sigma_{\|}e} = 0 \tag{6}$$

where $\sigma_{\parallel}$ is the plasma conductivity along the magnetic field. Since the wave-electron friction is weak, the first equation gives the relation between the electric drift and the ordered velocity of ions in a quasistationary state:

$$F_{\mathrm{fr},i}^{(0)} \sim U_{\parallel i} + \frac{v}{\Theta} - \frac{cT_i}{eB_\vartheta}\frac{d\ln n}{dr} = 0 \qquad (7)$$

Obviously, the state of quasistationary diffusion is quickly reached because of the small change of the plasma potential up to the level obeying eq. 7. Only after that do the effects violating the ambipolarity of the diffusion lead to a rather slow joint change of the ion velocity and the electric field. Rosenbluth et al. (5) considered the finite Larmor radius correction as one of the effects violating the diffusion ambipolarity and found the equation for the longitudinal ion velocity. They have shown that the ions slow down in the case of uniform ion temperature, just as in the case of plasma diffusion across straight magnetic field lines. In a plasma with nonuniform temperature the shear in the ion motion gives rise to considerable tension between trapped and transit ions.† As a result the ion velocity is quite large and corresponds to an electric field of the order $e\phi_0 \approx \epsilon^{1/2} T_i$. The latter value is maximal and can be reached only in the dilute plasma limit.

The second of eqs. 6 gives the value of the additional current maintaining the diffusion equilibrium

$$j_{\parallel} = -\frac{\tau_{\mathrm{M}}}{\tau_b}\frac{c}{B_\vartheta}\,\epsilon^{1/2}\frac{dp}{dr}, \qquad \tau_{\mathrm{M}} \ll \tau_b \qquad (8)$$

Diamagnetic drift of the resonant particles leads to a plasma flux across magnetic surfaces. This flux does not depend on collisions (compare with the Landau damping rate):

$$\langle nv_r \rangle = \left\langle c\int \mu \frac{[\nabla B \times \mathbf{B}]_r}{eB^2} f_{\mathrm{res}}\, d^3v \right\rangle_\vartheta \qquad (9)$$

$$\approx \frac{\pi^{1/2}}{2}\frac{r_{\mathrm{ce}}}{\theta r}\frac{c}{eB_0}\frac{dp}{dr}$$

Let us now consider the rare collision limit ("banana" regime):

$$\tau_{\mathrm{M}} \gg \tau_b$$

In this case nonlinear distortion of the particle distribution occurs, and the particle tends toward a particular distribution for which the damping rate is zero. The collisions try to make the distribution take the form of a local

† Equation 7, found by equating the friction force between transit and trapped ions to zero, obviously is insufficient for avoiding the tension.

Maxwellian and thereby maintain the damping. A good estimate for the wave particle friction is the interpolation formula given by Vedenov, Velichov, and Sagdeev (6):

$$F_{\mathrm{fr}} = \frac{F_{\mathrm{fr}}^{(0)}}{1 + (\tau_{\mathrm{M}}/\tau_b)} \tag{10}$$

The use of this formula leads again to the estimate of eq. 5 for the diffusion.

A more accurate approach is based on the solution of the drift kinetic equation

$$\frac{\partial f}{\partial t} + (\Theta v_\parallel + V_0)\frac{\partial f}{r\,\partial \vartheta} - \frac{\mu B_0}{m}\,\epsilon \sin\vartheta \left(\frac{1}{\omega_c}\frac{\partial}{\partial r} + \Theta\frac{\partial}{\partial v_\parallel}\right) f = Stf \tag{11}$$

In the rare collision limit we use here a Fokker–Planck type of collisional term, derived by Landau for the special case of Coulomb collisions. Then the highest derivative enters the kinetic equation with a small parameter; the ratio of the mean free path to the connection length of the magnetic field, $L = r/\Theta$. The mathematical methods for solving such problems are well developed and have been applied to the problem of Landau damping of a finite-amplitude wave in a rarefied plasma (7). It was found that the main contribution to the damping rate came from the boundary layer separating the phase spaces of the trapped and transit particles. Because of this analogy we have used the results of Ref. 7 to calculate the plasma transport coefficients in a torus (2).

Simple arguments can be used to qualitatively describe the solution of eq. 11. Let us note, first, that the toroidal drift of trapped particles in a non-uniform plasma causes an additional slope of their distribution in velocity space. As a result the ions with positive velocity come to point O (see Fig. 1) of the magnetic surface from the inner region, where the particle density is larger, and the ions with negative velocities come from the outer region. Therefore the number of ions with positive velocities is larger by an amount $\Delta f = -(df/dr)\,\Delta r_t$. The resulting velocity gradient of the distribution function,

$$\frac{\Delta f}{\Delta v_\parallel} \approx -\frac{1}{\omega_c \Theta}\frac{\partial f}{\partial r}$$

has different signs for ions and electrons and is sufficient to suppress the resonant interaction of the trapped ions and electrons with a "wave" (see Fig. 3). This slope of the trapped ion distribution spreads quickly over the transit ions, and the hole distribution appears to be moving with velocity $U = (v_{\mathrm{th}}^2/\omega_c \Theta n)(dn/dr)$.† The transit electrons are carried along by the

† For the sake of simplicity we are using the coordinate system where the electric field is zero.

ions and therefore are moving with the same speed. The trapped electron distribution corresponds, then, to a portion of the Maxwellian distribution, drifting in the opposite direction. As a result an additional current arises. We

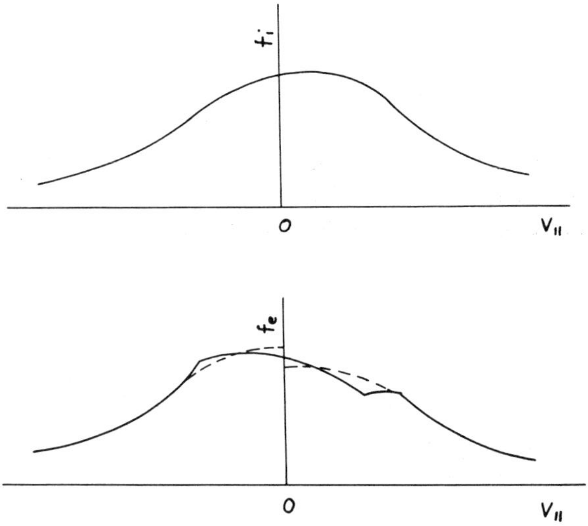

Fig. 3. Particle distribution in a rarefied plasma.

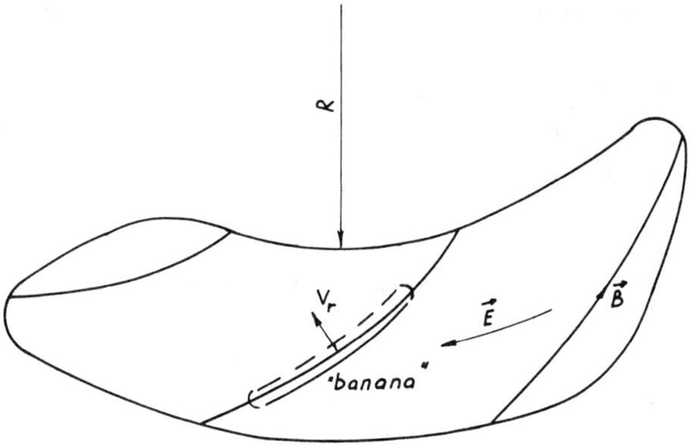

Fig. 4. "Banana" geometry.

give it the name "banana current" by analogy with Larmor current. The banana current value can be estimated as the product of the trapped electron surplus, $\Delta n_e \approx (dn\epsilon^{1/2}/dr)\,\Delta r_{\mathrm{te}}$, and its velocity, $\Delta v_\parallel \approx \epsilon^{1/2} v_{\mathrm{th\,e}}$. Then we obtain

$$I_{\mathrm{bt}} \approx -\frac{c}{B_\vartheta}\,\epsilon^{3/2}\,\frac{dp}{dr} \tag{12}$$

This current causes friction between the trapped electrons and the transit electrons and ions. The latter leads to a banana-drift-like guiding center drift (see Fig. 4):

$$\langle nv_r \rangle = \frac{c}{eB_\vartheta} F_{\mathrm{fr}}$$

$$= -\left(\nu_{ei} + \nu_{ee}\right) \epsilon^{1/2} \frac{r_{ce}^{\;2}}{T_e \Theta^2} \frac{dp}{dr} \tag{13}$$

Though the electron-electron collision frequency evidently enters into the diffusion coefficient expression, the diffusion is still ambipolar because the friction between transit and trapped electrons is finally balanced by transit

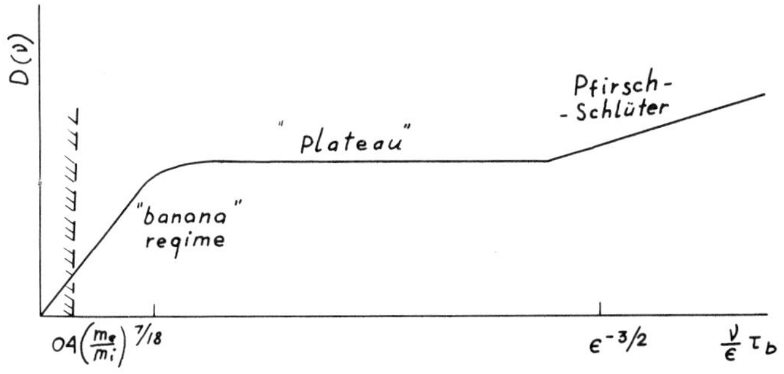

Fig. 5. Diffusion coefficient dependence on collision frequency.

electron-ion friction. Equations 9 and 13, together with the Pfirsch-Schlüter result, define the well-known dependence of the diffusion coefficient on collision frequency (Fig. 5).

II. Electrodiffusion Phenomenon

We have already noticed that in the plateau regime the wave-particle friction can be balanced by the electron-ion friction only by generation of some additional current. An analogous effect takes place in the banana regime. The friction between the trapped and transit electrons arising because of the banana current of trapped electrons carries along the transit electrons and results in much greater additional current:

$$\nu_{ei} j_{\parallel}^{(1)} \approx \frac{\nu_{ee}}{\epsilon} I_{bt}$$

In addition, in a rarefied plasma there is another mechanism to establish the additional current, which does not have any relation to an additional friction force. It results from the diffusion-type spreading of the slope of the trapped electron distribution over the transit electron phase space. The distribution of the transit electrons caused by such diffusion is shown by the dotted line in Fig. 3. It consists of the left and right sides of the steady Maxwellian distribution, and the densities of the electrons moving in opposite directions differ sufficiently to make the electron distribution continuous. Therefore the transit electrons are stationary relative to the ions, and electron-ion friction does not arise. But the difference in densities of the electrons moving in opposite directions gives rise to an additional current of the order

$$j_\parallel^{(2)} \approx e\,\Delta n v_{\text{th}} \approx -\frac{c}{B_\vartheta}\,\epsilon^{1/2}\,\frac{dp}{dr}$$

Both mechanisms have the same order of magnitude in the banana regime, and we should include them simultaneously in Ohm's law:

$$j_\parallel + \frac{c}{B_\vartheta}\,\epsilon^{1/2} \sum_j T_j \frac{dn}{dr} = \sigma_\parallel \left(E_\parallel - D_{ee} \frac{B_\vartheta}{c} \frac{d \ln n}{dr} \right) \tag{14}$$

Using the Onsager principle,† we can easily obtain the contribution to the particle flux resulting from the presence of the toroidal electric field and Ohm's current:

$$\langle n v_r \rangle = -D_e \frac{dn}{dr} - n\epsilon^{1/2} \frac{c}{B_\vartheta} E_\parallel - \frac{D_{ee}}{T_e} \frac{B_\vartheta}{c} j_\parallel \tag{15}$$

The second term represents here the banana drift in a toroidal electric field [the trapped particle pinch effect (3, 8)]. Both this term and the second term on the left-hand side of eq. 14 are absent in the plateau regime when collisions destroy bananas. The third term of eq. 15 describes the banana drift under the influence of the friction between the "nailed" trapped electrons and the transit electrons moving with current velocity. In the plateau regime the presence of a current leads to a change in the slope of resonant particle distribution and therefore to a change in the wave-resonant particle friction. The additional friction is responsible for an increase of plasma resistivity by the amount $\Delta\eta_\parallel = 0.75\,\pi^{-1/2}\theta v_{\text{th}\,e}\epsilon^2 m_e/ne^2 r$.

For lack of space we do not consider here the calculation of the ion thermal conductivity, which can be estimated from eqs. 9 and 13 by using the ion parameters instead of the electron ones. Besides that, we have taken into

† The second term on the left-hand side of eq. 14 was omitted by an oversight of the author in Ref. 3 and led him to conclude that the Onsager principle can be violated in the rare collision limit. In a later paper by Rosenbluth, Hazeltine, and Hinton (4) this current was calculated from the trapped particle pinch effect (3, 8) with the help of the Onsager principle.

account only the magnetic trapping of particles (due to the nonuniformity of the magnetic field). However, an important electric field directed along the magnetic surface can exist even in quiescent plasma if the Larmor radius of an ion in a poloidal magnetic field is of the same order as the radius of the plasma column. We estimate the electric potential, taking into account that electrons have a Boltzmann distribution and compensate for surplus of trapped ions at the inner portion of a torus: $(dn\, \epsilon^{1/2}/dr)\, \Delta r_t \approx ne\phi_1/T_e$. We see that we are obliged to take into consideration the trapping due to the electric field non-uniformity if $r_{ci}/\Theta r > 1$. However, this effect does not change the order of value of the transport coefficients, since in the worst case the electric well has the same depth as the magnetic one. As a result the potential well for ions becomes more shallow in order to make its displacement from the magnetic surface smaller, and the potential well for electrons tends to be deeper by a factor of 2.

III. Bootstrap Tokamak and Pressure Limit

The presence of the additional current in toroidal traps creates some new and interesting possibilities for the confinement of a rarefied plasma. The reason is that the neoclassical diffusion of particles in a hot plasma is able to maintain the poloidal magnetic field even in the absence of the toroidal electric field and therefore to make possible plasma equilibrium in a toroidal magnetic field. Substituting the additional current found earlier into the Maxwell equation for the poloidal field,

$$\frac{1}{r}\frac{d}{dr}\, rB_\vartheta \approx \frac{4\pi}{B_\vartheta}\, \epsilon^{1/2}\, \frac{dp}{dr}$$

we find that this possibility arises in a plasma with $\beta_j = (8\pi p/B_\vartheta{}^2) > \epsilon^{-1/2}$ (10, 11). But the Joule heating by this additional current is $\epsilon^{-1/2}$ times smaller than the heat loss due to electron thermal conductivity, and therefore we need some other heating mechanism to reach this regime. As soon as we reach thermonuclear temperatures, the energy release exceeds the Joule heating, and plasma diffusion maintains the additional current continuously, if there is some particle source in the volume. In other words, we come to the idea of the stationary Tokamak. A more careful treatment shows that plasma diffusion does not generate the poloidal field but only pushes it out of the axes. In addition, the "plateau" regime of neoclassical diffusion takes place in the vicinity of the axes, and therefore the additional current is absent there. This means that the current and therefore the poloidal magnetic field at the axes must be maintained by external sources (11, 12).

At this point an interesting question arises: What is the limit for the pressure increase in a stationary Tokamak? This limit is given by the Kruskal-Shafranov criterion for stability (10, 11):

$$\frac{B_z r}{B_\vartheta(\nabla p)R} \approx \frac{\epsilon^{3/4}}{\beta^{1/2}} = q \geqslant 1 \tag{16}$$

More accurate calculation, done for the case of similar profiles of the electron density and temperature in a nonisothermal plasma with uniform current, gives the following value of the pressure limit:

$$\beta = \frac{8\pi p(0)}{B_0{}^2} \simeq 1.8\left(\frac{a}{R}\right)^{3/2} q^{-2} \tag{17}$$

IV. Pseudoclassical Diffusion

Neoclassical confinement times were first reached in Tokamaks (13, 14). However, the construction of Tokamaks is such that plasma density and temperature are limited by the Joule heating law, with plasma current creating the rotational transform of the magnetic field. Therefore one cannot change these parameters independently in order to measure the experimental dependence of plasma diffusion on density and to compare it with the theoretical value. The possibility of such measurements exists in the stellarator, however, and the variation of $D(\nu)$ was recently measured with the stellarator "Proto-Cleo" in the Pfirsch-Schlüter regime and the plateau regime (15). The results, shown are in Fig. 6, are in a good agreement with neoclassical theory.

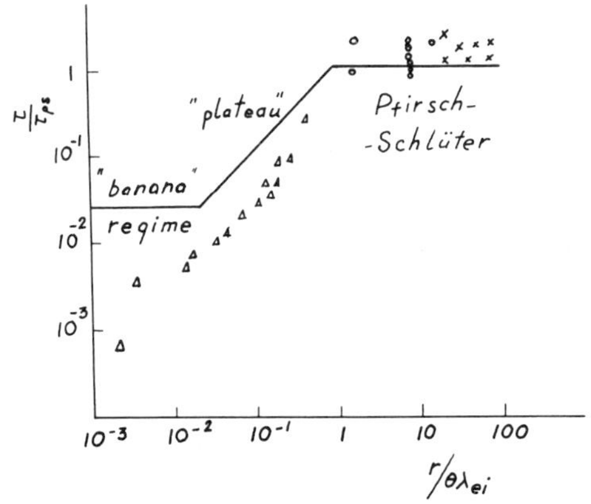

Fig. 6. Experimental points and theoretical dependence of plasma confinement time, in units of Pfirsch-Schlüter time, on the ratio of connection length to mean free path (stellarator "Proto-Cleo").

Artsimovich and his collaborators (16) have made a rather fine test of the applicability of the neoclassical theory for the explanation of ion energy losses. Considering that ions are heated because of pair Coulomb collisions with electrons (17) and lose energy through neoclassical ion thermal conductivity, they have plotted the expected temperature dependence on the parameter $\sqrt[3]{I_z B_z n R^2} \cdot A^{-1/2}$ (A = atom weight of the gas used in the experiments) for a uniform (I) and a parabolic (II) distribution of current and plasma density over the plasma column. This dependence is shown in Fig. 7, taking into account the last experimental data reported in Madison by the T-4 group. (18).

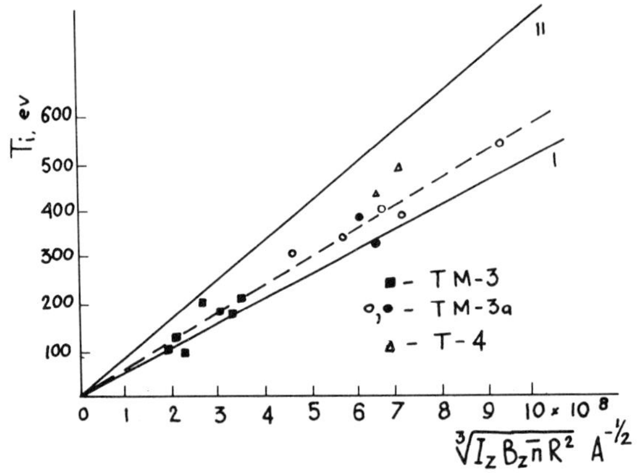

Fig. 7. Ion temperature dependence on plasma parameters (Tokamak).

The situation in regard to electron losses is more complicated. Artsimovich gives the following empirical formula for the electron thermal conductivity, named later as pseudoclassical:

$$\chi_e = \frac{5\nu_{\text{eff}} r_{ce}^2}{\Theta^2} \tag{18}$$

where ν_{eff} = effective collision frequency, calculated from the measured plasma conductivity, $\sigma_{\text{exp}} = ne^2/m_e\nu_{\text{eff}}$. This thermal conductivity is greater by a factor of 2 or 3 than the neoclassical thermal conductivity calculated for the case in which the anomalously large collision frequency in T-3 is due only to impurities (20). Therefore there is no reason as yet to consider speculations based on the concept of turbulent diffusion. Nevertheless the theory of turbulent diffusion in a very rare plasma has independent interest.

Yoshikawa and Christofilos (21) have considered the model of a plasma in which the external forces support turbulence with a known spectral density of the particle density fluctuations, $|n_k|^2$. Supposing that both the averaged and the fluctuating densities have a weak gradient, they have calculated the change in the amplitude and phase of electric-field and density fluctuations in the frame of a two-fluid approach. The particle flux across the magnetic field is expressed through the level of turbulence:

$$\langle nv_r \rangle = \frac{c}{B_0} \sum_k n_k^{(1)} E_k^{(1)*} = -\frac{1}{2} \sum_k \frac{k^2}{k^2} \frac{|n_k|^2}{|n_0|^2} \nu_e r_{ce}^2 \tag{19}$$

They reduced this formula to the pseudoclassical one, using speculation to find the nonlinear limitation of amplitude ($|n_k|^2 = n_0^2/k_\perp^2 a^2$) and defining $k_\| = a/\Theta$.

The particle distribution relaxation can lead to a self-suppressing of instability at a much lower level than in the fluid model used in Ref. 21. That is why we will not draw attention to a more self-consistent theory of turbulence developed a long time ago. The important example is the drift turbulence exhibited by resonant electrons. It is well known that because of quasilinear diffusion the slope of distribution in the resonant region becomes more steep, so that the enhanced Landau damping stabilizes the waves. The rare collisions try to make the distribution take the form of a local Maxwellian and thereby support instability at such a level that the turbulent diffusion appears to be proportional to the classical diffusion. Since the energy released because of quasilinear relaxation of electrons is sufficient to capture the resonant electrons in the wave potential,

$$\sum_k \frac{ne^2 |\phi_k|^2}{2T_e} \approx \tfrac{1}{2} m_e n_0 \left(\frac{\Delta v_\|}{v_{\text{th}\, e}}\right)^2 \Delta v_\|^2 \tag{20}$$

that is,

$$\sqrt{e\phi/m_e} \approx \Delta v_\|$$

the quasilinear theory gives an answer coinciding with the neoclassical diffusion in a monochromatic magnetostatic wave. The trapping takes place not in the magnetic, but the electric, well and we put $\epsilon = e\phi/T_e$. The particle excursion caused by the electric drift, $ck_\perp \phi/B_0$, during the bounce time in the electric potential well, $\tau_b = 1/k_\| \sqrt{e\phi/m_e}$, is bigger by a factor $k_\perp/k_\|$ than the value given by eq. 4. Therefore eq. 13 for the case of a drift wave takes the form

$$D \approx \sqrt{e\phi/m_e}\,(\nu_{ei} + \nu_{ee}) r_{ce}^2 \left\langle \frac{k_\perp^2}{k_\|^2} \right\rangle \qquad \nu_e \frac{T_e}{e\phi} < k_\| \sqrt{e\phi/m_e} \tag{21}$$

Using the value of the drift wave potential given by eq. 20, we reduce this equation to the well-known result of the quasilinear theory for drift wave turbulence (see, e.g., Ref. 22).

Pogutze has used formula 21 to explain the pseudoclassical diffusion in a Tokamak. But we should mention that it is valid in the form given here only for systems with straight magnetic field lines. Toroidal drift of "E-banana" form (Furth and Rosenbluth gave this name to the trajectory of particles trapped in an electric potential well) causes the creation of "superbananas" in the case of the toroidal system. Superbanana diffusion increases with the decrease of collision frequency in the region not far from the "plateau" (23, 24). The latter leads to a decrease of β_J and the formula of Artsimovich, derived under the condition β_J = const, is no longer valid. Diffusion decreases again with the decrease of collision frequency only in the rare collision limit.

Since we have started to discuss instabilities as a possible cause of anomalous electron thermal conductivity, it is useful to give a theoretical estimate of the particle and heat turbulent fluxes. The drift temperature instability of trapped electrons is the most probable candidate, since it is difficult to stabilize this mode by shear (see Ref. 4). The ratio of the heat flux to the thermodiffusional flux of particles (simple diffusion is absent in this case) can be estimated for this mode from the quasilinear equation

$$\frac{\partial f_e}{\partial t} = n \frac{\partial}{\partial r} \frac{\epsilon}{\nu_e(v)} \frac{c^2}{B_\vartheta{}^2} \left| \frac{1}{\tau_b} \int_0^{\tau_b} \frac{\partial \phi^{(1)}}{R\, \partial \zeta}\, dt \right|^2 \frac{\partial}{\partial r} \frac{f_e(v, r)}{n(r)}$$

and it is equal to 5. This is in agreement with measurements of the Artsimovich group, who found that the energy confinement time is shorter than the particle confinement time for electrons.

V. Applicability Restrictions for the Neoclassical Theory

Plasma parameters for which the diffusion is neoclassical are restricted by two modes of instabilities developing in the "banana" regime. These are the drift oscillations of "trapped" (25) and "transit" (26) ions. Both types of oscillations are excited by interaction with trapped electrons. We use the banana kinetic equation to estimate the disturbance of its distribution in a wave field:

$$\frac{\partial}{\partial t} \Delta f_{te}{}^{(1)} + \frac{\nu_e}{\epsilon}\left(\frac{v_{\mathrm{th}\, e}}{v}\right)^3 \Delta f_{te}{}^{(1)} \approx \frac{e}{T_e}\left(f_{te}{}^{(0)} \frac{\partial}{\partial t} + \frac{cT_e}{eB_\vartheta} \frac{\partial f_{te}{}^{(0)}}{\partial r} \frac{\partial}{R\, \partial \zeta}\right) \times \langle \phi^{(1)}(\vartheta, \zeta, t)\rangle|_{\zeta = q\vartheta} \qquad (22)$$

Here the banana drift velocity is expressed through the electric field averaged over the banana, and $\Delta f_{te}{}^{(1)}$ is the deviation of the particle distribution from Boltzmann one in a wave field. Collisions are taken into account in eq. 22

qualitatively in τ approximation. Leaving in the Fourier expansion only the main term

$$\phi^{(1)} = |\phi|\, e^{-i\omega t + im\vartheta + il\zeta}, \qquad m \approx lq$$

we find the following estimate for the trapped electron work in a wave field

$$\dot{W}_e = -\,\mathrm{Im}\frac{e^2}{T_e}\int d^3v\,|\phi|^2 \frac{\omega\left(\omega + \dfrac{lcT_e}{eRB_\vartheta}\dfrac{\partial}{\partial r}\right) f_{te}{}^{(0)}}{\omega + i\dfrac{\nu_e}{\epsilon}\left(\dfrac{v_{\mathrm{th}\,e}}{v}\right)^3} \tag{23}$$

Drift oscillations of trapped ions have low frequency, $\omega = -\,\epsilon^{1/2} lqv_*{}^e/r$ $[v_*{}^e = (cT_e/eB_0)\,(\nabla n/n)]$, and short wavelength along the magnetic field, $k_\| \approx (m - lq)\theta/r$. It is easy to see that such oscillations are excited even in the absence of a temperature gradient (25). However, a small portion of transit ions having velocities of the order $\epsilon^{1/2}v = \omega/k_\|$ can get into resonance with the wave and take its energy at the rate (27)

$$\dot{W}_i^{(\mathrm{res})} = -\epsilon^{1/2}\omega\left(\frac{\omega\tau_{bi}}{m - lq}\right)^3 W \tag{24}$$

In addition, the energy dissipation in the boundary layer separating trapped and transit ions [its width is $\Delta v_\| = \epsilon^{1/2}(\nu_{ii}/\epsilon\omega)^{1/2}$] gives the additional damping rate (28)

$$\dot{W}_i^{\mathrm{coll}} = -\,\epsilon^{1/2}\omega\left(\frac{\nu_{ii}}{\epsilon\omega}\right)^{1/2} W \tag{25}$$

The indicated effects lead to suppression of instability in a large fraction of the banana regime:

$$\frac{\nu_{ii}}{\epsilon}\tau_{bi} > 0.4\left(\frac{m_e}{m_i}\right)^{7/18}\left(\frac{T_e}{T_i}\right)^{7/6} |m - lq|\,\ln^{2/9}\left(\frac{4}{\sqrt{2\nu_{ii}/\epsilon\omega}}\right) \tag{26}$$

A second type of oscillation is exhibited in a plasma with nonuniform temperature under the condition

$$0.1\,\frac{\nu_e}{\epsilon} > \omega \approx -\frac{m}{r}\,v_*{}^e > |k_\||\,v_{\mathrm{th}\,i} \tag{27}$$

In the presence of shear, the longitudinal wave number grows when the waves are traveling from the initial magnetic surface, and therefore the waves damp because of resonance with ions. The maximal growth rate appears to be numerically small under the conditions of eq. 27, and this facilitates shear stabilization. A disturbance in the form of two waves traveling in opposite directions from the initial point is most difficult to stabilize. We obtain a stability criterion for such a mode by comparison of the growth rate with the damping rate resulting from energy flux to infinity of the order (29)

$$\dot{W}_\infty = -\left|\frac{\epsilon}{q}\frac{d\ln q}{d\ln T_e}\right|\omega W \tag{29}$$

More accurate treatment of numerical coefficients gives the criterion

$$\left|\frac{\epsilon}{q}\frac{d\ln q}{d\ln T_e}\right| > \frac{0.1\epsilon^{1/2}}{[1+(d\ln T_e/d\ln n)](T_i/T_e)+1} \tag{28}$$

Adam et al. (30) have drawn attention to the fact that variation of the toroidal drift with coordinates along the magnetic field lines leads to localization of some modes within a restricted part of the field line. These modes do not feel the shear. But the localization conditions seems to be in contradiction with the instability criterion eq. 27 if we take into account ion Landau damping due to diamagnetic drift. Thus, in the presence of shear (eq. 28), we get far into the banana regime.

VI. Application

This qualitative discussion of neoclassical theory can be concluded profitably by a summary of quantitative results for the particle and density fluxes and the generalized Ohm's law. We give here the results of Refs. 2 and 3, which differ only a little from the later results of Ref. 4 (see also the footnote on p. 93.

$$\langle nv_r\rangle = \langle nv_r\rangle^n + \langle nv_r\rangle^I + \langle nv_r\rangle^E$$

$$\langle nv_r\rangle^n = -\frac{\alpha_e D_e}{T_e}\sum_j nT_j\left[\frac{d\ln n}{dr} - \left(\frac{3}{2} - \gamma_j^n\right)\frac{d\ln T_j}{dr}\right] \tag{A.1}$$

$$\langle nv_r\rangle^I = -\frac{\alpha_I D_e}{T_e}\frac{I_\parallel B_\vartheta}{c}, \qquad \langle nv_r\rangle^E = -2\alpha_{ei}\epsilon^{1/2} n\frac{cE_z}{B_\vartheta}$$

$$q_j = T_j \sum_{s=n,I,E} \gamma_j^s\langle nv_r\rangle^s - K_j D_j n\frac{dT_j}{dr} \tag{A.2}$$

$$I_\parallel + \frac{2c}{B_\vartheta}\alpha_{ei}\epsilon^{1/2}\sum nT_j\left[\frac{d\ln n}{dr} - \left(\frac{3}{2} - \gamma_j^E\right)\frac{d\ln T_j}{dr}\right]$$

$$= \sigma_\parallel\left\{E_z - \frac{\alpha_I D_e}{T_e}\frac{B_\vartheta}{c}\sum_j T_j\left[\frac{d\ln n}{dr} - \left(\frac{3}{2} - \gamma_j^I\right)\frac{d\ln T_j}{dr}\right]\right\} \tag{A.3}$$

The coefficients here depend on collision frequency:
(*a*) in the banana regime:

$$\alpha_{ei} = 0.70, \quad \alpha_{ee} = 0.37, \quad \alpha_e = \alpha_{ei} + \alpha_{ee} = 1.07, \quad \alpha_I = 0.23, \quad K_e = 1.18, \quad K_i = 0.57$$

$$\gamma_i^{(n, I, E)} = 1.33, \quad \gamma_e{}^n = 1.11, \quad \gamma_e{}^E = \frac{5}{2}, \quad \gamma_e{}^I = 0.64$$

$$\sigma_\parallel = \frac{\sigma_{sp}(1 - 2\alpha_{ei}\epsilon^{1/2})}{1 + 0.35}, \quad \sigma_{sp} = \frac{ne^2}{0.51 m_e \nu_{ei}}$$

$$D_j = \frac{\nu_{jj} r_{cj}{}^2 \epsilon^{1/2}}{\Theta^2}$$

(*b*) in the plateau regime:

$$\alpha_{ei} = 0, \quad \alpha_{ee} = 1, \quad \alpha_I = 1.15, \quad \gamma_j{}^n = K_j = \gamma_i{}^I = 3, \quad \gamma_e{}^I = 3.8$$

$$\sigma_\parallel = \frac{\sigma_{sp}}{1 - 1.5\pi^{1/2} v_{\text{th}\,e} \theta \epsilon^2 / \nu_{ei} r}, \quad D_j = \frac{\pi^{1/2}}{2} \epsilon^2 \frac{r_{cj}}{|\theta| r} \frac{cT_j}{e_j B_z}$$

References

1. N. N. Bogoliubov and A. Mitropolskii, *Asymptotic Methods in the Theory of Nonlinear Oscillations*, Moscow, 1958.
2. A. A. Galeev and R. Sagdeev, *JETP*, **53**, 348 (1967); A. A. Galeev and R. Sagdeev, *Doklady*, **189**, 1204 (1969).
3. A. A. Galeev, *JETP*, **59**, 1378 (1970).
4. M. N. Rosenbluth, R. D. Hazeltine, and F. L. Hinton, *Plasma Transport in Toroidal Confinement Systems, Phys. Fluids*, to be published.
5. M. N. Rosenbluth et al., *Proceedings of the Fourth International Conference on Plasma Physics and Controlled Nuclear Fusion Research*, Madison, Wis., 1971, Paper No. CN-28/C-12.
6. A. A. Vedenov, E. P. Velikov, and R. Sagdeev, *Nucl. Fusion*, **1**, 82 (1961).
7. V. E. Shaksarov and V. M. Karpman, *JETP*, **43**, 490 (1962).
8. A. A. Ware, *Phys. Rev. Lett.*, **25**, 15 (1970).
9. T. Stringer, *Phys. Fluids*, **13**, 810 (1970).
10. A. A. Galeev and R. Z. Sagdeev, *JETP Lett.*, **13**, 162 (1971).
11. R. J. Bickerton, J. W. Connor, and J. B. Taylor, *Nat. Phys. Sci.*, **229**, 110 (1971).
12. B. B. Kadomtsev and B. D. Shafranov, *Proceedings of the Fourth International Conference on Plasma Physics and Controlled Nuclear Fusion Research*, Madison Wis., 1971, Paper No. CN-28/F-10.
13. L. A. Artsimovich et al., *Proceedings of the International Conference on Plasma Physics and Controlled Nuclear Fusion Research*, Vienna, (IAEA, 1969), p. 157.
14. G. A. Bodroviskii, N. D. Vinogradova, Z. I. Kuznetsov, and K. A. Razimova, *JETP Lett.*, **9**, 269 (1969).

15. Bolton et al., *Proceedings of the Fourth International Conference on Plasma Physics and Controlled Nuclear Fusion Research*, Madison, Wis., 1971, Paper No. CN-28/H-6.
16. L. A. Artsimovich, A. V. Glukov, and M. I. Petrov, *JETP Lett.*, **11**, 449 (1970).
17. L. A. Artsimovich, A. M. Anatsoni, E. P. Gordinov, D. P. Ivanov, and V. C. Strelkov, *JETP Lett.*, **10**, 130 (1969).
18. L. A. Artsimovich et al., *Proceedings of the Fourth International Conference on Plasma Physics and Controlled Nuclear Fusion Research*, Madison, Wis., 1971, Paper No. CN-28/C-8.
19. L. A. Artsimovich, *JETP Lett.*, **13**, 103 (1971).
20. A. A. Galeev and R. Z. Sagdeev, *Proceedings of the Fourth International Conference on Plasma Physics and Controlled Nuclear Fusion Research*, Madison, Wis., 1971, Paper No. CN-28/C-11.
21. S. Yoshikawa and N. C. Christofilos, *Proceedings of the Fourth International Conference on Plasma Physics and Controlled Nuclear Fusion Research*, Madison, Wis., 1971, Paper No. CN-28/F-1.
22. R. Sagdeev and A. A. Galeev, *Nonlinear Plasma Theory*, Benjamin, New York, 1969, p. 86.
23. A. A. Galeev, R. Z. Sagdeev, and H. P. Furth, *PMTP*, **6**, 3 (1968).
24. A. Gibson and D. W. Mason, *Plasma Phys.*, **11**, 121 (1969).
25. B. B. Kadomtsev and O. P. Pogutse, *JETP*, **51**, 1734 (1966).
26. B. B. Kadomtsev and O. P. Pogutse, *Doklady*, **186**, 553 (1969).
27. R. Z. Sagdeev and A. A. Galeev, *Doklady*, **180**, 839 (1968).
28. M. N. Rosenbluth and D. W. Ross, *Bull. Am. Phys. Soc., Ser. II*, **15**, 1400 (1970).
29. H. L. Berk and D. Pearlstein, *Phys. Rev. Lett.*, **23**, 220 (1969).
30. J. C. Adam, G. Laval, and R. Pellat, *Proceedings of the Fourth International Conference on Plasma Physics and Controlled Nuclear Fusion Research*, Madison, Wis., 1971, Paper No. CN-28/F-15.

Classical Plasma Diffusion

HAROLD GRAD

Courant Institute of Mathematical Sciences,
New York, New York, USA

Abstract

Macroscopic plasma diffusion as governed by the simplest form of Ohm's law for a small mean free path is a very complex phenomenon which is by no means described by the classical $1/B^2$ formula. Diffusion across a magnetic field is, in general, a nonlocal, nonlinear, and transient combination of skin effect, field diffusion, plasma diffusion, and singular convection. It turns out that several orders of magnitude are at stake in balancing these competing effects. There is probably at least as large an unknown factor to be determined in the equivalent calculation still to be done with a kinetic-physical model. The basic question—when can transport coefficients be calculated as averages over flux surfaces—is examined.

Introduction

We consider a macroscopic resistive model subject to Ohm's law in the form

$$\mathbf{E} + \mathbf{u} \times \mathbf{B} = \eta \mathbf{J} \tag{1}$$

The value of η is left open; it can be classical or "anomalous" [and some predictions turn out to be independent of the value of η (1, 2, 3)]. To complete the system, we can take conservation of mass, momentum, and flux in the form

$$\frac{\partial p}{\partial t} + \text{div}\,(p\mathbf{u}) = 0 \tag{2}$$

$$\nabla p = \mathbf{J} \times \mathbf{B} \tag{3}$$

$$\frac{\partial \mathbf{B}}{\partial t} + \text{curl}\,\mathbf{E} = 0 \tag{4}$$

The isothermal assumption and absence of mass source implicit in eq. 2, as well as the scalar pressure in eq. 3 and the scalar resistivity in eq. 1, are easily generalized. In order to decouple diffusion from magnetohydrodynamic wave motion, however, the inertia term, $\rho\, du/dt$, missing from eq. 3, must remain excluded.

This is a much simpler physical model than the ones treated in most of the recent literature, which couple energy flow with plasma diffusion, usually in a kinetic guiding center or drift model [e.g., (4, 5)]. But it was only recently recognized (1) that the standard calculation of diffusion in even the simplest macroscopic version (eq. 1) is woefully inadequate. It is based on a simplification, namely, curl $E = 0$ (the *special theory*), which turns out to eliminate from consideration all but a very special class of plasma equilibria. It can be shown (1, 2, 3) that this special class of equilibria *sometimes* takes over after a sufficient lapse of time (probably too long to be relevant in present experimental Tokamaks), but more often the special theory is not representative of the general case. Independently of the quantitative relevance of the macroscopic model (eq. 1) to a hot plasma with large mean free path, its qualitative understanding is crucial to the understanding of kinetic models for which the theory is still at the earlier stage of the standard macroscopic calculation with curl $E = 0$. The neoclassical orbit considerations supply some of the physics missing in the macroscopic model, but the additional physics that is missing because of the specialized profiles in the kinetic model is likely to yield equally important quantitative corrections.

The basic reason for the complexity of classical diffusion (also energy flow) is that what is called a "diffusion (heat transfer) coefficient" is not a traditional transport coefficient. It is not a local property of the plasma; rather it is a "black box" solution of a global boundary value problem. We can compare the elementary concept of a local viscosity coefficient with the drag coefficient for an airfoil, involving global interactions of wakes, boundary layers, shock waves, and so on, each one dependent on viscosity; plasma diffusion is more closely akin to the drag coefficient than to the viscosity coefficient. The special theory (curl $\mathbf{E} = 0$) can frequently be reduced to a calculation of local transport coefficients; the general theory almost never can.

This distinction is not mere semantics. For example, all practical numerical calculations of Tokamak behavior are done in a straight, cylindrical geometry with a diffusion (or heat transfer) coefficient which (among other ad hoc factors) is theoretically corrected for the toroidal curvature (e.g., (6, 7)]. But the correction factor is taken from a global, geometry-sensitive, convective calculation [as in Ref. 4 or 5]; moreover the theoretical calculation is valid only after complete relaxation of a transient—and it is the transient that is the object of study of the numerical calculation!

Without inertia in eq. 3, at least two time scales appear, separated by three or more orders of magnitude. With inertia, faster MHD time scales appear, increasing the spread by at least two more orders of magnitude. No numerical calculation that includes the fastest scale (MHD waves and compression) can include the slowest. But elimination of inertia makes the system very non-standard, with constraints replacing time derivatives; thus elaborate theoretical preparation is required before one can proceed with a numerical calculation (3).

Basic features of the general theory are described in Ref. 1; some specific Tokamak applications are given in Ref. 2; details of the general theory will appear (3). In this paper we briefly summarize these results and discuss the overall significance of constraints and averaging on transport properties.

I. Critique of the Special Calculation

Almost all diffusion calculations (both macroscopic and kinetic) are based on two postulates:

1. Small resistivity, implying slow passage through a family of quasistatic equilibria ($\nabla p = \mathbf{J} \times \mathbf{B}$ or an equivalent guiding center equilibrium).
2. Low β, implying curl $\mathbf{E} = 0$.

Although these seem eminently plausible, after examination they turn out to be unduly restrictive.

There are several consequences of these assumptions, of which we list three.

(*a*) The net flow can be calculated explicitly and very easily.
(*b*) The pertinent equilibria are restricted.
(*c*) The conclusions are very sensitive to slight changes in the model.

Implicit in postulate 1 is that there *exists* a family of neighboring equilibria (through which the plasma slowly diffuses). This is not necessarily so, however, because of the phenomenon of *marginal equilibrium* (8). For example, a mirror machine that is bent into a bumpy torus may require an abrupt change in the topology of the pressure surfaces (this phenomenon is distinct from the formation of *magnetic* islands by loss of symmetry). Similarly, in a straight circular screw pinch, certain helical perturbations *must* change the pressure topology. In addition to its effect on diffusion, this also requires reexamination of stability concepts such as resistive tearing "modes." Standard estimates of resistive growth rates are irrelvant since they do not describe an instability at all but instead a change in equilibrium topology; to follow the consequences of the change in topology requires a nonlinear calculation of an initial value problem rather than normal modes or growth rates.

With regard to point *a*, the calculation is miraculously able to ignore all the fine details of the global interplay between convection and diffusion, and it can always be carried out, both when it is correct and when it is incorrect. For example, it is known that neither scalar pressure nor guiding center equilibria exist in asymmetric geometries (9); nevertheless the calculation yields a diffusion rate (10, 11) (this is not an abstract objection–it is easily verified that the "solutions" do not satisfy the equations).

With regard to consequence *b* above, when there is no driving electromotive force the mean parallel current on each magnetic surface must vanish (12):

$$\langle \eta \mathbf{J} \cdot \mathbf{B} \rangle = 0 \tag{5}$$

More generally (3), in a Tokamak or multipole geometry with driving emfs, the restriction

$$\langle \eta \mathbf{J} \cdot \mathbf{B} \rangle = c_1 \psi_1'(V) + c_2 \psi_2'(V) \tag{6}$$

must be satisfied where

$$c_i = \oint_{C_i} \mathbf{E} \cdot d\mathbf{x} \tag{7}$$

are the two (poloidal and toroidal) constant, path-independent, emfs, and $\psi_1 = \psi$ and ψ_2 are the poloidal and toroidal fluxes.

The general static equilibrium is characterized by two arbitrary profiles, for example, $p(V)$ and $\psi(V)$, or $p(\psi)$ and $q(\psi) = \psi_2'(\psi)$ (q is the reciprocal rotation number or Tokamak "stability" parameter). The standard calculation, based on assumptions 1 and 2, is automatically limited to a *single* arbitrary profile through the condition of eq. 5 or 6. A natural question is whether these special profiles are representative of the general case. The answer is that the evolution in time (including diffusion properties) is even *qualitatively* different for the special and general profiles, both physically and analytically. The analytical features have necessitated development of an elaborate and entirely nonstandard mathematical structure (3). As just one example, the classical boundary condition that at a fluid interface the normal component of fluid velocity equals the velocity of the interface is violated in this theory.

Some of the physical conclusions arising in the general theory which may be more immediately relevant to Tokamak experiments have been summarized in Ref. 2.

1. Present Tokamaks are inherently transient since they do not persist long enough to settle down; this casts some doubt on the significance of scaling laws.

2. The fact that the plasma does not *appear* to be transient is probably due to the dominant effect of the neutral influx.

3. A noncircular plasma cross section can give a decided advantage with regard to diffusion rate and β [a stability advantage has been noted previously (13)].

This transient conclusion was seemingly disputed in a later calculation (14). There is actually no contradiction (15), however, since a steady state is *assumed* in Ref. 14 which is an ad hoc calculation based on several experimental inputs that would undoubtedly change if the experiment were extended sufficiently in time.

With regard to point *c*, the sensitivity of the standard calculation, we point to at least three distinct sources. The classical skin effect has a diffusion coefficient $D_0 \sim \eta/\mu_0$. The "classical" plasma diffusion coefficient is approximately $D_1 \sim \beta D_0$. Diffusion in a general equilibrium is characterized by a competition between these two processes. Since $\beta \sim 10^{-3}$ in a Tokamak and possibly 10^{-8} in an alkali plasma, even a very slight coupling with the much faster skin effect might raise D_1 enormously (it is surprising that factors of only one or two orders of magnitude have so far arisen in the theory).

Another cause of great sensitivity is the complexity of the convective flow. Comparing the two different weighted averages:

$$\begin{gathered} \langle u \cdot \nabla\psi \rangle \sim \text{net flow} \\ \langle r^2 \rangle \left\langle u \cdot \frac{\nabla\psi}{r^2} \right\rangle \end{gathered} \tag{8}$$

we should expect them to be approximately equal in a large-aspect torus ($\delta r/r$ is small on a flux surface), but instead we find that their ratio is a large number ($\sim q^2$, where $2.5 < q < 4$ in a Tokamak). In a finite-aspect torus, where the two expressions can be expected to be somewhat different, they are even found to scale differently. The point is that convection is strongly geometry dependent (whereas a true transport coefficient would not be).

Related to this phenomenon is the fact that D_1 for a straight circular cylinder is increased by a factor of $1 + 2q^2$ [Pfirsch and Schlüter (16)] in a torus of arbitrarily large aspect ratio, even though q itself is given the same value in the straight and toroidal geometries.

A similar calculation with resistivity $\eta \neq \eta(\psi)$ varying slightly (i.e., of the order of 1/aspect) on a flux surface gives a correction factor to D_1, which may be large compared to unity (3, 15), just as for the Pfirsch-Schlüter factor.

Another independent cancelation occurs in a Tokamak (or in any discharge with $\oint E \cdot dx \neq 0$). In a passive plasma, the net diffusion is directly related to the dissipation $\langle \eta J^2 \rangle$; in a discharge it is smaller by an order of magnitude. A similar sensitivity to the presence of a driving emf has been found for large mean free path in the form of trapped particle pinching (17).

II. The General Classical Calculation

With a general static equilibrium profile, the assumption curl $E = 0$ is not justified, even with arbitrarily small resistivity and β. One contributive factor is that the skin effect (field penetrating plasma) and plasma diffusion (plasma penetrating field) are coupled through the pressure balance, giving rise to a complex convective flow that modifies both primitive concepts. The skin effect is not the same in a deformable plasma as in a resistive solid conductor, since this would clearly violate the pressure balance. The plasma diffusion that accompanies the skin penetration is entirely nonstandard, even in scaling. There is a very strong geometrical dependence; circular sections are qualitatively different from noncircular ones; Tokamaks are qualitatively different from hard cores or multipoles; there are singular limits as the toroidal or poloidal field dominates the other; the skin effect can involve one field component or both; and in a θ-pinch or bumpy torus there is no skin effect at all on the scale D_0. Not only the physical conclusions, but also the mathematical structures, differ in these cases. Entirely different numerical schemes are required in order to "march" in different cases, and a physical variable (such as ψ) which is advanced because of a time derivative in one case $(\partial\psi/\partial t)$ will advance in virtue of an implicit constraint in another.

In a large number of cases a *standard sequence* (skin effect followed by "classical" diffusion) can be verified theoretically. For example, in a low-β, large-aspect Tokamak, the toroidal current will decay quickly in a time related to D_0. This process will approximate the standard skin effect if the plasma cross section is approximately circular; it will have the same scaling but will differ quantitatively from the standard skin effect if noncircular. During this process the pressure contours follow the changes in flux surface shape; the pressure remains approximately constant on a contour of fixed volume. The toroidal field cannot be said to diffuse; instead it fills in the gaps to maintain pressure balance. During the skin diffusion the plasma does diffuse slightly in relation to a fixed volume, but at a nonstandard rate. The diffusion rate is not local, and the process cannot be described in terms of mean values. After a number of skin times, the system settles down to one of the special profiles of the standard calculation. Provided that certain additional conditions are met, the system then diffuses slowly to a universal equilibrium with no adjustable parameters, independent of the initial profile [and even, to some extent, of the value of η (2)].

This standard sequence is not necessarily followed, for example, at finite β or even at low β at finite aspect. The ultimate equilibrium may not be reached because of instability (this is a new type of resistive instability), or it may not be reached because it does not exist. For example, a Tokamak equilibrium in a straight circular cylinder can be found with a suitable mass source (ionization of neutrals) taking the place of the discharge, that is, with $\oint E \cdot dx = 0$ (2). But no such equilibrium exists for arbitrarily small toroidal curvature (3). This is significant with regard to the "bootstrap Tokamak" concept (18, 19).

By varying the plasma cross section, the rate of approach to the universal equilibrium can be accelerated, and the plasma can be given more interesting parameters (e.g., higher β) (2).

A beginning of a transient (i.e., a general) theory can be found in the guiding center model as the "trapped particle pinch effect" (17).

The details of the diffusion theory for a general profile are quite elaborate (3); one qualitative reason for the vast difference between the special and the general theory can be described in terms of the relation of each to irreversible thermodynamics, which we now discuss.

III. Transport Coefficients, Symmetry, and Averaging

Perhaps one reason that the general theory of diffusion for the classical system (eqs. 1-4) remained undeveloped for so long is that is cannot be put into a standard "transport coefficient" framework, as can the special theory. It turns out, however, that even the special theory becomes standard only with further restrictions. This is best seen after we discuss several simpler examples.

All standard transport coefficients are subject to the Onsager symmetry relations. However, additional nonthermodynamic symmetries arise from quite a different source, as, for example, through a Green's function or topological symmetry. Entirely new problems arise when forces or fluxes are made subject to constraints or are averaged. Also, the extension of the symmetry principle to include skew symmetry for terms linear in B (Hall effect) may not hold for global or averaged quantities.

Consider first the classical problem of the flow of heat in a simply connected solid body (Fig. 1). We single out portions of the boundary, $S_1, S_2, \ldots$ which are held at temperatures $T_1, T_2, \ldots$. Solutions of the elliptic system

$$q = -\lambda \nabla T, \qquad \operatorname{div} q = 0 \tag{9}$$

can be described by a symmetric Green's function. Thus the net heat flow

$$Q_i = \int_{S_i} q \cdot dS \tag{10}$$

is related to the set of temperatures T_i by a symmetric heat conductivity matrix:

$$Q_i = \Lambda_{ij} T_j \tag{11}$$

The original problem has a single local transport coefficient λ and consequently has no Onsager identities. The discrete formulation in terms of T_i and Q_i gives

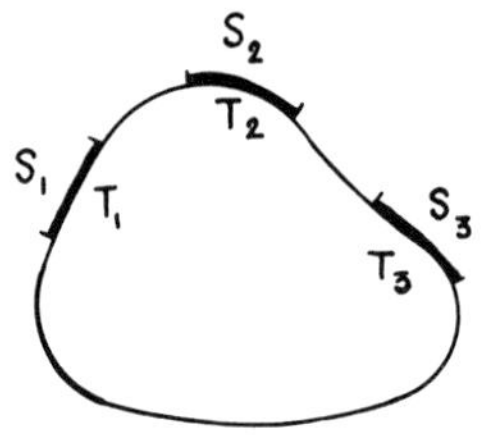

Fig. 1. Simply connected domain.

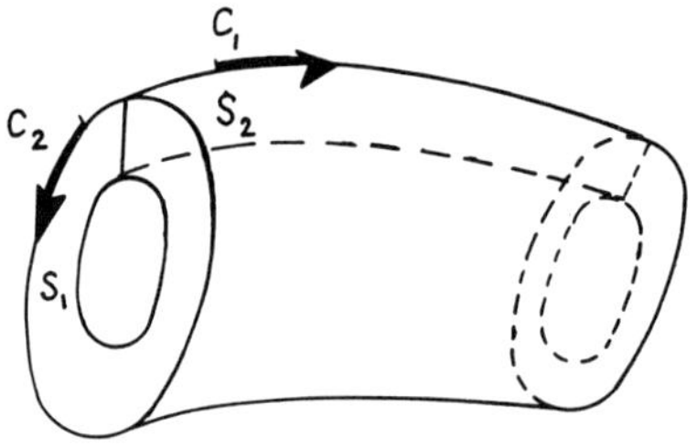

Fig. 2. Toroidal domain.

an array of forces and fluxes, and the symmetric relation (eq. 11) arises from the symmetry of the Green's function.

Consider next the dissipative relation:

$$E = \eta J \tag{12}$$

in a hollow toroidal copper shell (Fig. 2). We suppose that $\eta(x)$ is given and that

$$\text{curl}\, E = 0, \qquad \text{div}\, J = 0 \tag{13}$$

and $J_n = 0$ on the boundary. This is a classical elliptic system, and the manifold of solutions is known to be a two-dimensional linear vector space (20). A unique solution is determined by specifying a pair of periods, either emf's:

$$\oint_{C_i} E \cdot dx = c_i \tag{14}$$

or currents:

$$\int_{S_i} J \cdot dS = I_i \tag{15}$$

(see Fig. 2), which are related by the symmetric resistance matrix,

$$c_i = R_{ij} I_j \tag{16}$$

As in the previous example, a single local transport coefficient $\eta(x)$ proliferates to an array of forces and fluxes when examined globally.

In an abstract formulation,

$$v_i = A_{ij} u_j \tag{17}$$

where v_i are the "fluxes," u_j are the "forces," and A_{ij} is any nonsingular matrix, we can symmetrize trivially by choosing appropriate new fluxes $\bar{v}_i$. Given any symmetric nonsingular L_{ij}, we simply set

$$\bar{v}_i = L_{ij} u_j \tag{18}$$

where the new fluxes $\bar{v}$ are defined in terms of the old by $\bar{v} = LA^{-1}v$ (only minor modifications are necessary if the forces and fluxes are not independent and A and L are singular). Thus one can *always* symmetrize. The problem is whether this occurs through a natural or evident choice of variables in a given physical problem.

The proper abstract formulation of irreversible thermodynamics is to take forces and fluxes as dual linear vector spaces (in the standard terminology of linear algebra) with the entropy production as the defining inner product. The question of whether an abstract inner product (u, v) is symmetric has no meaning. The symmetry of a specific matrix representation of (u, v) depends on the choice of base vectors in u and in v.

Similarly, we have just seen that an abstract mapping from u to v, such as is represented by eq. 17 or 18, can be symmetric or not, depending on the choice of base vectors. However, the symmetry of a map from u to v *relative* to a given inner product (u, v) does have an invariant meaning. This is the content of the Onsager principle: that the relation between forces and fluxes is symmetric to a specific bilinear form, the entropy production.

If in the examples above the medium is anisotropic and there is a matrix relating the three components of J and of E, or if there are several coupled vector fluxes J^α and forces E^α, then we may have local Onsager identities at a given point x of the medium. The global symmetry, relating emf to total current, is a consequence of two distinct symmetries, one thermodynamic and local, and the other primarily topological.

The geometrical symmetry can be separated out and is also related to an inner product. In a domain of arbitrary topology, the linear vector space of closed curves C_i is isomorphic to the linear vector space of cross cuts S_i (Alexander duality theorem; cf. Ref. 20). A reciprocal basis can be found so that

$$(C_i, S_j) = \delta_{ij} \tag{19}$$

where (C_i, S_j) is the algebraic number of intersections of the curve C_i with the surface S_j.

Given two vector fields X and Y satisfying

$$\text{curl } X = 0, \qquad \text{div } Y = 0 \tag{20}$$

in the domain and $Y_n = 0$ on the boundary, an easy calculation gives

$$(X, Y) = \int X \cdot Y \, dV = \sum_1^n \xi_i \eta^i \tag{21}$$

where

$$\xi_i = \oint_{C_i} X \cdot dx, \qquad \eta^i = \int_{S_i} Y \cdot dS \tag{22}$$

(Many generalizations and extensions of this formula, Hodges' Theorem, are given in Ref. 20.) The system of eq. 20 is incomplete and does not determine X or Y. To complete the system we need a linear relation between X and Y. The simplest, $X = Y$, determines a manifold of solutions which is an n-dimensional linear vector space spanned by either ξ_i or η^i. If Z is a specific solution, its dual representation

$$Z = \Sigma \xi_i X^i = \Sigma \eta^i Y_i \tag{23}$$

where X^i and Y_i are reciprocal bases relative to the inner product (eq. 21)

$$(X^i, Y_j) = \delta_j{}^i \tag{24}$$

gives the symmetric matrix

$$\xi_i = (Y_i, Y_j)\eta^j \tag{25}$$

Taking $\xi_i = c_i$ and $\eta_j = I_j$, we have proved that $R_{ij} = (Y_i, Y_j)$ is symmetric.

Any local symmetric linear relation between X and Y that is used to complete the system will combine with the topological symmetry to give a globally symmetric result.

In a simple domain, dropping the boundary condition $Y_n = 0$ and setting $X = \nabla\phi$, we have

$$(X, Y) = -\int \phi Y_n \, dS \tag{26}$$

This symmetric inner product between boundary values of ϕ and Y_n [which is the source of the symmetry (eq. 11)], coupled with a symmetric local (Onsager) relation between Y and $\nabla\phi$, gives a globally symmetric structure.

To illustrate the effect of averaging, we consider an abstract mathematical model of forces $u_i(x)$ and fluxes $v_i(x)$, related by a symmetric matrix $L_{ij}(x)$:

$$v_i(x) = L_{ij}(x)u_j(x) \tag{27}$$

If u_j = const (independent of x), averaging clearly preserves the symmetric structure,

$$\int v_i \, dx = \left(\int L_{ij} \, dx\right) u_j \tag{28}$$

The same is true if v_i = const in terms of the reciprocal matrix, L^{-1}. The condition u_i = const or v_i = const is natural and not arbitrary. For example, minimizing the quadratic form

$$(u, v) = \int v_i L_{ij}{}^{-1} v_j \, dx \tag{29}$$

subject to fixed *total* flux:

$$\int v_j \, dx = V_j \tag{30}$$

gives, by the Lagrange multiplier rule,

$$\delta\left(\int v_i L_{ij}{}^{-1} v_j \, dx - \lambda_i \int v_j \, dx\right) = 0 \tag{31}$$

the natural minimization condition:

$$2u_i = \lambda_i = \text{const} \tag{32}$$

It is easy to see that there is no natural averaging which preserves the linear structure when both forces and fluxes depend on x.

We can now understand the successful averaging in the first two examples given above. The heat flow problem (eq. 9) gives a simple, symmetric averaged structure (eq. 11) when the T_i are taken to be constant on each boundary segment; otherwise it does not. Similarly, the electrical conduction problem (eqs. 12 and 13) gives a simple averaged result because curl $E = 0$ guarantees that the forces c_i are constant (path-independent).

Averaging occurs in a more complex way if J is constrained to lie in given surfaces, or if, because of highly anisotropic heat conductivity, T is constrained to be constant on specified surfaces. The standard transport formulation must be modified when there is a constraint, and, as we might expect, the inner product gives the clue.

If T is constrained to be constant on a given set of surfaces, $T = T(\psi)$, the inner product is

$$\int \nabla T \cdot q \, dV = \oint T q \cdot dS - \int T \operatorname{div} q \, dV \tag{33}$$

With $T = T(\psi)$, the volume term vanishes if $\langle \text{div } q \rangle = 0$ on each ψ surface. Thus restricting the admissible T forces us to expand the admissible q from $\text{div } q = 0$ at each point to $\langle \text{div } q \rangle = 0$ on each surface.

If J is constrained to lie in a specified surface, $\nabla\psi \cdot J = 0$, we can write

$$J = \nabla\psi \times \nabla\omega \tag{34}$$

and evaluate

$$\begin{aligned} \int E \cdot J \, dV &= -\int E \cdot \text{curl } (\omega \nabla\psi) \, dV \\ &= -\int \omega(\nabla\psi \cdot \text{curl } E) \, dV + \text{boundary terms} \end{aligned} \tag{35}$$

Thus the restriction on J requires us to relax E from $\text{curl } E = 0$ to $\nabla\psi \cdot \text{curl } E = 0$; E is a surface gradient on each ψ surface. Note that taking the conventional equation $\text{div } q = 0$ or $\text{curl } E = 0$ is incorrect (and overdetermines the system) when there is a constraint such as $T = T(\psi)$ or $\nabla\psi \cdot J = 0$. (In the plasma diffusion problem we are permitted to retain $\text{curl } E = 0$ rather than $\nabla\psi \cdot \text{curl } E = 0$ only because Ohm's law is relaxed to include an adjustable term, $u \times B$.)

The constrained current problem is formulated as

$$\begin{aligned} E &= \eta J, \qquad \nabla\psi \cdot \text{curl } E = 0 \\ \text{div } J &= 0, \qquad \nabla\psi \cdot J = 0 \end{aligned} \tag{36}$$

On each surface we have an elliptic system and eqn. 16 generalizes to

$$c_i(\psi) = R_{ij}(\psi) I_j'(\psi) \tag{37}$$

where $I_j(\psi)$ is the total current within the ψ surface. A simple symmetric structure obtains on each ψ surface, since the forces c_i are constant on a surface. But no such structure exists when averaged over ψ unless $\text{curl } E = 0$, rather than $\nabla\psi \cdot \text{curl } E = 0$.

Up to now we have taken a restricted Onsager principle without a magnetic field. With Hall effect, the relation between forces and fluxes is no longer symmetric. It is conventional to include the skew symmetry of terms linear in B (this is not a physical principle at all, but a consequence of the fact that B is not a vector) as an extension of the "symmetry" principle. This convention is not tenable, however, when we turn to global considerations. A Hall term in Ohm's law, $\mathbf{E} = \eta\mathbf{J} + \sigma\mathbf{J} \times \mathbf{B}$, or in heat conduction, $\mathbf{q} = -\lambda\nabla T + \tau\mathbf{q} \times \mathbf{B}$, makes the differential system non-self-adjoint, destroys the symmetry of the Green's function, and usually destroys the global symmetry between forces and fluxes. Symmetry cannot be restored by any expedient related to reversal of part of the matrix, since, for example, various magnitudes and orientations of $\mathbf{B}$ may be averaged globally.

The simplest example showing Hall effect is two-dimensional heat flow; we take the same problem as before (Fig. 1). We have

$$q = -\lambda \nabla T + \tau \mathbf{q} \times \mathbf{B} \tag{38}$$

where **B** is a constant vector perpendicular to the plane of interest, and λ and τ are constant. Using div **q** = 0, we can set

$$\mathbf{q} = \mathbf{n} \times \nabla \psi \tag{39}$$

where **n** is the transverse unit vector. We have

$$\mathbf{n} \times \nabla \psi = \nabla \phi \tag{40}$$

where

$$\phi = \tau B \psi - \lambda T \tag{41}$$

From eq. 40 we see that ϕ and ψ are conjugate harmonic potentials related by a symmetric Green's function. In particular, ϕ = const on each segment S_i of the boundary gives a symmetric global relation between ϕ_i and Q_i. But ϕ = const on S_i is inconsistent with T = const (ψ must vary if there is any heat flow through S_i). In fact, we have symmetrized by *changing bases* in eq. 41 (as we have shown, this can always be done). Even without making the problem discrete, that is, without averaging, we can verify that the boundary values $q_n(S)$ are *not* symmetric in terms of $T(S)$, although they are symmetric in terms of $\phi(S)$.

It should be pointed out that, when a Hall term such as **q** x **B** points in an innocuous direction, the desymmetrizing effect may disappear. For example, take

$$\mathbf{E} = \eta \mathbf{J} + \sigma \mathbf{J} \times \mathbf{B} \tag{42}$$

together with the remainder of eq. 36 (note that B is a given constraining field with $\mathbf{B} \cdot \nabla \psi = 0$ and, in this context, has no relation to J). The surface equation is the same as before, and the normal component (eq. 42) serves merely to determine the component $E \cdot \nabla \psi$ after J has been found. Note that this system is not self-adjoint, yet eq. 37 remains symmetric.

More generally, a set of fluxes q^α and forces $\nabla T^\alpha(\psi)$ gives a symmetric averaged matrix even with Hall effect. The same is true of a set of fluxes J^α, each constrained by $\nabla \psi \cdot J^\alpha = 0$. However, a system with mixed constraints, coupling (q, T), $T = T(\psi)$, with (J, E), $J \cdot \nabla \psi = 0$, is *not* symmetric and does *not* allow averaging in the presence of Hall effect.

Turning to the plasma diffusion problem, since there is only one local transport coefficient, η, any global symmetries that arise are not basically thermodynamic, With the special equilibrium of eq. 6, the dissipation takes the form

$$\int \eta J^2 \, dV = c_1 I_1 + c_2 I_2 = \int (u \cdot \nabla p) \, dV \tag{43}$$

where I_1 and I_2 are the total poloidal and toroidal currents. An unusual, symmetric "matrix" arises (3) relating the forces

$$c_1, c_2, -p'(V) \tag{44}$$

and the fluxes

$$I_1, I_2, \oint_V u \cdot dS \tag{45}$$

Note that c_1, c_2 and I_1, I_2 are global (averaged over the entire domain), whereas $p'(V)$ and $\oint u \cdot dS$ are local to a surface. The symmetry under averaging is preserved because c_1 and c_2 are constant (curl $E = 0$) and p is constant on a flux surface. Variation of η with ψ or even over a flux surface is irrelevant to the structure (but can change the numerical values considerably).

It is immediately evident that inclusion of heat flow, either by itself or coupled to diffusion, can yield a standard structure after averaging over flux surfaces only when $T = T(\psi)$. This restriction is inadmissible in the present theory, but it may be valid at very large mean free path. Coupled heat flow and diffusion yields an innocuous Hall effect (both are of type q^α above). But in a discharge, $\oint \mathbf{E} \cdot \mathbf{dx} \neq 0$, symmetry is lost and there is no longer any simple relation between averaged forces and fluxes (since the system then has mixed constraints).

Similarly, with fluid flow (steady-state rather than static equilibrium) and Hall effect, since the nondissipative equilibrium has distinct flux and pressure surfaces, $p \neq p(\psi)$ (21), even diffusion by itself cannot be described in terms of net flow and other flux surface averages. In a guiding center model, the equilibrium can sometimes be written as (22)

$$\begin{gathered} \mathbf{K} \times \mathbf{B} = \left(\frac{\partial p}{\partial \psi}\right) \nabla\psi \\ \mathbf{K} = \text{curl}\,(\sigma \mathbf{B}) \\ \sigma = 1 - \frac{(\partial p/\partial B)}{B} \end{gathered} \tag{46}$$

where $p = p(\psi, B)$ is the parallel pressure. Here $\mathbf{K}$ takes the place of $\mathbf{J}$ and is constrained to lie in ψ surfaces. In Ohm's law,

$$\mathbf{E} + \mathbf{u} \times \mathbf{B} = \ldots \tag{47}$$

multiplication by $\mathbf{K}$ gives an averaged term $\langle \mathbf{E} \cdot \mathbf{K} \rangle$ which can be evaluated in terms of global emfs, but

$$\langle \mathbf{u} \times \mathbf{B} \cdot \mathbf{K} \rangle = -\left\langle (\mathbf{u} \cdot \nabla\psi) \frac{\partial p}{\partial \psi} \right\rangle \tag{48}$$

is not simply interpretable. Specializing to

$$p = p_1(\psi)p_2(B) \tag{49}$$

we have

$$\langle \mathbf{u} \times \mathbf{B} \cdot \mathbf{K} \rangle = -p_1' \langle (\mathbf{u} \cdot \nabla\psi) p_2 \rangle \tag{50}$$

which indicates that the force p_1' may be associated with a *weighted* plasma flow. The net plasma flow occurs in an averaged calculation only to the extent that B = const on flux surfaces (see also Ref. 15).

In a kinetic model, there are no local transport coefficients; it requires flux surface averaging to introduce this concept at all. There is the same distinction between a general and a special theory (the latter with curl $E = 0$ and restricted profiles). As in the small-mean-free-path theory, there are no averaged transport coefficients in the general theory, and strict limitations govern the use of averaging in the special theory. The consequence is that there is no widely applicable theory of transport coefficients, either local or averaged over surfaces. In other words, each problem is solved globally as a special boundary value problem for a special profile, and there is no direct carryover to another problem with different shape or profile (as there would be for a legitimate transport coefficient).

To elucidate the general theory, we consider the relation between conservation equations and dissipation. Conservation of energy in a static fluid can be written as

$$C_V \frac{\partial T}{\partial t} + \operatorname{div} \mathbf{q} = 0 \tag{51}$$

which together with the transport relation

$$q = -\lambda \nabla T \tag{52}$$

gives a simple dissipative diffusion equation for T. Note that the force is a gradient of the quantity that is conserved. The isothermal conservation of mass

$$\frac{\partial p}{\partial t} + \operatorname{div}(p\mathbf{u}) = 0 \tag{53}$$

together with "classical" plasma diffusion,

$$\mathbf{u} = \frac{-D_1 \nabla p}{p} \tag{54}$$

gives a similar diffusion equation for p. Conservation of flux,

$$\frac{\partial \mathbf{B}}{\partial t} + \operatorname{curl} \mathbf{E} = 0 \tag{55}$$

together with the transport relation $\mathbf{E} = \eta \mathbf{J}$, gives a similar diffusion equation for flux:

$$\frac{\partial \mathbf{B}}{\partial t} = -\text{curl}\left(\frac{\eta}{\mu_0} \text{ curl } \mathbf{B}\right) \tag{56}$$

There is a complete analogy of flux, mass, and energy conservation. But, curiously, the analogy holds only if **E** is called the flux and **J** the force; also the heat *conductivity* λ takes the same position as electrical *resistivity* η, namely, as a diffusivity coefficient. This has a number of interesting consequences that we shall not pursue here.

In all these time-dependent formulations the transport relation itself does not depend explicitly on time. This is the crucial difference, in the present context, between the special plasma diffusion theory and the general theory. In the latter, although the local transport relations (when they exist) are always time-independent, the averages over flux surfaces can *never* be separated from the global transient solution of the entire system. There is no doubt that the situation will be found to be the same in all finite or large-mean-free-path models when the general time-dependent case is attempted. We conclude that the special solutions, with curl $\mathbf{E} = 0$, give limited information about plasma transport behavior in general. And even within the special theory, only in certain well-defined special cases do the transport properties reduce to a calculation of flux surface averages.

IV. Conclusion

In a small-mean-free-path plasma, local transport coefficients exist at each point of the plasma. Transport coefficients in a finite-mean-free-path system can exist only as a consequence of averaging over flux surfaces. The condition curl $\mathbf{E} = 0$ follows, not from low β, but from a specialization of profiles (and sometimes from aging the system). In the special theory, with curl $\mathbf{E} = 0$, in addition to the fact that this postulate represents only a small part of the physical world, averaging leads to a conventional system of transport coefficients only after additional specialization. Even when there do exist averaged transport coefficients, they frequently cannot be found directly from an entropy production quadratic form because of the interference of Hall effect, which may destroy the averaged symmetry.

The general theory, with curl $\mathbf{E} \neq 0$, exhibits many new dissipative properties and requires new analytic techniques. It turns out that numerical calculations of transients must be done in an actual toroidal geometry. Theoretical results have thus far been obtained only for the macroscopic case, and even here mostly for diffusion, uncoupled to energy flow. It has been predicted (1, 2) that corrections of up to two orders of magnitude should be expected to arise when this theory is generalized to finite or large-mean-free-path problems.

References

1. H. Grad and J. Hogan, *Bull. Am. Phys. Soc.*, **14**, 1017 (1969).
2. H. Grad and J. Hogan, *Phys. Rev. Lett.*, **24**, 1337 (1970).
3. H. Grad and J. Hogan, article to appear.
4. A. A. Galeev and R. Z. Sagdeev, *JETP*, **26**, 233 (1968) [Russian: **53**, 348 (1967)].
5. M. N. Rosenbluth, R. Hazeltine, and F. L. Hinton, article to appear.
6. Y. N. Dnestrovski, D. P. Kostamarov, and N. L. Pavlova, *Fourth European Conference on Plasma Physics*, Rome, 1970.
7. D. F. Düchs, H. P. Furth, and P. H. Rutherford, Radial Transport in Tokamak Discharges, IAEA Conference, Madison, Wis., June, 1971.
8. H. Grad, Plasma Containment in Closed Line Systems, IAEA Conference, Madison, Wis., June 1971.
9. H. Grad, *Phys. Fluids*, **10**, 137 (1967).
10. E. Frieman, *Phys. Fluids*, **13**, 490 (1970).
11. J. L. Johnson and S. von Goeler, *Phys. Fluids*, **12**, 255 (1969).
12. M. D. Kruskal and R. M. Kulsrud, *Phys. Fluids*, **1**, 265 (1958).
13. T. Ohkawa and H. G. Voorhies, *Phys. Rev. Lett.*, **22**, 1275 (1969).
14. H. P. Furth et al., *Phys. Fluids*, **13**, 3020 (1970).
15. J. Nührenberg, Comparison Between Stationary Tokamak Models (to appear).
16. D. Pfirsch and A. Schlüter, Max Planck Inst., Munich, Reprt. No. MPI/PF/7/62, 1962.
17. A. A. Ware, *Phys. Rev. Lett.*, **25**, 916 (1970).
18. R. J. Bickerton, J. W. Connor, and J. B. Taylor, *Nature*, **229**, 110 (1971).
19. B. B. Kadomtsev and V. D. Shafranov, Stationary Tokamak, IAEA Conference, Madison, Wis., June, 1971.
20. A. A. Blank, K. O. Friedrichs, and H. Grad, Notes on MHD-V, NYO-6486, New York University, 1957.
21. D. Dobrott and H. Grad, Steady Flow with Hall Effect (to appear).
22. H. Grad, *Trans. N.Y. Acad. Sci.*, II, **33**, 163 (1971).

Present Status of the Theory of Plasma Instabilities

A. B. MIKHAILOVSKII

Kurchatov Institute of Atomic Energy, Moscow, USSR

Abstract

The paper considers the present status of the theory of plasma instabilities. The role played by stability theory in the general balance of plasma theoretical studies is pointed out, and possible trends in the theory of plasma stability are considered. A review of some experimental results is given as evidence that instabilities exist.

Introduction

Studies of instabilities occupy a primary place in plasma theory, as the program of the present conference convincingly demonstrates. It should be kept in mind that instabilities are also discussed even in papers whose titles do not include the word "instability." In any event one has to deal with the problem of instability when considering transport processes in mirror machines. When talking about anomalous resistivity (or anomalous conductivity), we mean certain nonlinear effects accompanying the development of some kind of instability. The main concern of plasma turbulence theory is the study of a nonlinear stage of instability. Investigations of instabilities also occupy a considerable place in works on interactions between a high-frequency field and plasma.

Such a "dominance" of instabilities is a characteristic feature not only of the present conference. The number of journals publishing papers on theoretical and experimental studies of instabilities is growing. Now not only general physics journals or periodicals on nuclear fusion research publish papers concerned with plasma instabilities. Instabilities are widely discussed also in connection with magnetohydrodynamic energy conversion. The presence of instabilities is revealed in solid-state plasma. Instability concepts are widely used to interpret geophysical and astrophysical phenomena. All this sufficiently demonstrates that instabilities are "omnipresent". It should be added that all review papers in a recent issue of *Reviews of Plasma Physics* (Vol. 6) were also concerned with instabilities.

I. "Thermal" Instabilities

Thus the theory of plasma stability is being developed. What is the main trend in this development? It involves the fact that now the conservation of energy principle must be more carefully analyzed, whereas in the past theoreticians exercised more care in analyzing the law of conservation of momentum. For the simplest case of a hydrodynamical approximation this means that the equations of heat balance should be written along with the equations of motion. Practically, this should be understood as a certain displacement of the interests of the theoreticians from one class of instabilities, which may be referred to as "force" instabilities, to another class: the so-called "thermal" instabilities.

The two-stream instability first investigated by Akhiezer and Feinberg can be included in the class of "force" instabilities. In deriving the dispersion relation for the two-stream instability, we use such concepts as deceleration and acceleration of particles, while the energy balance is considered only if one has to take into account the square amplitude of the waves. The drift instability may serve as another example of force instability. The dispersion relation for this instability can be written without the equations of heat balance.

"Thermal" instability includes, for example, the acoustic instability of a weakly ionized plasma. This instability is due to transfer to the neutral atoms of the energy obtained by electrons during Joule heating of the plasma. Thermal instability may also be represented by the so-called instability of entropy waves. It is associated with the drift-convective heat transfer in a nonuniformly heated plasma.

The tendency toward development of a theory of thermal instabilities is associated with a departure from the traditional lines of theoretical studies, which were concentrated for a relatively long time on collisionless low-pressure plasma in a uniform magnetic field. The "thermal" effects appear also in such a plasma. However, they declare their presence immediately as we start to study a finite-β plasma, or a plasma in a curved magnetic field, or a plasma that is under the conditions of frequent particle collisions.

One class of thermal instabilities inherent to a finite-β plasma has been known for a long time. This consists of the electromagnetic instabilities of an anisotropic plasma. When these instabilities develop, redistribution of energies connected with the motion of particles along and transverse to the magnetic field takes place. The theory of electromagnetic anisotropic instabilities remained static for a long period of time. A few years ago, however, it acquired a new meaning due to a simple but rather fruitful idea about a two-component character of the velocity distribution of particles. In other words, one began to take into account the fact that, in addition to a group of fast particles with anisotropic velocity distribution, a plasma could contain a group of "cold" particles of the same kind (electrons or ions). Such a situation proves to exist very often in practice—in particular, in the radiation belts of the Earth and in adiabatic traps as well. Under such assumptions it is possible to interpret a rather broad range of phenomena. According to Tverskoy in particular, the synchrotron X-radiation from the Crab nebula can be explained on this basis.

Another kind of thermal instabilities of a collisionless finite-β plasma is observed when the temperature gradient is transverse to the magnetic field. These instabilities were first pointed out by Bogolubov and Tzerkovnikov. As is true of the above instability of entropy waves, the effect of the drift-convective heat transfer across the magnetic lines of force forms the basis for the mechanism of these instabilities.

In recent years great attention has been paid to entropy wave instabilities (known also as magneto-drift instabilities). In particular, this interest is associated with possible thermonuclear utilization of θ-pinch installations. According to modern ideas, such instabilities are the basis of the potentially possible micro-instabilities of a θ-pinch. Investigations carried out in recent years have shown that these instabilities are not insurmountable obstacles to the prolonged confinement of a high-temperature θ-pinch plasma.

Thermal effects appear also in a low-β plasma in a curved magnetic field. In this case the convective heat transfer is due to the particle drift under the influence of the curvature of the magnetic lines of force. Most favorable conditions for devoloping the magneto-drift instabilities appear when the lines of force are of variable curvature (i.e., in the stellarator-type systems).

Thermal effects of another kind can be observed under conditions of frequent collisions between the particles when heat exchange between the electrons and ions, thermal force, and finite thermal conductivity are of great importance. It has been shown by Moiseev that even at small β a number of new nontrivial instabilities of an inhomogeneous plasma can be found by taking temperature disturbances into account. However, a finite-β collisional inhomogeneous plasma is especially rich in "thermal" effects. In this case, in addition to the effect of spatial heat transfer, of great importance are the collisional redistribution of heat between different degrees of freedom (a hydrorelaxation effect) and the so-called "cross" effects: the effect of viscosity on thermal conductivity and a few other effects that are not taken into account in generally known equations of two-fluid hydrodynamics. As a whole, it is evident from an analysis of such effects that a plasma which is heated less than is required for being collisionless should be more unstable than a collisionless plasma.

One more important result obtained from the study of a finite-β inhomogeneous collisional plasma is that at finite β some "force" instabilities are not very important, whereas they are essential for $\beta \to 0$. In particular, it appears that the Moiseev-Sagdeev drift-dissipative instability is of no importance for finite β. This is quite understandable if one remembers that this instability is due to friction between the electrons and ions, that is, it depends on the current velocity disturbances, and if one takes into account that the disturbances of the current velocity in a dense plasma should be small (otherwise the magnetic field would be very greatly disturbed).

Accounting for the "thermal" effects is necessary also in considering certain kinds of two-stream instability in a collisional plasma. It was believed long ago that the current flowing through a fully ionized plasma should not give rise to collisional excitation of the ion sound. This was associated with the fact that in the case of Coulomb collisions the conductivity should not depend on the density and hence does not vary when the density is disturbed.

Kuckes has shown, however, that both the density and the temperature are disturbed in ion-sound oscillations. Thus a ban on the collisional excitation of ion sound in a fully ionized plasma was removed.

One more kind of thermal instability of a collisional fully ionized plasma is observed if the Joule heating is accounted for. An example of such an instability is an overheating instability considered by Kadomtsev in the early 1960s. Recently the theory of overheating instabilities has been further developed by Rosenbluth, Furth, and others. An interesting feature of their work is an indication of possible excitation of disturbances in a nonuniform-temperature plasma due to impurity radiation.

The Forslund paper (from Los Alamos) also suggests very interesting ideas concerning the "thermal" effects. We are used to the instabilities caused by a current flowing through the plasma. Current instabilities are known to result from the positive derivative of the longitudinal velocity electron distribution function, $\partial f/\partial v_\parallel > 0$. Even if the current is absent, the temperature gradient and an electric field balance the thermal force and the pressure gradient. Forslund noticed that in this case also $\partial f/\partial v_\parallel > 0$, that is, instabilities may be present even in the absence of the current. They may have quite real meaning if the ratio between the electron free path length and the characteristic temperature gradient [greater than $(m_e/m_i)^{1/2}$] is not very small. In this connection a few questions arise. In particular, what will be the result of these instabilities? How will the true expressions for the transfer coefficients—thermal conductivity, thermal force, and others—look under these conditions?

II. Some Other Aspects of the Theory of Plasma Stability

Though the conservation of energy principle is almost omniponent", the development of the stability theory is not bound only to an analysis of that law. Understanding of the fact that, in addition to the laws of mechanics, one should also take into account the law of electrodynamics following from the Maxwell equation is growing. For example, it has become more and more evident recently that nonuniformity of a magnetic field should be taken into account when studying the instabilities of a current-carrying plasma. As a result we have to rule out some false instabilities and introduce new, true ones. The "false" instabilities include both the kinetic instability of ion-acoustic waves, which is due to the current flowing along the magnetic field (this is described in *Reviews of Plasma Physics*, Vol. 4), and the collisional instability of electromagnetic waves, which is caused by the current when there is no magnetic field (JETP, 1965).

Among the new "true" instabilities is an instability described by Ivanov and Parailov of nonpotential waves of a helicon kind in a slightly ionized recombination plasma. As has been shown by Garry and Sanderson, the nonuniformity of a magnetic field can also give rise to excitation of the Bernstein modes.

The "quasiclassical approximation" is a powerful weapon of theoreticians. In recent years, in connection with developments in geophysics, the quasiclassical approximation has been enriched with new ideas. In particular, "new" instabilities were discovered when the nonquasiclassical terms are included in the quasiclassical dispersion equation.

III. Some Applications of Stability Theory

Plasma instabilities are the most active kind of collective interactions between particles. They often play a determining part in plasma dynamics. This accounts for the fact that theoreticians who work in the field of plasma stability theory may influence experimental researches more appreciably than theoreticians working in other areas of plasma physics. Certain conclusions drawn from the stability theory have been recognized by not only the experimentalists but also a wider circle of the scientific public. Thus current-convective instability was registered as a discovery, while the discovery of flute instability was granted the State Prize. It should be expected that the discovery of beam stream instability, that is, the number 1 instability in plasma physics, would gain similar recognition.

Recent laboratory studies have shown that, in addition to such instabilities as stream, flute, and current-convective instabilities, many others also exist in practice.

Studies of a hot electron plasma confirmed the fundamental conclusions of the theory of electron-cyclotron instabilities, which are due to anisotropy and transverse nonequilibrium in the electron velocity distribution function. Morse from the Culham laboratory carried out a special experiment for studying an electrostatic anisotropic instability. He revealed good agreement between the experiment and theory. Experimental work carried out by Perkins and Barr resulted in the confirmation of $\partial f/\partial v_\perp > 0$ as responsible for the cyclotron instability. Very interesting experiments on the study of maser-type cyclotron instabilities were performed by Blanken and Kuckes.

A large number of experiments carried out in the laboratories of many countries provided support for the theory of electromagnetic electron-cyclotron instabilities. In this connection it is of interest to point out that the electromagnetic instabilities (which are sometimes referred to as the Sagdeev-Shafranov instabilities) obtained their "right to citizenship" under laboratory conditions only after their important contribution to the dynamics of the Earth radiation belts had been demonstrated. Before that the experimentalists "preferred" the electrostatic instability (known as the Harris type of instability). However, measurements of the dispersion characteristics and polarization of oscillations excited in plasma revealed that in many cases the electromagnetic instabilities were responsible for the phenomena observed.

Cooperation between theory and experiment has been very clearly demonstrated by investigations of electron-ion collective interactions. First of all, this involves two cases: the one in which electrons move relative to ions (plasma current), and the problem of plasma confinement in the loss-cone adiabatic traps. In the former case anomalous resistivity was produced and the plasma was effectively and rapidly heated. In the latter case it was possible to find additional arguments against adiabatic traps. The best understanding has been gained of the current instabilities, largely because of a wide range of experimental and theoretical studies, though the opinions of the experimentalists and the theoreticians are sometimes quite opposed to each other. As to the problem of loss-cone instabilities, the final answer will not be obtained so soon. One cause is the dismantling of experimental facilities (e.g., the OGRA device) just before physical results might have been obtained. Another cause is Dr. Rosenbluth's move from Livermore to Princeton.

One of the "oldest" instabilities, that is, the electromagnetic instability due to ion anisotropy, has been found in the laboratory. It is not very easy to observe instabilities of this type, since for their appearance to be clearly seen it would be necessary to produce an anisotropic finite-ion-β plasma. These instabilities could be observed only after such a plasma had been produced in θ-pinches. The axisymmetrical disturbances of a θ-pinch were identified with the anisotropic instability.

The slightly ionized plasma instabilities due to the density gradient (the so-called drift or universal instabilities) are studied well experimentally, and it is possible to deal with them rather successfully, including their feedback stabilization. As to more dangerous instabilities, they have been studied with considerably less success.

Interesting phenomena had been theoretically predicted to exist in a strongly inhomogeneous plasma. According to the theory, oscillations having frequencies of the order of the ion-cyclotron frequency and even higher should be excited in such a plasma. Experiments carried out in recent years have confirmed this prediction of stability theory. Nezlin and his followers from the Lebedev Institute found a family of so-called beam-drift instabilities. Marinin (Kharkov), followed by Suprunenko, Stepanov (Kharkov), and Alexeff (Oak Ridge), has profoundly studied both the high-frequency and ion-cyclotron instabilities, which are caused by centrifugal ion drift in a radial electric field. Vdovin and Rusanov investigated the excitation of a high-frequency ion sound due to a high-density gradient.

Plasma becomes strongly inhomogeneous also when the electrons are heated by a high-frequency field. In experiments of the type described by Dubovoy, the electron Larmor radius turned out to be finite in comparison with a characteristic plasma inhomogeneity. Under these conditions, in addition to the usual acoustic instability due to the longitudinal current, instabilities excited by the electron Larmor current (caused by the transverse density gradient) may be of importance. These instability increments are greater than those of the ion-acoustic instability, and the oscillations are excited not only at $T_e > T_i$ but also at $T_e \simeq T_i$. The possibility is not excluded that anomalously strong ion heating, which was observed in the experiments with high frequency discharges, can be attributed to gradient instabilities.

IV. Conclusion

As is clearly seen, the theory of instabilities is not "a thing in itself." It is widely used for interpreting experimental results. On the other hand, it is the development of experimental studies that causes the theoreticians to write new dispersion relations and to calculate the increments of new instabilities. In regard to the development of the instability theory itself, the above considerations show that it should be extensive and fruitful for a long time.

The Propagation and Stability of Finite-Amplitude Waves in Plasmas

V. N. ORAEVSKY

Institute for Nuclear Research, Academy of Sciences of the Ukrainian SSR, Kiev, USSR

Introduction

The peculiar character of plasma physics and, in particular, its distinction from ordinary gas dynamics are closely connected with wave and oscillatory processes. For a neutral gas information concerning its state is communicated to other regions by particle collisions and the information transfer velocity is connected with the velocities of particles. For plasmas the picture differs substantially. Because of Coulomb forces the entire ensemble of particles often "feels" very quickly any local changes in plasma state and begins participating in collective motions—oscillations and waves.

The most widespread situation in plasma is one in which the characteristic space-time scales of wave and oscillation motions are substantially smaller than collision ones; hence it is natural that in such cases plasma can abolish thermodynamical nonequilibrium in a most rapid way, through the excitation of waves and oscillations. This leads, in particular, to transport phenomena that are not connected with particle-particle collisions, to the so-called abnormal transport phenomenon, and to its attendant effects—the rapid departure of particles from the confinement volume, so-called collisionless heating and collisionless shocks, and so on.

In the light of the facts mentioned above, it is not surprising that there are many papers on the theory of wave processes in plasmas. In this survey a brief account of the main ideas and results of the theory of finite-amplitude wave propagation and stability in plasmas is given.

I. Wave Propagation in Plasmas; Finite-Amplitude Steady-State Waves

In the processes of wave propagation in plasmas dispersion and polarization play an important role. It is known (1–25) that steady-state waves of finite amplitude exist in plasmas with absolute neglect of dissipation. Generally speaking, this neglect cannot be justified in the study of steady-state motion in ordinary gas. It is just the dissipative effects which limit the growth of steepness of the front of an initially cosine wave (that is seen clearly in the solution to Burger's equation; see, e.g., Ref. 24). The picture differs in plasmas. Here there exist, for example, Alfvén waves and helicons, which can propagate without "breaking" because cross-polarization (nonlinear) terms of the type $(v \cdot \nabla)v$ responsible for "breaking" are absent in equations describing these waves. In cases in which the tendency to "break" does exist, dispersion effects can limit the growth of steepness of the wave front (see, e.g., Ref. 6). Indeed, increasing the wave front steepness means the appearance of multiple harmonics. The rapid appearance of multiple harmonics takes place when the "driving" force (arising in the equations on account of nonlinear terms) is in resonance with the linear oscillations of the system. Such a force has the form (in the second approximation) $\alpha_1 A^2 \exp[-i(2\omega t - 2kx)]$, where A, ω, and k are the amplitude, frequency, and wave vector of the initial wave, respectively. Obviously, if the dispersion law differs from the linear one (i.e., $\omega \neq ks$), the resonance will not take place, and steady-state waves can exist in rare plasma (where collisions can be neglected).

As an example, let us consider magnetosonic wave propagation in low β plasma ($p \ll H_0{}^2/8\pi$). The equation describing the magnetosonic waves can be obtained when

$$\alpha^2 \gg \frac{k^2 c^2}{\omega_{0i}{}^2}, \qquad h = H_2 - H_0 \sin\alpha \ll H_0 \sin\alpha$$

where α is the angle between the magnetic field and the direction of propagation, $\omega_{0i}{}^2 = (4\pi n e^2/m_i)$:

$$h_t + \left(v_A + \frac{3}{2} v_A \sin\alpha \frac{h}{H_0}\right) h_x + \beta h_{xxx} = 0 \tag{1}$$

where

$$v_A = \frac{H_0}{\sqrt{4\pi\rho_0}}, \qquad \beta = v_A \frac{c}{2\omega_{0i}{}^2}\left(\frac{m_e}{m_i} - \cot^2\alpha\right)$$

This equation coincides with the Korteveg-de Vries equation (26).

Let us make use of perturbation theory (small parameter = wave amplitude) and find the solution in the form $h = h_1 + h_2 + \cdots$. Then the linear equation for the quantity h_1 has the form

$$h_{1t} + v_A h_{1x} + \beta h_{1xxx} = 0 \tag{2}$$

A cosine wave is the solution $h_1 = A \exp i(-\omega t + kx)$, and the dispersion equation can be written as $\omega = kv_A - \beta k^3$. Let us write down the equation for h_2 and its formal solution:

$$h_{2t} + v_A h_{2x} + \beta h_{2xxx} = -3_{ikv_A} \frac{\sin \alpha}{H_0} A^2 \exp [i(-2\omega t + 2kx)] \tag{3}$$

$$h_2 = \frac{3kv_A A^2 \sin \alpha \exp(-2i\omega t + 2ikx)}{H_0[2\omega(k) - 2kv_A + \beta(2k)^3]} \tag{4}$$

It is seen from eqs. 3 and 4 that for a linear law of dispersion ($\beta = 0$) the multiple harmonics get into resonance with the first harmonic beats. But dispersion limits the harmonic growth.† With the help of eq. 1 it is a simple matter to find the steady-state wave profile: $h = h(x - ut)$. Then $h_t = -uh_x$, and eq. 1 reduces to the conventional differential equation, easily integrated once:

$$\beta \frac{d^2h}{dx^2} = -\frac{\partial w}{\partial h}; \qquad w = -b\frac{h^2}{2} + c\frac{h^3}{3} \tag{5}$$

where $b = v_A - u$, and $c = (3/2)v_A (\sin \alpha/H_0)$. Note that more complex equations than eq. 1, for cases in which laminar steady-state waves of low amplitude exist, may be reduced, in fact, to equations like eq. 5. The solution of eq. 5 takes the periodic form:

$$h(x - ut) = \hbar_1 + (\hbar_2 - \hbar_3) \operatorname{sn}^2 \left(\sqrt{\frac{\hbar_1 - \hbar_2}{3\beta}} \frac{x - ut}{2}; \gamma \right) \tag{6}$$

where $\hbar_1$, $\hbar_2$, and $\hbar_3$ are roots of the equation $bh^2 - (c/3)\, h^3 + \text{const} = 0$, and $(\hbar_1 \geqslant \hbar_2 \geqslant \hbar_3)$; the solitary wave solution is

$$h(x) = ch^{-2} \left(\sqrt{\frac{b}{\beta}} \frac{x - ut}{2} \right) \tag{7}$$

For a qualitative consideration of the steady-state wave problem in plasma and of the influence of dissipation on these waves, it is convenient, following Sagdeev, to make use of a mechanical analogy (10). Equation 5 is the equation of motion of a material point with mass equal to β in a potential well w (h serves

† The nonlinearity parameter is seen from eq. 4 to be: $1/4 (\sin \alpha A^2/H_0 \beta k^2)$.

Case	Linear Dispersion	Dispersion Size	Form of Solutions	Critical MACH Number	Instabilities and Turbulent Shock Sizes	Ref.
$u \perp H;\ \beta \ll 1$ $nmc^2 \ll \frac{h^2}{8\pi} \ll nMc^2$	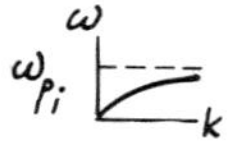	$\frac{c}{\omega_{p_2}}\sqrt{\frac{H^2}{nmc^2}}$		2		1 5 7
$u \perp H;\ \beta \ll 1$ $\frac{H^2}{8\pi} \ll nmc^2$	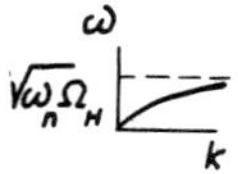	$\frac{c}{\omega_{p_e}}$		2	Beam instability $M > 1 + \frac{1}{3}\left(\frac{8\pi nT}{H^2}\right)^{1/3}$ $\Delta \sim \frac{c}{\omega_{p_e}}\left(\frac{m_i}{m_e}\right)^{1/6}\left(\frac{c}{v_A}\right)^{1/3}$	2 3 4
$u \perp H;\ \beta \ll 1$ $e_1 M_1 n_1$ $e_2 M_2 n_2$		$\frac{c}{\omega_i}$ $M_1 \ll M_2$				
$H = 0$ $T_e \gg T_i$	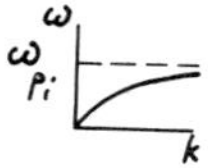	r_{D}		$-1, 5$		6 10
$u \perp H$ $\theta > \sqrt{m/M}$	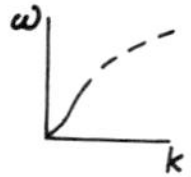	$\frac{c}{\omega_{p_i}}\theta$			Decay instability of Buneman	9 10

$u_{\perp}H$ $\beta_i \geqslant 1$	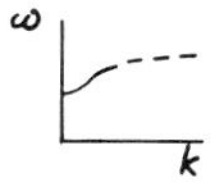	r_{H_i}			Decay instability	12
$H = 0; H \neq 0$ $\langle \epsilon \rangle = T_e \gg T_i$		r_D	Depend on distribution junction of the particles			13
$H \neq 0; T_e \gg T_i$ $\nabla T_e \neq 0; \nabla n \neq 0$ $\tan\theta > s/v^*$	Drift waves	$\sqrt{r_D{}^2 + r_{H_i}{}^2}$	 $\dfrac{d \ln T_e}{d \ln h} < 0$  $\dfrac{d \ln T_e}{d \ln h} > 0$			15 20
$T_e \gg T_i$	Electrosonic waves	$\dfrac{c}{\omega_{p_i}}$		< 1		21
$H = 0; T_e \gg T_i$ $\nabla T_e \neq 0; \nabla n \neq 0$ $\tan\theta < s/v^*$	Ion-sonic and slow magnetic-sonic waves	$\sqrt{r_D{}^2 + \tan^2\theta(r_D{}^2 + r_{H_i}{}^2)}$				22

as coordinate, x as time). Motion taking place near the well bottom, $h = 2b/c$, is almost a harmonic oscillation. The corresponding solution is

$$h = 2\frac{b}{c} + h_0 \exp\left[i\sqrt{\frac{b}{\beta}}(x - ut)\right] \tag{8}$$

The maxima (or minima) move apart as the wave amplitude increases. For the case under consideration, when h is increased, the "elasticity" increases too; hence the particle would jump more quickly to larger values of h. The maxima would move apart until the value $h = 0$ became possible. In this case the solution has the form of solitary waves (infinitely long "rolling up" of corresponding particles, reflection in the point $h = 3(b/c)$, and infinitely long "raising" up to the point $h = 0$). Taking account of dissipation changes the character of the motion. Independently of dissipation type (collisional or collective), the motion can be of two types: (*a*) damped oscillations (when dissipation is small)–in accordance with this the solutions belong to laminar shock waves with oscillatory structure (1, 7, 10); and (*b*) shocks (when the particle rolls up to the well bottom at once). If the dissipation character is a collective one, we deal with so-called collisionless shock wave turbulent structure. The accompanying table (the composition scheme is taken from Ref. 14) reflects the main results of steady-state and collisionless shock wave investigations in plasma. In addition to the data reported in the table, it is desirable to point out some other details.

In Ref. 13 (see also Refs. 20 and 22), the steady-state wave character (compression or rarefaction) is shown to be dependent on the type of distribution function.

A highly interesting wave type, the so-called electrosonic, is described in Ref. 21. The initial wave pressure was shown to change the density in a periodic manner upon incidence of a wave with frequency $\omega < \omega_0$ on the plasma. This results in capture of the wave field between density humps, which these propagate into the plasma with acoustic velocity.

After finding solutions of the types in eqs. 7 to 9, the following natural questions arise: can laminar steady-state waves exist in real plasma, and, if so, for how long? To answer these questions the stability of the solution must be investigated first of all. To do this not only waves belonging to the same dispersion branches as the initial waves, but also all possible perturbations, must be accounted for.

II. Wave Packets: Self-Focusing and Self-Compression Effects

Let us now discuss the effects to which deviation from one-dimensional wave propagation leads. Here the wave self-influence effects, self-focusing

(27, 28) and self-compression (29), should first be pointed out. These effects do not result in the excitation of other types of oscillation; they only change the dynamics of the initial waves.

First let us consider the self-focusing phenomenon (27, 28). One can understand this effect on the basis of rather simple considerations. Let the nonlinear dependence of the refractive index be such that it is increasing in regions with high field strengths. Then, if the wave packet has maximum strength in the center, the central part of the wave front will move more slowly than the peripherical portions. It is as if a lens, focusing the wave front, forms.

Obviously, in studying the self-focusing problem the nonlinear effects that can destroy self-focusing must be considered. Among them one must include the decay processes (see Section III) and the process of wave breaking. In the preceding section nonlinear dispersion was shown to stop the breaking.

There can exist also conditions in which decay processes either are prohibited or are going slowly. The self-focusing in plasma was apparently observed in the work described in Ref. 30.

The self-focusing effect can also exist because of saturation of the wave packet by harmonics. In Ref. 31 self-focusing is shown to arise even for one solution.

The self-compression phenomenon (29) is no less interesting than self-focusing. This effect is connected with the dependence of the phase velocity on wave amplitude and of the group velocity on the wave vector k. For example, let the phase velocity increase with amplitude. Then at small modulations the waves from the high-amplitude region will tend to escape into the region of less amplitude, thus increasing the gradient in the modulation wave forefront and hence the effective wavenumber k. Obviously, if the group velocity decreases as the wavenumber increases, the energy from the wave packet regions with high k will roll into the region with smaller k, that is, into the high-amplitude region. In such a way self-compression of the wave occurs.

III. Instabilities Leading to Wave Oscillation Excitations of New Types

Let us proceed now to discuss periodic wave stability in plasma. The main reason for instability of such waves is the appearance of positive feedback between fluctuating waves, which are propagating on the finite-amplitude wave "background." This can be shown most easily for waves of small amplitude.

For this purpose it is convenient to make use of perturbation theory (33, 39). We decompose the vectors into natural vectors of zeroth approximation; then, projecting in the direction of the chosen natural vector and averaging over rapid oscillations, we obtain the following system of equations for c_{ki}, the interacting wave amplitudes:

$$\frac{dc_k}{dt} = i \sum V_{kk'k''} c_{k'} c_{k''} \tag{9}$$

Equations 9 lead to the decay instability of finite-amplitude waves [discovered by Sagdeev and the present author in studying Langmuir wave stability (33)]. At this instability the waves connected by decay conditions with the initial

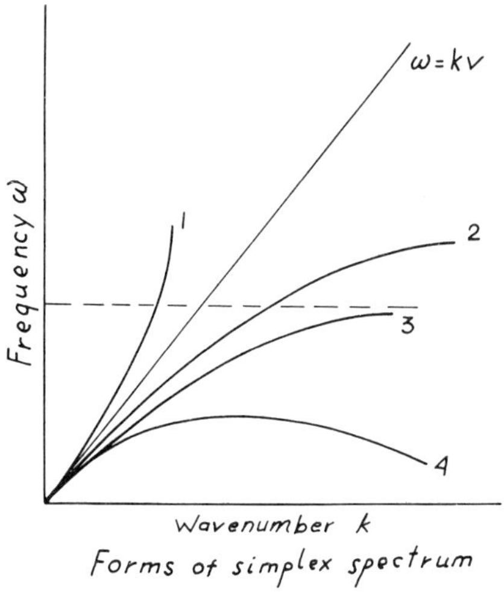

Fig. 1. Simplest possible spectral shapes.

wave (in nonlinear optics the equivalent terminology is "conditions of space-time synchronism") are excited:

$$\omega = \omega' + \omega'', \qquad k = k' + k'' \tag{10}$$

With the help of eq. 9 the growth rate of decay instability is easily shown to be equal to

$$\gamma = \sqrt{N_0}\,|V_{kk'k''}|, \quad \text{where} \quad N_0 \equiv |c_0|^2 = \frac{\epsilon_k}{\omega_k} \tag{11}$$

Knowing the spectrum shape (33), we can predict the presence or the absence of decay instability. Let there be only one branch of oscillations. The simplest possible spectral shapes are shown in Fig. 1. The decay conditions are easily shown to be fulfilled only for spectra 1 and 4. The oscillations having spectra analogous to 2 and 3 are stable as regards decay. However, because of the presence of various branches in the oscillation spectrum, the oscillations characterized by spectra of types 2 and 3 could be unstable in regard to the decay of oscillations, even though one of them did not belong to the given branch. More characteristic for plasma are the decay instabilities that are connected with the excitation of other types of waves (33–45, 48–99).

The wave dissipations have not been accounted for in the considerations above. It is necessary, however, to take dissipation into account in order to determine the threshold amplitude above which the waves are unstable to decay. The inclusion of weak damping or increase of the waves can be made also with the help of perturbation theory. Equation 9 is easily shown to be changed: on the right-hand side the quantities $\gamma_k c_k$ must be added, that is,

$$\frac{dc_k}{dt} = \gamma_k c_k + i \Sigma V_{kk'k''} c_{k'} c_{k''} \tag{12}$$

and this in turn leads to the following expression for the threshold values of the wave energy:

$$N_0 | V_{012} |^2 = \gamma_1 \gamma_2 \tag{13}$$

It must be noted that, although decay instabilities were shown to be a threshold process in the first work on these instabilities 33, theorists and experimenters did not pay any attention to this question for a long time. Later, however, Silin, Aliev, Gorbunov, and others investigated the threshold and near-threshold behavior of the growth rates of some decay instabilities of high-frequency waves.

If eq. 9 is used directly for the study of decay instability in fixed-phase approximation in the case of random relative wave phases, the kinetic equation for interacting waves is obtainable from the same equations, following Galeev and Karpman (39); this leads to a slower excitation of decay-interacting waves with growth rates, proportional to the initial wave amplitude squared (40, 44, 67).

Let us distinguish the peculiarities of decay interactions of waves in plasma. First of all, to them must be attributed the possibility of successive excitation of a number of oscillations. Therefore, on one hand, the solution of the classic problem of nonlinear optics about three-wave dynamics often loses meaning; on the other hand, the conversion problem becomes purely academic on account of the large increase in amplitude within a period (96). If the decay calculations are carried out with accuracy up to $\Delta\omega < \gamma$, it is a simple matter to show the existence of wave excitation, which is, however, somewhat slower. Therefore in actual conditions sufficiently broad wave packets could be excited and lead to random-phase behavior (see, e.g. Ref. 98).

It is convenient to illustrate the next peculiarity with an example (66). Let us consider nonisothermal plasma ($T_e \gg T_i$), in which Langmuir oscillation has been excited by an electron beam. We are interested in the intensity of radiation from such plasma with frequencies near the double Langmuir frequency. This radiation is connected with the process

$$l + l' \rightarrow t \tag{14}$$

In the absence of processes leading to a notable change in the initial turbulence spectrum, the calculation of radiation intensity could be carried out within the theory of Langmuir wave transforming into transverse waves on the thermal fluctuations of plasma (46, 47, 49). This is due to the fact that the process of eq. 14 can take place only when l and l' are propagating in nearly opposite directions. In nonisothermal plasma the waves l and l' participating in eq. 14 could simultaneously participate in the decay:

$$l \rightarrow l' + s \tag{15}$$

(so-called conjugation of three-plasma processes, eqs. 14 and 15). Now, first, both waves l and l' are substantially superthermal; second, the velocity of Langmuir wave merging is defined by the velocity of the more rapid process of Langmuir wave decay (eq. 15). All this leads to an increase in radiation power from plasma by a few orders, as compared with the radiation power connected with the transformation on thermal fluctuations (see Refs. 66 and 67).

Therefore the conjugation of three-plasma processes could lead to a notable increase in the velocity of one process at the expense of the others, and also to the appearance, because of decay instability, not only of red (l' and s) waves, but also of violet (t-wave) satellites (50, 51, 54).

It should be noted that the conjugated three-wave process takes place only when the beats formed in the spectrum by neighboring high-frequency waves get into resonance with the low-frequency wave (equidistant high-frequency spectrum). Then the problem regarding the interaction of arbitrary numbers of high-frequency waves could be solved exactly.

The existence of waves with negative energy leads to very interesting consequences, to which attention was paid for the first time in the paper of Dikasov, Rudakov, and Rjutov (57). Coppi, Rosenbluth, and Sudan (75) have shown that, if a wave with negative energy interacts with other waves, in three-wave approximation the wave amplitudes grow infinitely within a finite time—so-called explosive instability (see also Ref. 62). In a real situation this growth is limited by nonlinear processes, resulting in such a change of dispersion features of the plasma that the initial wave energy becomes positive. The explosive instabilities can develop both in turbulent and in laminar plasma (92, 93). Moreover, the time of explosive instability development is dependent on the initial conditions in laminar plasma.

Extension of the general calculation procedure of wave interaction to plasma systems with sufficiently sharp boundaries has become an important step in the wave interaction theory. Such a procedure has been proposed by

Rosenbluth, Cavalieri, and Sudan (63) and by Karpljuk, Kolesnichenko, and Oraevsky (73, 74, 87). This procedure makes possible to calculate the interactions of both types of waves, volume and surface.

The characteristic features of wave interaction in finite systems are connected with the following. First, there exist oscillation types (e.g., surface waves) which are not accounted for in the approximation of infinite plasma. Therefore new types of wave interactions, leading to new channels for decay instabilities (73, 74, 87), appear. Second, the dispersion relation differs from that existing in infinite plasma. Third, the absence of translation invariance in the direction the normal to the plasma boundary leads to the impossibility of introducing the notion of corresponding quasipulse projection. Nevertheless, on interaction of the short-wavelength volume waves, the laws of conservation of some values, which have a sense analogous to the quasipulse one, are approximately fulfilled. Hence for cylindrical systems (73) the short-wavelength mode interaction increases sharply for

$$\omega_0 = \omega_1 + \omega_2, \qquad k_{0z} = k_{1z} + k_{2z}, \qquad m = m_1 + m_2 \tag{16}$$

$$|T_0| = |T_1 \pm T_2| \tag{17}$$

In this case the space wave profile is described by the functions

$$\exp\,(ik_z z + im\,\varphi)\,J_m(Tr)$$

(Here z, r, and φ are of the mth order.) Conditions 16 and 17 are analogous to the decay conditions in infinite plasma. Moreover, eqs. 16 are satisfied by interactions of any wave types, whereas eq. 17 is satisfied only by interactions of short wavelength volume waves. The condition linking the transverse wave numbers, T_i, does not exist for interactions involving average and large wavelength volume and surface waves.

Decay processes of the first order were considered above. Second-order processes were shown by Zakharov (61) to lead to decay instabilities, namely, to the second harmonic decay of oscillations for oscillation branches for which the first-order processes are prohibited (see also Refs. 14, 41, and 43).

Shapiro and Shevchenko (81) have reported a finite-amplitude wave instability, connected with the interaction of the wave with captured particles. This leads to the appearance in the oscillation spectrum of satellites, shifted, in regard to the main frequency, by a value proportional to the square root of the potential amplitude. The mechanism of decay instability connected with the appearance of positive feedback between fluctuation waves propagating on the cosine wave background is expected to operate for waves of arbitrary amplitude. The natural mathematical difficulties do not allow investigation of the stability of a periodic wave with arbitrary amplitude and shape. Therefore the stability of waves with profiles for which the stability problem could be solved for an arbitrary value of wave amplitude was considered first.

The Alfvén wave stability with plasma-shaped profile of magnetic field lines has been investigated by Galeev and the present author (42, 45). The problem reduces to the solution of equations with periodically discontinuous coefficients, which are constant between discontinuity points. The joining of the solutions at the discontinuity points makes it possible to obtain eigenvalues, ω, that is, to solve the stability problem. The Alfvén wave of above-mentioned shape is shown to be unstable at arbitrary amplitude. The instability mechanism—the rise of positive feedback between given waves—is analogous to the instability mechanism of cosine waves with low amplitude.

In works by Silin and his collaborators (58–60, 68, 69, 78, 80, 83–86, 90, 95, 97) the stability of high-frequency oscillations has been considered (the wave vector of the initial wave was chosen to be zero). The case $k_0 = 0$ is the simplest one. When collisionless dissipation effects are neglected (this is valid for a large amplitude initial wave), the coefficients in equations describing the small perturbation dynamics of waves that are propagating on the initial wave background depend only on time. Therefore the differential equations could be reduced to conventional differential equations of Hill's type. In general, the stability problem can seldom be solved for arbitrary wave amplitude.

At this point some general discussion may be helpful. The processes considered in this section could be attributed to the parametrical processes in plasma, in systems both with distributed parameters and with different oscillation types. The generalized condition of parametrical excitation is

$$n\omega_0 = \omega_1 + \omega_2, \qquad nk_0 = k_1 + k_2 \tag{18}$$

At $n = 1$ this represents the decay of the first harmonic; at $n = 2, 3, \ldots$ it represents the decay of upper harmonics into waves $\omega_1 k_1$ and $\omega_2 k_2$, which can propagate in the system plasma + wave (see Refs. 14, 41, 43).

Therefore the decay instability of harmonics is a concrete display of parametrical wave excitation.† The width of the instability region in which the nth harmonic is unstable is proportional to the instability growth rate and hence, for waves with not too large amplitude, is proportional to the nth degree of the wave amplitude. It is clear that the decay conditions can be carried out with accuracy precisely up to the instability region width (eq. 18). The nonlinear wave interaction leads not only to the decay instability, but also to a frequency shift (33). As was shown by Silin, such a frequency shift could transfer the initial wave to the instability region. In considering the interaction of high- and low-frequency waves, the frequency shift could be of the same

† The term "parametrical instability" is often used (58–60, 68, 69, 78, 80, 90). By this term the authors of the indicated works mean the instability of waves only of infinitely large wavelength: $k = 0$. For this case it is characteristic that decay of either harmonic results in the excitation of waves that are propagating in opposite directions, $K_1 = -K_2$ (see also eq. 18).

order as the low frequency itself. This results in the so-called aperiodical instabilities (58, 68, 78, 91, 95)†, when the shifted low-frequency ω_2 is of the order of the instability growth rate. Therefore, in the case under consideration, the frequency condition of eq. 18 takes a simple form: $n\omega_0 \simeq \omega_1$.

The wave processes considerated in this survey play an important role in the dynamics of collisionless plasma. They lead to the collisionless heating of plasma (see, e.g., Refs. 33, 83, 99); determine, very often, the turbulence spectra that appear because of the development of plasma instability; lead to collisionless shock wave formation, and so on. Many problems in the theory of wave propagation and stability in plasma await solution. This is especially true of the wave processes in finite systems.

Acknowledgment

The author would like to express his deep gratitude to R. Z. Sagdeev for discussion of questions touched upon in this survey, and to V. P. Silin for consideration of the general scheme of the paper and valuable discussions.

References

1. R. Z. Sagdeev, *Plasma Physics and the Problem of Controlled Thermonuclear Fusion*, Vol. 4, Iz-vo Akad. Nauk SSSR, Moscow, 1958, p. 454.
2. A. Adlam, *Proceedings of the Second U.N. International Conference on the Peaceful Uses of Atomic Energy*, Iz-vo State Com. Inst. At. Energy SSSR, Moscow, 1959, p. 156.
3. L. Davis, R. Lüst, and A. Schlüter, *Natürforsch. Z.*, 916 (1958).
4. C. Gardner, H. Goertzel, H. Grad, C. Morawetz, M. Rose, and A. Rubin, *Proceedings of the Second U.N. International Conference on the Peaceful Uses of Atomic Energy*, Iz-vo State Com. Inst. At. Energy SSSR, Moscow, 1959, p. 94.
5. R. Z. Sagdeev, *Report on Uppsala Conference on Phenomena in Ionized Gases*, 1959.
6. A. A. Vedenov, E. P. Velihov, and R. Z. Sagdeev, *Nucl. Fusion*, **1**, 82 (1961).
7. R. Z. Sagdeev, *Zh. Tekh. Fiz*,, **31**, 1955 (1961).
8. A. A. Galeev and V. J. Karpman, *Zh. Eksp. Teor. Fiz.*, **44**, 592 (1963).
9. K. W. Morton, *Phys. Fluids*, 7, 1801 (1964).
10. R. Z. Sagdeev, *Problems of the Plasma Theory*, Vol. 4, Atomizdat, Moscow, 1964, p. 20.
11. P. V. Polovin, *Nucl. Fusion*, **4**, 10 (1964).
12. C. F. Kennel and R. Z. Sagdeev, Preprints IC/66/68 and IC/66/88, ICTP, Trieste, 1966.
13. I. A. Akhiezer and A. Borovik, *Zh. Eksp. Teor. Fiz.*, **51** (1966); *Ukr. Fiz. Zh.*, **13** (1968).
14. R. Z. Sagdeev and A. A. Galeev, Lectures on the Nonlinear Theory of Plasma, Preprint I/66/78, ICTP, Trieste, 1966.

† Aperiodical instabilities were considered for the first time by T. F. Volkov (32), who investigated the dispersion features of plasma in the field of strong electromagnetic waves.

15. V. N. Oraevsky, H. Tasso, and H. Wolig, *Plasma Phys. Controlled Nucl. Fusion Res.*, **1**, 671 (1969).
16. A. V. Schutko, *Zh. Eksp. Teor. Fiz.*, **55** (1968); **57**, 452 (1969).
17. N. J. Zabusky and M. D. Kruskal, *Phys. Rev. Lett.*, **17**, 966 (1966); C. S. Gardner, J. M. Green, M. D. Kruskal, and R. M. Miura, *Phys. Rev. Lett.*, **19**, 1095 (1967).
18. T. K. Chu et al., *Plasma Phys. Controlled Nucl. Fusion Res.*, **1**, 611 (1969).
19. T. Stix, Princeton Plasma Physics Laboratory, MATT-558, 1968.
20. A. A. Zhmudsky, V. V. Lisitchenko, and V. N. Oraevsky, *Nucl. Fusion*, **10**, 151 (1970).
21. V. J. Karpman, *Plasma Phys.*, **13**, 477 (1971).
22. V. V. Lisitchenko, V. N. Oraevsky, and S. N. Reznik, *Report on the Conference of Plasma Theory*, Kiev, 1971.
23. R. Igithanov and B. B. Kadomtsev, *Zh. Eksp. Teor. Fiz.*, **60**, 477 (1971).
24. B. B. Kadomtsev and V. I. Karpman, *Usp. Fiz. Nauk*, **103**, 193 (1971).
25. H. Alfvén, *Cosmic Electrodynamics*, Iz-vo IL, Moscow, 1952.
26. D. Korteweg and G. de Vries, *Phys. Mag.*, **39**, 442 (1895).
27. G. A. Askarjan, *Zh. Eksp. Teor. Fiz.*, **42**, 1567 (1962).
28. V. I. Talanov, *Izv. High School Ser. Radiophys.*, 7, 564 (1964).
28a. R. Chiao, F. Gardmire, and C. H. Townes, *Phys. Rev. Lett.*, **13**, 479 (1964).
29. M. J. Lighthill, *Proc. Roy. Soc (London)*, **A299**, 28 (1967).
30. B. G. Eremin, Letter in *Zh. Eksp. Teor. Fiz.*, **13**, 603 (1971).
31. B. B. Kadomtsev and V. I. Petriashvili, *Dokl. Akad. Nauk SSSR*, **192**, 753 (1970).
32. T. F. Volkov, *Plasma Phys. and Controlled Nucl. Fusion Res.*, **4**, 98 (1958).
33. V. N. Oraevsky and R. Z. Sagdeev, *Zh. tekh. Fiz.*, **32**, 1291 (1962).
34. V. N. Oraevsky, *Zh. Prikl. Method. Tech. Fiz.*, **5**, 39 (1962).
35. A. A. Galeev and V. N. Oraevsky, *Dokl. Akad. Nauk SSSR*, **147**, 71 (1962).
36. V. N. Oraevsky, *Zh. tekh. Fiz.*, **33**, 251 (1963).
37. L. A. Ostrovsky and H. C. Stepanov, *Zh. Eksp. Teor. Fiz.*, **45**, 1475 (1963).
38. A. N. Kondratenko, *Zh. Eksp. Teor. Fiz.*, **33**, 1395 (1963).
39. A. A. Galeev and V. I. Karpman, *Zh. Eksp. Teor. Fiz.*, **44**, 592 (1963).
40. A. A. Vedenov, *Problem of Plasma Theory*, Vol. 3, Atomizdat, Moscow, 1963, p. 203.
41. V. N. Oraevsky, *At. Energy*, **16**, 441 (1964).
42. A. A. Galeev and V. N. Oraevsky, *Akad. Nauk SSSR*, **154**, 1069 (1964).
43. V. N. Oraevsky, *Nucl. Fusion*, **4**, 263 (1964).
44. B. B. Kadomtsev, *Problem of Plasma Theory*, Vol. 4, Atomizdat, Moscow, 1964, p. 188.
45. A. A. Galeev and V. N. Oraevsky, *Report of the Second All-Union Congress of Mechanics: Thesis of the Reports*, Iz-vo Akad. Sci. USSR, Moscow, 1964, p. 61.
46. A. G. Sitenko, *Electromagnetic Fluctuation in the Plasma*, Iz-vo Karkov State University, Karkov, 1964.
47. A. I. Ahiezer, N. L. Danelija, and N. L. Tsintsadze, *Zh. Eksp. Teor. Fiz.*, **46**, 300 (1964).
48. Ju. L. Klimontovich, *Statistical Theory of Nonequilibrium Processes in the Plasma*, Iz-vo Moscow State University, Moscow, 1964.
49. R. E. Aamodt and W. E. Drummond, *Nucl. Energy*, Part C, **6**, 147 (1964).
50. L. M. Kovriznih and V. N. Tsitovich, *Zh. Eksp. Teor. Fiz.*, **47**, 1454 (1964).
51. A. A. Vedenov and L. I. Rudakov, *Dokl. Akad. Nauk SSSR*, **159**, 767 (1964).
52. R. E. Aamodt and W. E. Drummond, *Ann. Phys.*, **51**, 278 (1965).
53. V. N. Tsytovich, *Zh. Tekh. Fiz.*, **34**, 773 (1965).
54. D. D. Rjutov, *Dokl. Akad. Nauk SSSR*, **164**, 1273 (1965).
55. A. A. Ivanov and D. D. Rjutov, *Zh. Eksp. Teor. Fiz.*, **48**, 5 (1965).
56. Ju. M. Aliev and V. P. Silin, *Zh. Eksp. Teor. Fiz.*, **48**, 901 (1965).

57. V. M. Dikasov, L. I. Rudakov, and D. D. Rjutov, *Zh. Eksp. Teor. Fiz.*, **48**, 913 (1965).
58. V. P. Silin, *Zh. Eksp. Teor. Fiz.*, **48**, 1679 (1965).
59. L. M. Gorbunov and V. P. Silin, *Zh. Eksp. Teor. Fiz.*, **49**, 1973 (1965).
60. Ju. M. Aliev, V. P. Silin, and X. Wotson, *Zh. Eksp. Teor. Fiz.*, **50**, 944 (1966).
61. V. E. Zaharov, *Zh. Eksp. Teor. Fiz.*, **51**, 1107 (1966).
62. A. V. Timofeev, Letter in *Zh. Eksp. Teor. Fiz.*, **4**, 48 (1966).
63. M. Rosenbluth, A. Cavaliere, and R. Sudan, Preprint, ICTP, Trieste, 1966.
64. I. S. Danilin, *Zh. Tekh. Fiz.*, **35**, 266 (1966).
65. K. S. Karpljuk and V. N. Oraevsky, Letter in *Zh. Eksp. Teor. Fiz.*, **5**, 451 (1967).
66. V. N. Oraevsky and V. N. Tsytovich, *Zh. Eksp. Teor. Fiz.*, **53**, 1116 (1967).
67. I. S. Danilkin, *Zh. Tekh. Fiz.*, **36**, 667 (1967).
68. L. M. Gorbunov, *Zh. Eksp. Teor. Fiz.*, **55**, 2298 (1968).
69. V. P. Silin, Letter in *Zh. Eksp. Teor. Fiz.*, 7, 242 (1968).
70. K. S. Karpljuk, V. N. Oraevsky, and V. P. Pavlenko, *Ukr. Fiz. Zh.*, **13**, 1114 (1968).
71. V. N. Oraevsky and Ju. P. Judkin, *Ukr. Fiz. Zh.*, **13**, 460 (1968).
72. A. S. Bakai, *Zh. Eksp. Teor. Fiz.*, **55**, 266 (1968).
73. K. S. Karpljuk and V. N. Oraevsky, *Zh. Tekh. Fiz.*, **38**, 1214 (1968).
74. K. S. Karpljuk et al., *Plasma Phys. Controlled Nucl. Fusion Res.*, **1**, 753 (1969).
75. B. Coppi, M. Rosenbluth, and R. Sudan, *Plasma Phys. Controlled Nucl. Fusion Res.*, **1**, 771 (1969); *Ann. Phys.*, **55**, 207 (1969).
76. V. N. Oraevsky and V. P. Pavlenko, *Zh. Tekh. Fiz.*, **39**, 1799 (1969).
77. N. S. Erohin, V. E. Zakharov, and S. S. Moiseev, *Zh. Eksp. Teor. Fiz.*, **56**, 179 (1969).
78. N. E. Andreev, Ju. A. Kirii, and V. P. Silin, *Zh. Eksp. Teor. Fiz.*, **57**, 1024 (1969).
79. M. I. Rabinovich and S. M. Fainshtein, *Zh. Eksp. Teor. Fiz.*, **57**, 1298 (1969).
80. Ju. M. Aliev and D. Zünder, *Zh. Eksp. Teor. Fiz.*, **57**, 1324 (1969).
81. V. D. Shapiro and V. I. Shevchenko, *Zh. Eksp. Teor. Fiz.*, **57**, 2066 (1969).
82. A. N. Kondratenko and V. G. Shaptala, *Zh. Tekh. Fiz.*, **39**, 2229 (1969).
83. V. P. Silin, *Zh. Eksp. Teor. Fiz.*, **57**, 183 (1969).
84. Ju. M. Aliev and E. Ferlensh, *Zh. Eksp. Teor. Fiz.*, **57**, 1263 (1969).
85. N. E. Andreev, *Zh. Tekh. Fiz.*, **39**, 1560 (1969).
86. L. M. Gorbunov and V. P. Silin, *Zh. Tekh. Fiz.*, **39**, 3 (1969).
87. K. S. Karpljuk, Ja. I. Kolesnichenko, and V. N. Oraevsky, *Nucl. Fusion*, **10**, 3 (1970).
88. A. S. Bakai, *Nucl. Fusion*, **10**, 53 (1970).
89. K. S. Karpljuk and Ja. I. Kolensnichenko, *Ukr. Fiz. Zh.*, **14**, 1459, 1468 (1970).
90. N. E. Andreev, Ju. A. Kirii, and V. P. Silin, *Radiophysics (Izv. Vus'ov)*, **13**, 1321 (1970).
91. Ju. A. Kirii, *Zh. Eksp. Teor. Fiz.*, **58**, 1002 (1970).
92. K. S. Karpljuk, V. N. Oraevsky, and V. P. Pavlenko, *Ukr. Fiz. Zh.*, **15**, 857 (1970).
93. H. Wilhelmsson, *Phys. Scripta*, **2**, 113 (1970).
94. N. S. Erohin and S. S. Moiseev, *Zh. Tekh. Fiz.*, **40**, 1144 (1970).
95. A. Ju. Kirii, *Zh. Eksp. Teor. Fiz.*, **60**, 955 (1971).
96. V. E. Zakharov and V. S. Lvov, *Zh. Eksp. Teor. Fiz.*, **60**, 2066 (1971).
97. N. E. Andreev and A. Ju. Kirii, *Zh. Tekh. Fiz.*, **41**, 1080 (1971).
98. A. A. Ivanov and V. V. Parail, *Zh. Eksp. Teor. Fiz.*, **60**, 2113 (1971).
99. A. I. Akhiezer, V. F. Aleksin, and V. Khodusov, *Report on the Conference of Plasma Theory*, Kiev, 1971.

Present Status of the Theory of Nonlinear and Stochastic Plasma Processes

B. B. KADOMTSEV

Kurchatov Institute of Atomic Energy, Moscow, USSR

Initially my report was to be devoted to the narrower topic of the turbulent processes in a plasma. Since a number of the conference papers are concerned with nonlinear processes, I shall concentrate in the main on this narrower subject in order not to encroach on the topics of other surveys. I shall begin with some general aspects well known to plasma theoreticians which, however, should be mentioned for the benefit of those participants of the conference whose activities are not directly concerned with studies of plasma turbulence.

It is customary to assume that plasma turbulence is a state in which strong noise and oscillations are spontaneously excited in the plasma. In other words, plasma turbulence corresponds to excitation of many degrees of freedom to a very high level, which is considerably above the thermal level. Such degrees of freedom may include, for example, slow motions; in such cases plasma turbulence looks like the turbulence in common fluids. More often, however, different, near-harmonic oscillations and waves are excited in the plasma because of a certain "elasticity" of the system of interacting charged particles. If there are many excited degrees of freedom, then, as a result of the complicated nature of their interactions, the entire process may be considered as being stochastic, and statistical methods have to be used for its description.

I. Weak Turbulence

Among various turbulent processes in the plasma there is a sufficiently large and rather important class of so-called weak turbulent processes. Weak turbulence corresponds to rather a slight excitation of the plasma in which the characteristic times of interactions between the waves (or relaxation of waves due to particle interactions) are found to be very large as compared to the periods of oscillation. Therefore, in the roughest approximation, weak turbulence is a set of many waves in a plasma.

Because of the weak interaction the theory of perturbations may be used to describe weak turbulence, that is, the scheme of expansion of the solutions of nonlinear equations into series, with consequent averaging over random phases and retention of only secular terms. Within the past 10 years the corresponding theory has been developed. At present a number of processes occurring in a slightly turbulent plasma have already been analyzed within the framework of this theory.

The general structure of the weak turbulence theory is very simple. It is assumed that weak turbulence should be simply a set of a large quantity of waves which, in the lowest approximation of the perturbation theory, satisfy the linear system of equations. In other words, they are either standing waves in finite systems or traveling waves in systems large in comparison with the relaxation length of the waves. For the sake of convenience or, rather, for unification of the terms with those used in other fields of physics, these waves are referred to as plasmons.

To describe the plasmon interactions the small nonlinear terms should be taken into account in the next approximation of the perturbation theory. When taking into account only the amplitude square terms, we obtain the so-called quasilinear approximation suggested by Vedenov, Velikhov, and Sagdeev (1) and by Drummond, and Pines (2). The quasilinear approximation takes into account the wave-resonance particle interactions. In other words, only the Cherenkov radiation (to be more exact, the induced Cherenkov radiation) is taken into account, as well as the absorption of waves by particles.

The field of application for the quasilinear theory is very limited (the amplitudes are very small). However, the theory is extremely simple; therefore, there is a natural desire to extend it over as wide a range of phenomena as possible, sometimes even beyond its formal applicability. This often proves to be successful, and the quasilinear theory has made it possible to explain qualitatively a very broad range of phenomena in a rarefied plasma, in particular, the phenomena developing in plasma-beam interactions. As to the experimental confirmations of the quasilinear theory, a recent experiment carried out by Robertson, Gentle, and Nielson (3) should be mentioned. This experiment has shown that, if all the approximations of the quasilinear theory are satisfied, the theory is in good agreement with experiment.

It should be pointed out, however, that the quasilinear theory is rather approximate, and the range of its applicability is narrow. Moreover, the processes described by this theory have no right, strictly speaking, to be called turbulent processes. In fact, let us remember what we mean when talking about tubulence in common fluids. We mean a strongly irregular flow of fluid in which a great number of vortex motions are excited. Interactions between vortices of various scales should result in a spread of the energy of irregular motion over all degrees of freedom. However, as the smallest vortices damp very rapidly because of viscosity, a constant-energy flux is generated from large vortices to smaller ones.

The quasilinear approximation scheme does not allow for any energy flux over the spectrum—we simply neglect this. If the oscillation amplitude is sufficiently large, however, this cannot be done, and we have to retain the next approximation terms of the perturbation theory which take into account the processes quadratic in the wave intensities (the intensity is proportional to the square of the amplitude). This approximation takes into account the induced scattering of waves by the particles and three-wave processes: coalescence of two waves into one wave, and disintegration of one wave into two waves (we may talk about plasmons instead of waves). Such processes cause transport of energy over the spectrum.

However, in contrast to the usual hydrodynamic turbulence, in which the energy flux is directed from large scales to smaller ones, the quasilinear approximation allows many variations. For example, if plasmons are scattered by the particles, it is easy to see from the quantum-mechanical analogy that the number of plasmons remains constant whereas their energy and hence their frequency decrease. Therefore the spectral transfer takes place in the direction of smaller frequencies, that is, smaller wave numbers. The transfer process itself is of a more complicated nature. As a matter of fact, the induced processes of the weak turbulence scheme have a clear tendency to form rather complicated localized or oscillating distributions of energy in the wave vector space. Roughly, the plasmons tend to converge in the **k** space. All these factors make the scheme of weak turbulence very complicated. Nevertheless a number of situations have been theoretically analyzed in which a stationary weak turbulence with constant energy flux through the spectrum could be produced in the plasma. The oscillations are excited for a certain class of waves due to instability. Then this energy is transferred by nonlinear processes to the region of damping. The ion-acoustic, Langmuir, helicon, and magnetodynamic turbulences, in which only one of the possible branches of oscillations (4–7) is excited, were thus analyzed. In spite of the high sophistication of particular theoretical constructions the degree of their reliability is still not great. Therefore experimental verification of the foundations and conclusions of the theory of plasma weak turbulence seems to be even more desirable than that of the quasilinear theory.

Some progress is evident in this direction. First, a number of experiments have recently been carried out on the "elementary" processes. These studies inquire into the foundations of weak turbulence, that is, the induced scattering of waves (sometimes known as the nonlinear Landau damping) and disintegration of waves. The results of these experiments are in good agreement with theory. Step by step, more complete theoretical schemes of weak turbulence begin to be borne out.

Thus Hamberger and Jankarik (8) have measured the spectrum of the ion-acoustic turbulence in a plasma to which a very strong electric field is applied. Such a field generates a strong electric current, and because of the instability of ion-acoustic waves the plasma becomes turbulent and exhibits high resistivity. The shape of the spectrum measured in their work when the frequency was changed by two orders of magnitude is in good agreement with theoretical predictions made on the assumption that the principal nonlinear process should be the scattering of waves by ions. Tsytovitch (6) pointed out that under similar experimental conditions the three-wave processes should also play almost the same role with regard to the broadening of the frequency spectrum of the wave due to interactions, while the shape of the spectral energy distribution remains the same in this case. A similar spectrum was also observed in shock wave experiments for a smaller range of the wavenumbers (9). These results arouse hope that the concepts of plasma weak turbulence can be applied to the interpretation of a broad range of phenomena and that the theoreticians did not "plough sand" when making comprehensive investigations of particular constructions.

At the same time it is already quite clear that the nature of the plasma stochastic processes is more complicated than it appears from the point of view of the weak turbulence theory, that is, in terms of the collective excitations. Hence the stochastic theory will need further development before making more profound and complete descriptions of the phenomena occurring in a plasma. We have to arrive at this conclusion even within the framework of the weak turbulence scheme itself. The trouble is that the matrix elements of the terms of interactions between the waves and the particles and between themselves contain singular factors with resonance denominators of the kind $\omega - kv$. The higher the expansion terms, the more resonance factors are involved. To eliminate the singularities it is sometimes necessary to take a selective summing of the perturbation theory. This procedure is quite usual and probably lawful, but as the amplitude of the turbulent noise increases we become uncertain whether the interactions between resonance particles and waves are taken into account correctly and sufficiently. In order to obtain a qualitative or semiqualitative answer to this question, it is necessary to picture in detail what occurs with the resonance particles in their interactions with the waves.

II. Trapped Particles

This problem has been considered in many theoretical studies; in particular, some of them are presented at this conference. The picture is quite clear qualitatively. It is especially simple in the case of a single monochromatic wave. In such a wave, all particles naturally separate into two groups—the trapped particles and the transit particles. The trapped particles (i.e., those trapped by a wave) exhibit oscillations between neighboring crests of the wave, whereas the energy of the transit particles is sufficiently great to pass over the crests. The presence of the trapped particles influences the wave amplitude. Instead of monotonically growing due to instabilities or monotonically decreasing according to the linear Landau solution, in the case of stability the oscillation amplitude oscillates because of the trapped particle-wave energy exchange. The frequency of these oscillations is of the same order of magnitude as that of the trapped particles near the "valley" of the wave. As is easy to see, the oscillation frequency of the trapped particles is proportional to the square root of the wave amplitude. These features of the Landau damping for nonlinear waves were not only theoretically predicted but also observed experimentally, first by Malmberg and Wharton (10) and then by other experimentalists.

An interesting feature appears in interactions between a low-density beam and a wave. Initially the wave amplitude increases because of plasma beam instability; then, as the amplitude grows and becomes sufficiently high, the beam is trapped. The largest fraction of the beam electrons form a blob, which then oscillates near the valley of the wave, giving rise to oscillations in its amplitude. This blob acts as a macroparticle. The ideas of the particle-bunching mechanism were suggested in early papers on the collective interactions of particles in a plasma. Lately, however, they have been replaced by the more formal concepts of the quasilinear theory.

It is obvious that similar processes of production of particle blobs correlated in the phase space, which look like macroparticles, may also take place in more complicated situations, not only with a monochromatic wave but also with many waves, that is, in the stochastic processes. Before passing to consideration of such phenomena let us consider a simpler case—namely, let us begin with the works of Lynden-Bell (11) on collisionless relaxation in a system of Coulomb interacting particles.

Lynden-Bell investigated relaxation to the state of thermodynamical equilibrium in a system of Coulomb interacting particles when the pair collisions may be completely neglected, that is, at a very high Debye number. Under such conditions the evolution of the system is described by a Vlasov equation with a self-consistent field. It is assumed that there are no collective disturbances, for example, in the fields with wavelength smaller than the

Debye length (in particular, for stellar systems or galaxies). Lynden-Bell's consideration is as follows. Assume that initially the configuration was strongly nonequilibrium, so that some areas in the phase space were occupied by the particles and other areas were empty. Because of self-consistent field interactions some intermixing of the distribution functions should take place in the phase space. If this intermixing is strong enough, the distribution function will reach a certain final equilibrium state. This state should be the most probable one for the given energies and number of particles. In addition, it is necessary to take into account the fact that, according to the Vlasov equation, the flow in the phase space is incompressible. Therefore a certain principle of exclusion should act: the distribution function in a given region of the phase space should vanish if a cell without particles enters there, and it should be equal to the initial value in an occupied cell if such a cell enters there. If initially the distribution function was unity within the area where the particles were present, we arrive at the Pauli exclusion principle, and the final equilibrium function proves to be exactly similar to the Fermi distribution.

This very interesting result was obtained from purely statistical arguments, and it was not completely clear under what conditions the same final result could be reached in practice. To elucidate this problem numerical calculations were started (12). These calculations showed that under certain simple initial conditions the quasistationary state would be achieved after a few plasma periods, with the function close to the Fermi distribution but with an additional high-energy tail. For more complicated initial conditions the final state will deviate from the Fermi distribution.

These results are easily understood qualitatively if one again applies a method of the quasilinear type, and instead of collective perturbations one takes into account disturbances of the macroparticle kind (13). In the process of relaxation the correlation length decreases–the effective charge of the corresponding macroparticles decreases with time. The relaxation process itself may be considered as a result of pair collisions of such macroparticles. The quasilinear approximation naturally gives a collisional term of the Landau kind, but a rather more complicated one, so that it gives rise to relaxation toward the Fermi distribution. This term is proportional to the square of the effective charge of the macroparticles. If this charge is sufficiently large (this corresponds to "simple" initial conditions with broad regions of correlation), the distribution function will really have time for relaxation to the Fermi function. The presence of the "tail" at high energies may be associated with plasmon excitation, that is, with usual quasilinear effects.

Now we may return to consideration of a role played by the trapped particles in the weak turbulence. When the trapped particles are interacting with a single wave, we again observe correlated particle blobs–an analog of the macroparticles. In plasma-beam interactions such blobs are especially clearly seen. They have been observed experimentally by Gabovich and Kovalenko. These experiments are well described in terms of bunching a beam which is weakly modulated at the entrance to the plasma (ϵ is negative in the plasma for frequencies somewhat smaller than ω_0, so that attraction of the beam particles, followed by their coalescence into a blob, takes place). These experiments confirm the earlier ones carried out by Merill and Webb on the bunching of a beam in the plasma.

The effect of macroparticle production (i.e., correlated blobs in the phase space) may be of appreciable importance if the beam density is sufficiently high (14, 15). As the density decreases, the part played by these new excitations diminishes, and finally we come to the common quasilinear theory.

It is seen from this qualitative consideration that the quasilinear theory is approximate in the sense that we neglect new, higher harmonics of the disturbed distribution function and retain only the first of them. In practice, the trapping of particles and their consequent mixing due to oscillations in the well give rise to a motion in the phase space that is similar to the motion of incompressible fluid. In chaotic intermixing, the dimensions of the correlation regions will decrease with time, while the spectral function for the square of the perturbed distribution function will acquire a spectrum flux toward the higher wavenumbers. In other words, even for a weak small-scale turbulence in the phase space there appears a picture similar to that of strong turbulence. This picture is complicated by oscillations of the trapped particles whose frequency is proportional to the fractional power of the amplitude (this is the square root of the amplitude for a monochromatic wave). Since the fractional power appears in the situation considered, this means that if the turbulence is rather moderate it is not possible to use expansion into series, that is, strictly speaking, there is no analyticity. This conclusion is very interesting from the general theoretical point of view. Even if the macroparticles rapidly disintegrate with time, and they turn out to be not very important for a description of the experimental results, their evolution is very interesting from a theoretical point of view. This may elucidate the mathematical structure of the collective excitation scheme, and it may be useful in another field of physics. These aspects have not been yet investigated. This is quite understandable because the absence of the analyticity does not allow the application of conventional methods of the perturbation theory and greatly complicates progress in this direction.

In conclusion it may be pointed out that the range of application of the theory of weak turbulence developed so far is rather broad if we do not accept all its conclusions completely, in particular, those concerning very fine effects, like the components in the turbulence spectrum.

A rougher quasilinear theory is even more preferable in certain respects because it is deliberately more approximate and consequently less pretentious.

Development of a more exact theory encounters difficulties. It is clear that the next step will give us, instead of higher-order terms, some new concepts and approaches that are more closely related to strong turbulence, Such results certainly will be of interest for general theory. Moreover, there are a number of problems, such as anomalous resistivity and anomalous diffusion, that are not entirely clear from the theoretical point of view and are also awaiting solution. I have not mentioned them previously because here we have a broad investigation to carry out, taking into account many experimental details. It should also be remembered that a full theory of strong turbulence has not been developed as yet for common fluid, and consequently the elucidation of various kinds of strong plasma turbulences is still awaiting the development of adequate mathematical methods.

References

1. A. A. Vedenov, E. P. Velikhov, and R. Z. Sagdeev, *Nucl. Fusion*, **1**, 82 (1961).
2. W. E. Drummond and D. Pines, *Nucl. Fusion, Suppl.*, **3**, 1049 (1962).
3. C. Robertson, K. W. Gentle, and P. Nielson, *Phys. Rev. Lett.*, **26**, 226 (1971).
4. B. B. Kadomtsev, *Plasma Turbulence*, Academic Press, New York, 1965.
5. V. N. Tsytovitch, in *Nonlinear Effects in Plasma*, ed. by S. M. Hamberger, 1970.
6. V. N. Tsytovitch, *Theory of a Turbulent Plasma*, Atomizdat, Moscow, 1971.
7. A. A. Galeev and R. Z. Sagdeev, *Nonlinear Plasma Theory*, New York, 1969.
8. S. M. Hamberger and L. Jankarik, Preprint CLM-P269, Culham Laboratory, 1971.
9. C. C. Daughney, L. S. Holmes, and L. W. Paul, *Phys. Rev. Lett.*, **25**, 479 (1970).
10. J. H. Malmberg and C. B. Wharton, *Phys. Rev. Lett.*, **17**, 175 (1966).
11. D. Lynden-Bell, *Mon. Notic. Roy. Astron. Soc.*, **136**, 101 (1967).
12. F. Holi, *Bull. Amer. Phys. Soc.*, **13**, 1745 (1968).
13. B. B. Kadomtsev and O. P. Pogutse, *Phys. Rev. Lett.*, **25**, 1155 (1970).
14. T. H. Dupree, *Phys. Rev. Lett.*, **25**, 789 (1970).
15. B. B. Kadomtsev and O. P. Pogutse, International Centre for Theoretical Physics, Rept. 1C/70/45, 1970.

Anomalous Resistivity of Plasma

R. Z. SAGDEEV

Institute of High Temperatures,
Academy of Sciences of the USSR, Moscow, USSR

The anomalous resistivity of plasma occurs when the electric current flowing through the plasma exceeds some critical value. This critical value above which the plasma resistivity increases sharply is very small. It is convenient that the current density be expressed in terms of the so-called drift velocity $\bar{V}$. If the electron distribution function is characterized by a velocity $\bar{V}$ relative to the ion distribution function that exceeds some critical value, instability occurs. As a result of such instability, the electrons loose momentum by generating oscillations and waves, in addition to the usual loss due to binary collisions.

It is convenient to start with the consideration of a table of instabilities occurring when the critical value of velocity is exceeded. Listed in this table are

TABLE I

CURRENT-DRIVEN PLASMA INSTABILITIES

Type of Instability	Threshold of Instability	Frequency	Growth Rate
Buneman	$\bar{V} \gtrsim V_{\text{th}\,e}$	$\sim \Omega_0$	$\sim \Omega_0$
Ion-sound ($T_e \gg T_i$)	$\bar{V} > c_s$	$\leqslant \Omega_0$	$< \Omega_0 \dfrac{\bar{V}}{V_{\text{th}\,e}}$
Drummond-Rosenbluth	$\bar{V} > c_s$	$\sim \Omega_H$	$\sim \Omega_H \dfrac{\bar{V}}{V_{\text{th}\,e}}$
Electron-sound	Very low, sometimes $< V_{\text{th}\,i}$	$\leqslant \omega_{H_e}$	$< (\omega_H \Omega_H)^{1/2}$
Bernstein mode		$n\omega_H$	$\omega_{H_e} \dfrac{\bar{V}}{V_{\text{th}\,e}}$

all the main instabilities discussed in connection with the problem of anomalous resistivity in plasma. The simplest instability, known for some time, is the Buneman instability (1), which now is sometimes called the Buneman-Budker instability, since Budker (2) actually studied analogous phenomena several years ago in relation to stabilized electron beams. This instability is quite simple when there is an electron and an ion distribution funcion in the form of two delta functions shifted relative to each other by the drift velocity $\overline{V}$. This instability is the excitation of longitudinal electrostatic oscillations of the plasma with a growth rate of the order of the plasma ion frequency. Another example which can practically be regarded as being of the same mode is the ion-sound instability (3). These oscillations occur at the electron drift velocity that may be substantially lower than the thermal velocity. The growth rate of ion-sound oscillations (the imaginary part of the frequency) is the plasma ion frequency reduced by the drift-to-thermal-electron velocity ratio. In the extreme case, $\overline{V} \to \overline{V}_{the}$, the ion-sound instability changes almost smoothly into Buneman instability.

In the presence of a magnetic field instabilities of other types may occur. One of these also occurs because of the imaginary part of the electron contribution in ion-cyclotron waves, the so-called Drummond-Rosenbluth instability (4). This occurs when the current passes along the magnetic field, whereas the former two instabilities are in a sense invariant to the presence of a magnetic field if the magnetic field is not very high ($\omega_0 \gg \omega_H$). The Drummond-Rosenbluth instability has been discussed less extensively in connection with the problem of anomalous resistivity, inasmuch as it leads to low fluctuation growth rates (the imaginary part is comparatively small) and, apparently, can be easily suppressed by simple quasilinear effects such as plateau formation.

Also, there exists an instability called electron sound (5). These are waves in which the wave vector along the magnetic field is substantially less than the transverse component of the wave vector, and the frequencies are considerably lower than the electron Larmor frequency. This particular mode reminds one of the mode known to occur in the presence of the loss cone, and even of common drift instabilities. Such an instability is observed when the current flows across the magnetic field. This instability leads to very low growth rates and manifests itself clearly only when the currents are comparatively low, so that other, stronger instabilities, such as Buneman or ion-sound ones, do not enter the picture.

Finally, a Bernstein mode (6) instability has recently been under discussion. This instability features a relatively high growth rate and occurs when the current flows across the magnetic field.

Up to now, instabilities of the first two types, namely, Buneman and ion-sound instabilities, have received primary attention. Buneman suggested a heuristic formula for the nonlinear stage of instability in his first paper. He assumed the effective electron collision frequency to be on the order of the imaginary part of the frequency in the linear theory of instability, that is, on the order of the plasma frequency of ions. This simple formula, in which ν_{eff} in Ohm's law is replaced by the plasma frequency of ions, is called Buneman's conductivity formula. Obviously, this formula cannot claim to provide anywhere near an exhaustive description of experiments; however, for some simple experiments it yields a correct order of magnitude. Next, however, the following takes place: the emerging fluctuations slow down the electrons, the electron drift velocity, $\bar{V}$, decreases, and Buneman instability should cease.

At this point the remaining possibility is the ion-sound instability. The ion-sound instability is suitable for investigation by means of the weak turbulence method. In this case the imaginary part of the frequency is considerably smaller than the real part, since the drift velocity may be much lower than the mean thermal velocity of electrons. Quite a number of papers devoted to the nonlinear theory of ion-sound instability and the calculation of anomalous resistivity are available. We shall discuss this problem in greater detail. The energy density $\mathcal{E}_k$ of the oscillation increases exponentially at low amplitudes. Then, at high amplitudes, nonlinear saturation effects should occur, and one can expect the emergence of a stationary or quasistationary state. Furthermore, one can neglect the left-hand part of the equation and find the spectrum of $\mathcal{E}_k$ by simply equating the linear growth to some effect connected with nonlinear saturation. Nonlinear effects can be presented symbolically as follows:

$$\frac{\partial \mathcal{E}_k}{\partial t} = 2\gamma_k \mathcal{E}_k - \hat{A}\mathcal{E}^2 - \hat{B}\mathcal{E}^3 - \cdots \tag{1}$$

There are quadratic effects, which are proportional to the squares of wave amplitudes, and cubic effects, For ion-sound oscillations three-wave interaction resonances are forbidden; therefore, the only effect yielded by the term on the order of $\mathcal{E}^2$ can be that of nonlinear wave-ion scattering (7). This effect occurs because of the presence of nonlinear denominators of the $\omega_1 - \omega_2 + (\mathbf{k}_1 - \mathbf{k}_2)\cdot\mathbf{v}$ type. In fact, these are Landau resonance oscillations from each randomly selected pair of waves. These oscillations get into resonate with the ions, absorption occurs, part of the energy is absorbed and part passes over to a wave with lower frequency.

Actually, the quadratic term is a rather complex integral expression in which the integration is performed with respect to all the wave vectors. The magnitude of this term can be evaluated as follows. Inasmuch as this effect is connected with the thermal motion of ions, the operator A will contain a small quantitative factor, T_i/T_e. This is a small value, since we are discussing ion sound, and it is required for propagation that this value be small. The other effect, the cubic one, relates to the four-wave interaction. The four-wave interaction is allowed and, when it is accounted for, leads to a rather cumbersome nonlinear operator containing the wave energy density to the third power. In weak turbulence theory this effect is weaker than nonlinear ion scattering. Thus the main problem to be solved is the balance between the linear rise and the first (quadratic) nonlinear term.

A problem of this type was first solved by Kadomtsev (8), who reduced the complex integral operator to differential form while bearing in mind that in such a nonlinear interaction the frequency varies but slightly. While solving the balance equations, Kadomtsev discovered, in the region of wavenumbers substantially lower than the Debye wavenumber (in other words, of wavelengths considerably greater than Debye length), a simple energy density dependence proportional to k^{-3}. It is unknown what happens at $k \to k_D$. In this case the integral operators (or the wave–wave collision integral) do not reduce to a simple form. However, it is possible to investigate the opposite extreme of very large wave vectors, that is, wavelengths shorter than Debye length. Here the dispersion law for the ion sound is very simple, $\omega \approx \Omega_0$, and the spectrum appears to fall very rapidly ($\mathcal{E}_k \sim k^{-13}$), as shown by Galeev and the present author. The Kadomtsev spectrum features a logarithmic divergence (the total wave energy diverges for small wave vectors). However, this logarithmic divergence is not dangerous, since we are interested in the loss of momentum by electrons rather than in the energy density. The loss of momentum by the electrons can be represented in the following manner:

$$nm\bar{V}\nu_{\text{eff}} = \int \frac{\mathcal{E}_k}{\omega_k} k_{\parallel} \gamma_k \, dk \tag{2}$$

($k_{\parallel}$ is the projection of the wave vector on **j**).

The momentum lost by the electrons is transferred to the waves. It is thus seen that another integral is used in this case. For long waves k and ω are approximately proportional to each other. However, in this case there emerges an additional factor, an imaginary part γ_k, a value proprotional to the frequency—hence k enters once more. Thus no divergence is now observed at low values of k; on the contrary, a contribution to the integral is provided by the region of large values of k. An assumption follows naturally that the cutoff should be effected at wave vectors on the order of the Debye one (further on, the Kadomtsev spectrum is invalid and a sharp damping commences). Calculation of this integral leads to the following formula for the effective collision frequency (10):

$$\nu_{\text{eff}} \approx 10^{-2} \frac{T_e}{T_i} \frac{\bar{V}}{V_{\text{th}\,e}} \omega_0 \tag{3}$$

The numerical factor 10^{-2} emerges during calculations and is something like $1/32\pi$. Then follow the temperature ratio (and this is quite natural because the nonlinearity itself contained a small T_i/T_e ratio and, finally, the drift-to-thermal-electron velocity ratio contained in the instability increment. Therefore, if a current could be passed through the plasma that would exceed considerably the critical value, so that electrons would lose momentum because of coherent phonon radiation, that is, ion-sound oscillations, some stationary or, rather, quasistationary spectrum would be established and ν_{eff} could be found from such a simple formula. This formula has a more profound meaning than Buneman's formula because it reflects the specific features of the nonlinear saturation of the instability.

Tsytovich (11) called attention to the fact that, although the spectrum of ion-sound oscillations is a nondecay one, nevertheless, in the region of low wavenumbers where the dispersion law is almost linear, the three-wave resonance condition can be met owing to the small imaginary part of the frequency occurring because of nonlinear line broadening. Then he introduced such a three-wave resonance, that is a decay process, into the kinetic equation for waves as a nonlinear term bringing about the saturation of oscillations. After that he solved this kinetic equation. In this case a spectrum is obtained close to the Kadomtsev spectrum, since the nonlinearity here is likewise quadratic; however, an additional small parameter is lacking which is present in Kadomtsev's theory, namely, the ion-to-electron temperature ratio. Naturally, a somewhat different value is obtained for Ohm's law. The Tsytovich spectrum can occur only at sufficiently low wavenumbers, where a small non-linearity is sufficient for resonance overlapping and three-wave interaction can be thereby effected. However, the major contribution to conductivity (as distinct from the oscillation energy) is given by short waves. In short waves, at frequencies on the order of the ion plasma frequency, the deviation from the linear dispersion law is rather great, and a high nonlinearity is required to meet the condition of three-wave interaction. Therefore this formula does not yet have a region of applicability, especially in view of the fact that in the case of high nonlinearity we have to shift to the strong turbulence region.

There exist indirect experimental data that are not quite reliable but that, in appropriate extreme cases, corroborate formula 3 for resistivity. However, one should handle these experimental data with great caution. Indeed, in what does the main problem consist? The answer is that none of the four quantities included in this formula, $\overline{V}$, $V_{\text{th}\,e}$, T_i, T_e, can have such a simple meaning in the actual plasma in the absence of real pair collisions and in the presence of only fluctuation scattering. Let us start with the electron temperature. If there are no pair collisions, one can hardly expect the distribution function to be a Maxwellian one. But we should not necessarily require that the distribution function of, say, electrons, be a Maxwell function; at least, however, this function should behave properly, that is, so that it can be characterized by some mean thermal spread, so that it can feature some quite rapidly converging tails, and so that one can speak of single-temperature electrons. The same holds true for ions, although the ion situation is still more complicated. It is clear from the beginning that the ion distribution function will behave rather exotically if the ions interact with waves alone and there are no pair collisions. In regard to the mean drift velocity, the picture is usually as shown in Fig. 1*a*. In this diagram can be observed the ion distribution function and the electron distribution function shifted relative to the former function in the tacit assumption that the electron distribution is shifted relative to the ion distribution as a whole. Nothing, however, prevents us in principle from imagining

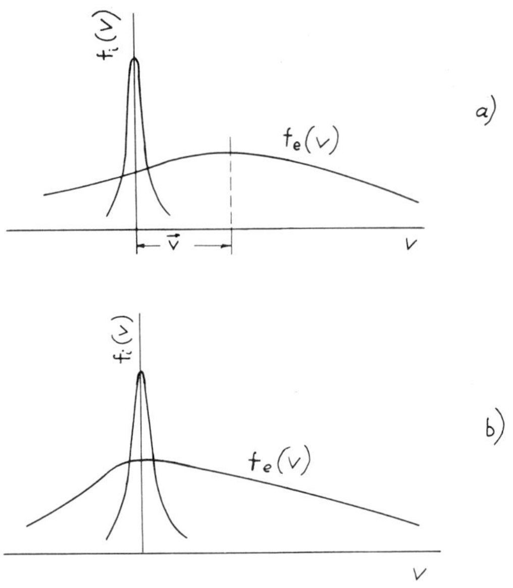

Fig. 1. Distribution functions in current carrying plasma. (*a*) Usual assumption. (*b*) With coinciding electron and ion maxima.

a situation (cf. Fig. 1*b*) in which the electron distribution preserves its maximum in the same position as the ion one, while some portion of the electron distribution deforms considerably so that there occurs an asymmetric electron distribution. As for the possible appearance of the ion and electron distribution functions, the complexities of the emerging phenomena can be well seen with the aid of the two-dimensional diagram presented in Fig. 2. Plotted on the x axis is the particle velocity component along the direction of the current flow; on the y axis, the transverse component. Let us assume that there is initially present a normal Maxwell distribution of electrons and ions. In this Maxwell distribution the equivalent lines of the distribution function in this plane are circles.

The wave-particle interaction is especially strong when Landau resonance is realized. Imagine that we have some wave with vector **k**, traveling at a phase velocity ω/k. This wave interacts with particles located near line A in Fig. 2. It is for such particles that the resonance is realized.

Finally, we should consider waves of diverse directions and different phase velocities. Thus all the particles in the portion where one can draw a line corresponding to the Landau resonance condition are affected by a random wave field. Waves with velocities lower than a certain value are absent; roughly speaking, there are no waves with velocities below that of sound. Therefore particles having a velocity, say, less than the sound velocity, in the quasilinear approximation do not interact with waves. In this case the interaction appears

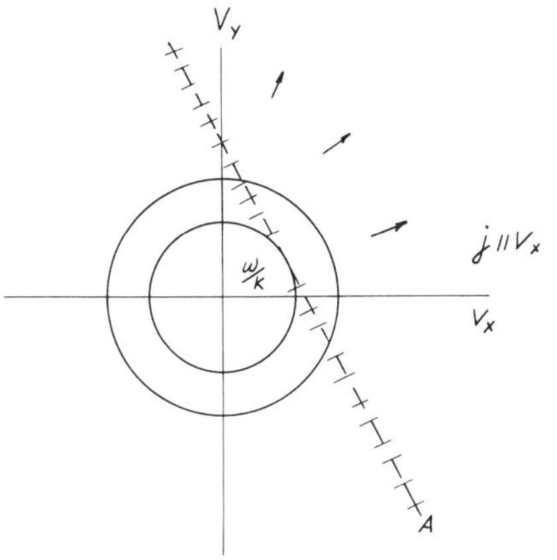

Fig. 2. Distribution function contours, illustrating wave-particle resonance.

to be much weaker; it is due to the nonlinear effects of the next approximation. The number of ions in the region of interaction with waves is rather small, therefore, but a small fraction of ions is subjected to the powerful effect of the waves. The distribution function in the main region essentially is not deformed in zero approximation, whereas in the resonance region a strong deformation occurs. From the viewpoint of the quasilinear theory, this deformation is nothing but the diffusion of particles in velocity space. Quite a number of intermediate experiments in which such diffusion takes place rather slowly, and pair collisions of particles still succeed in creating something like a Maxwellian distribution, are known. We refer here to cases in which the plasma resistivity exceeds but slightly the classic value.

We shall discuss the most extreme cases, in which the pair collisions are of no importance whatsoever. It is here that such a distortion of the ion distribution function is essential. It greatly changes the ion imaginary part, the ion residue being proportional to the number of ions that may be present at the resonance. Although this is a very small number as a rule, it is seen to be very sensitive to what is taking place on the tail of the ion distribution. At present there is no theory that can self-consistently describe the development of the ion distribution function at sufficiently long times. Certain evaluations can be made, however; in particular, the use of the two-temperature approximation is rather effective on the assumption that the ions can be divided into two groups: cold ions, to which nothing or something slow happens; and hot ions, which are on the tail of the distribution function and have quite a high temperature. We shall discuss this later.

The following takes place with the electrons: the region of forbidden velocities, within which there is no resonance between the particles and the waves, is relatively small, since the sound velocity is $\sqrt{m/M}$ times lower than the mean thermal velocity of electrons. In principle, what takes place inside this small circle can be neglected. However, another complexity appears. It is hard to imagine a situation in which the current flowing in this direction should generate waves almost transverse to the direction of the current. Indeed, it is obvious from the theory of ion-sound instability that waves with a large vector across the direction of current have a small imaginary part and can be regarded as practically stable. Therefore we arrive at the conclusion that in this case a small cone is formed in the velocity space, in which there are no waves for resonance with electrons. These electrons are freely accelerated by the electric field generating the current in the plasma. The contribution made by these electrons may greatly reduce the resistivity of plasma. Hence how large is the fraction of electrons that gets into this "cone of losses" and is freely accelerated thereafter?

This problem can be split into two extreme cases. It is expedient that from the overall problem there be separated the simpler part involving the presence of a magnetic field, if only a weak one (a weak magnetic field $\omega_0 \gg \omega_H$ is more convenient), in the plane perpendicular to **j**. Such a magnetic field slowly rotates the electrons (slowly in comparison to the plasma oscillation frequency). But this rotation can be rather fast in the time scale in which the escape of electrons into the loss cone would be pronounced. In this manner, on the average all the electrons during one Larmor revolution appear to be interacting with the waves. No additional difficulties arise in this problem for the electrons. Although such a problem is not reduced to the Maxwellian distribution of electrons, nevertheless in this case we can speak of the mean electron temperature. Moreover, with rather general assumptions, assuming only that the phase velocity of the oscillations appearing is much less than the mean thermal velocity of electrons, and assuming nothing about the oscillation spectrum, one can obtain a simple formula for the electron distribution function. After a certain period of time, when the energy of electrons becomes substantially higher than their initial thermal energy, there occurs a universal distribution according to the law e^{-V^5} (12). A similar distribution is observed in some experiments (13). In the case of such a distribution one can speak of mean temperature, and all the calculations for electrons are carried out in a manner that is practically the same as that for a Maxwell distribution, a slight distinction being that somewhat different numerical coefficients are obtained. Therefore one can believe that, if a weak transverse magnetic field is present, all the results for the effective number of collisions can be transferred to even remote moments of time when a distortion of the electron distribution (its deviation from Maxwellian distribution) could make itself felt.

It is such a situation that takes place in experiments with collisionless shock waves across a magnetic field. In such a shock wave the current flows across the magnetic field. However, the complexities accompanying the ion distribution are present in this case as well. The ion distribution function has no time to become mixed under the effect of the magnetic field, and therefore the ion distribution may assume a rather exotic form and, in the end, differ greatly from the Maxwellian distribution. One can expect the major fraction of ions to be cold and some fraction, starting from velocities on the order of the sound velocity, to become heated.

In all probability, it will be impossible in the near future to obtain such a complex ion distribution without resorting to numerical methods. At the same time, general considerations can lead us to expect the following. In case no interaction with the walls is observed in the course of the current flow and the growing increase in electron and ion energy, that is, if no heat losses to the outside take place, we should arrive at some self-similar form for distribution functions and for turbulence spectra. All should grow according to some relatively simple universal law. The assumption concerning the existence of a self-similarity solution was checked by Vekstein, Ryutov, and the present author (14). Self-similarity variables were found in which the equations of the weak turbulence theory had assumed a simpler form; however, these equations have yet to be solved.

Nevertheless, the following approach can be attempted. For the case in which the current flows across the magnetic field and the wave spectrum is three-dimensional, the ion distribution is divided into two groups, cold and hot ions. In such two-group approximations quantities characterizing the flow of the current have been found. It has been observed that the group of ions which come into resonance with ion-sound oscillations and are then accelerated constitutes a relatively small quantity, the order of magnitude of the concentration of such hot ions being defined by a simple expression, $(m/M)^{1/4}$. The mean energy of these hot ions appears to be close to that of the bulk of the electrons. And, finally, the cold ions have a temperature considerably lower than T_e and make practically no contribution to the emerging phenomena. In this case, Ohm's law for electric fields that are not very large has a simple form: if the electric field is not large, so that the force of friction due to coherent radiation of ion-sound oscillations slows down the electrons and prevents them from acquiring a mean velocity greater than the critical velocity at instability, the system appears to be permanently at the threshold of instability. It appears in this case that finally, after a certain period of time when the system assumes the self-similar mode, the mean drift velocity turns out to be on the order of the sound velocity times the factor $(M/m)^{1/4}$. This value is in satisfactory agreement with experimental data for the current flowing in a collisionless shock front (15).

Imagine now that the magnetic field either is altogether absent or is present, say, in the direction of the current flow. In that case the mechanism serving to mix all the electrons disappears, and the problem of the electron distribution function becomes more complicated. What shall we do in this case? Let us assume that here, as well, a self-similarity distribution will set in after a certain period of time. The electron distribution function assumes some universal form, and there follow a further heating of the electrons and increase in their mean velocity, although the form of the function remains self-similar. Nothing can be done with these self-similarity variables; it does not seem possible to introduce a two-temperature electron distribution because of the absence of a clearly marked division of electrons into two groups, as was the case with ions. On the contrary, there occurs a smooth transition from slower electrons to faster ones; and finally, with the passage of time, as shown by qualitative analysis in self-similarity variables, a considerable fraction of the electrons, on the order of unity, gets into the region of velocities where waves are practically absent. This phenomenon reminds one of the electron escape in gas with Lorentz collisions (when the collision frequency drops with velocity as V^{-3}). The interaction of electrons with ion-sound waves shows this particular property, as revealed by Korablev and Rudakov.

What, then, finally happens? This problem is still open to discussion, and various points of view exist. The present author, for instance, adheres to the following opinion. A considerable fraction of the electrons gets into the escape mode, the electron distribution function projected onto the parallel velocity appears to be greatly prolonged in the direction of current, and the mean drift velocity of the electrons, divided by their mean thermal velocity (some mean spread), can become of the order of unity. It is hard to predict the precise numerical value; this may be, say $\frac{1}{2}$ or $\frac{1}{3}$. The conclusions arrived at by Coppi are quite similar. Data obtained from laboratory experiments are still quite inadequate. The difficulty is that the majority of experiments on anomalous resistance relate to discharges with so-called open ends, that is, to discharges in which the plasma electrons can freely escape along the force lines of the magnetic field. Obviously, no long-range self-similar mode can be attained under such conditions; continuous heat removal is observed, with the fast particles being the first to escape since the path length of particles with high velocities is great. Therefore, there occurs a continuous cutoff of the electron distribution tail. The problem becomes extremely complicated, depends greatly on the boundary conditions, and is no longer of universal interest. In this case, the $\overline{V}/\overline{V}_{\text{th}\,e}$ ratio may appear to remain much less than unity.

There are some idealized extreme cases in which the equations of weak turbulence in the problem of anomalous resistance can be solved accurately. These are one-dimensional cases. Just as in statistical thermodynamics there is a class of one-dimensional soluble models, in the weak turbulence theory the one-dimensional models turn out to be considerably simpler. In some cases,

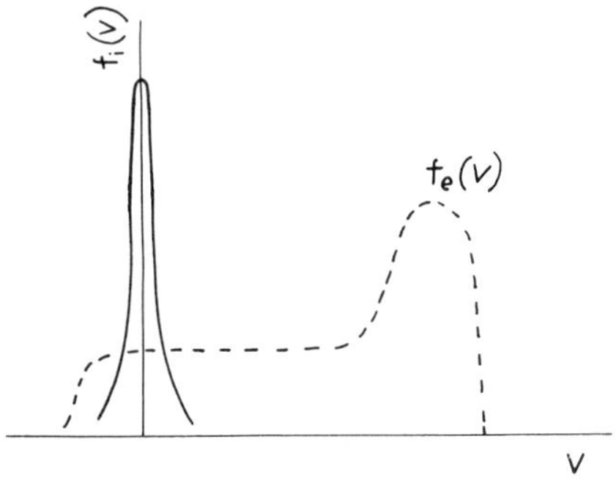

Fig. 3. Distribution function in the one-dimensional case similarity solution.

one-dimensional models can have real physical meaning. For example, if the magnetic field is so high that the Larmor frequency of the electrons exceeds considerably their plasma frequency, electron motion in the direction across the force lines of the magnetic field is practically forbidden, and we deal with purely one-dimensional motion. In this case, the one-dimensional theory can be expected to provide an adequate description of what is taking place. Self-similarity equations in one-dimensional formulation are solved exactly (14). The universal self-similar form of electron distribution, which finally sets up and evolves in a self-similar manner, turns out to be the following (cf. Fig. 3). There is a plateau region from $V \approx 0$ up to the free acceleration velocity acquired by the major fraction of the electrons.

Numerical experiments in the one-dimensional case exhibit approximately such a form of the distribution function (16). Some difference is observed, between the distribution function in numerical experiments and the self-similar form. In particular, in the self-similarity theory the number of electrons in the plateau region is quite small (several per cent), whereas in the computer experiments about half of the electrons are in the plateau region. A simple reason why the self-similarity solution underestimates the number of electrons in this region has been revealed. As a matter of fact, we believed the system to be at the threshold of instability—in other words, the imaginary part of the frequency to be equal to zero with a great degree of accuracy. In any real experiment there takes place a noise increase from some thermal equilibrium level, and we should have some finite value of the imaginary part of the frequency in excess of zero. If some reasonable finite value γ, such that the noise could manage to increase during a finite time, is assumed a development of a plateau immediately occurs. We do not know yet to what extent such improved theory can be brought closer to the experimental data.

Attempts have been made to consider the problem of anomalous resistivity within the framework of the strong turbulence theory. Of course, the strong turbulence theory provides more room for speculations. There exist a series of other variants of nonlinear saturation whose authors try to depart from the quasilinear consideration of electrons, in particular, to study the part played by trapped electrons. As shown by Kadomtsev and Pogutse (17), who introduced macroparticles trapped in potential wells, these "particles" are preserved for a sufficiently long time only if the wave packet is sufficiently narrow apart from being simply one-dimensional. Problems of anomalous resistivity lead to two- and three-dimensional spectra, and it is therefore hard to expect that in real cases trapped macroparticles play any part. At the same time, the existing numerical experiments can still relate to a case in which the trapped particles are important, since in view of the finiteness of the number of points the number of harmonics is small and the trapped particles can play a significant role (there may be no overlap between the areas of influence on the particles interacting with different waves).

Let us turn once more to the table on p. 153. The Drummond-Rosenbluth instability leads to relatively weak imaginary parts of the frequency; moreover, from the standpoint of the quasilinear approximation, this is a one-dimensional instability. The wave-electron interaction takes place in a one-dimensional manner, and therefore an emerging electron plateau similar to that described above should quickly arrest the development of instability. The current may increase, while in a small region of velocity space the electron distribution will develop a plateau and there will be no instability. As to the electron-sound instability, it has small imaginary parts of the frequency and brings about a relatively low anomalous resistivity; however, in some cases in which stronger instabilities are absent this resistivity is well pronounced and important. This problem has been adequately studied in the papers by Stepanov and Rudakov (5).

Of late, considerable interest has been shown in the so-called Bernstein mode instability (6). This rather interesting instability, which escaped the earlier attention of theoreticians, consists in the following. Bernstein modes are oscillations with a wave vector that is strictly or almost strictly transverse to the magnetic field; the frequencies are grouped near the $n\omega_H$ harmonics. Imagine purely electron oscillations; their frequency is rather high, and there is no interaction with ions. Let a current flow through the plasma. Thus, in the reference system moving together with the electrons, such oscillations are present. However, because of the Doppler effect, in the laboratory reference system where the ions are resting the oscillation frequency is shifted to $\omega' = n\omega_H - k\bar{V}$. If the drift velocity is adequately high, at sufficiently great values of k (which may be selected up to the Debye wave vector), we can substantially reduce the frequency in the ion reference system so that these oscillations interact with ions at $\omega' \sim kV_i$.

Then there appears a negative energy of wave instability. If we take into account the imaginary part of the ion interaction with Bernstein modes due to Landau resonance, the normal Maxwellian ion distribution yields an instability. This instability appears to have a rather great imaginary part of the order of the electron Larmor frequency reduced by the drift-to-thermal-electron velocity ratio. Moreover, this instability exhibits little sensitivity to the temperature ratio. As opposed to the case of the ion sound, this instability does not require that the electron temperature exceed substantially the ion temperature. A higher anomalous resistivity could be expected. It turns out, however, that a very slight nonlinearity—a very low effective collision frequency occurring in the course of the instability development—proves sufficient to fully suppress this instability. Indeed, the electron inertia is essential to these oscillations. This means that the electron collisions will make a considerable contribution to the imaginary part. Obviously, ν_{eff} will be subtracted from the instability increment. It appears that by equating these two quantities one can readily find ν_{eff}. In reality, however, the instability yields a still greater contribution. One should remember that the oscillating part of the distribution factor has the term $e^{ik_\perp V_\perp/\omega_M}$; therefore ν_{eff} will enter with the "Pitayevskii factor," $k_\perp{}^2V^2/\omega_H{}^2$. Inasmuch as we speak of very great values of k for short waves, this factor is of great importance. As a result, a small level of nonlinearity is sufficient to suppress the Bernstein mode instability.

References

1. O. Buneman, *Phys. Rev.*, **115**, 503 (1959).
2. G. I. Budker, *At. Energy*, **5**, 9 (1956).
3. G. V. Gordeyev, *Zh. Eksp. Teor. Fiz.*, **27**, 19, 24 (1954).
4. W. E. Drummond and M. N. Rosenbluth, *Phys. Fluids*, **5**, 1507 (1962).
5. V. L. Sizonenko and K. N. Stepanov, *Nucl. Fusion*, **7**, 131 (1967); V. I. Arefiev, I. A. Kovan, and L. I. Rudakov, *Letter in Zh. Eksp. Teor. Fiz.*, **8**, 286 (1968).

6. Collision-Free Shock Meeting, Los Alamos, 1971.
7. A. A. Galeev, V. I. Karpman, and R. Z. Sagdeev, *Nucl. Fusion*, 1965.
8. B. B. Kadomtsev, *The Problems of Plasma Theory*, Vol. 4, Atomizdat, Moscow, 1964.
9. R. Z. Sagdeev and A. A. Galeev, *Nonlinear Plasma Theory*, Benjamin, New York, 1969.
10. R. Z. Sagdeev, Symposium on Applied Mathematics, New York, 1965.
11. V. N. Tsytovich, Preprint, Culham Laboratory, 1970.
12. See Ref. 9.
13. Montgomery, Shock Symposium at ESRIN, 1969.
14. G. A. Vekstein, D. D. Ryutov, and R. Z. Sagdeev, *Zh. Eksp. Teor Fiz.*, **6**, 2142 (1971).
15. V. G. Eselevich, A. G. Eskov, R. Kh. Kurtmulayev, and A. I. Malyutin, *Zh. Eksp. Teor. Fiz.*, **60**, 1658 (1971).
16. D. Biskamp and A. Chodura, Preprint, 1971.
17. B. B. Kadomtsev and O. P. Pogutse, *Doklady*, **186**, 553 (1969).

Some Aspects of the Theory of Strong Plasma Turbulence

SETSUO ICHIMARU

Department of Physics, University of Illinois, Urbana, Illinois, USA and Department of Physics, University of Tokyo, Bunkyo-ku, Tokyo, Japan

Abstract

The theory of plasma turbulence based on a solution of the Bogoliubov-Born-Green-Kirkwood-Yvon hierarchy is reviewed. The salient feature of the theory is a recognition of the fact that the effective interactions between two particles may be drastically modified because of the presence of strong correlations in a turbulent plasma. Fluctuations and physical consequences arising therefrom are discussed; these include a wave kinetic equation, size-dependent effect, similarity law, and the possible existence of a universal spectrum. Related topics such as anomalous transport and radiation phenomena are also discussed.

Introduction

For a plasma in thermodynamic equilibrium, the density fluctuations associated with the collective modes are at the level of the thermal fluctuations; the motion of single particles is not affected appreciably by the presence of such weak fluctuations. The plasma parameter, $g \equiv 1/n\lambda_D{}^3$, where n is the number density of the particles and λ_D is the Debye length, provides a measure of the ratio between the fluctuation energy and the kinetic energy under these circumstances. The Liouville equation or, equivalently, the Bogoliubov-Born-Green-Kirkwood-Yvon (BBGKY) hierarchy (1, 2) may be solved for such a quiescent plasma by means of a systematic expansion in powers of the plasma parameter.

In most cases, the plasmas found in nature are far from thermodynamic equilibrium. A collective mode may even become unstable in some cases; fluctuations are then greatly enhanced over the thermal level. The plasma may eventually go over to a state of turbulence. The resulting state may be either weakly or strongly turbulent, depending on the deviation of the plasma from equilibrium. Clearly, the BBGKY hierarchy should be able to produce solutions applicable to any of the three plasma states, namely, the quiescent, weakly turbulent, and strongly turbulent states, although the truncation scheme of a quiescent plasma is no longer applicable to plasmas in a turbulent state.

A quantity of primary interest in the theory of turbulence is the energy spectrum $\mathcal{E}_\mathbf{k}$ contained in the wave vector **k** mode of the fluctuations. For a plasma in thermodynamic equilibrium at temperature T, $\mathcal{E}_\mathbf{k}$ in a given collective mode takes on an equipartition value κT, where κ is the Boltzmann constant. Generally, in a quiescent plasma, $\mathcal{E}_\mathbf{k}$ is written in a form proportional to the discreteness parameters, such as the electric charge (e), the mass (m), the characteristic volume ($1/n$), and an average kinetic energy (κT) per particle. We may portray the fluctuations under these circumstances as arising essentially from the random motion of the discrete particles; hence $\mathcal{E}_\mathbf{k}$ vanishes in the fluid limit (2), in which those discreteness parameters approach zero in such a way that en, mn, and $n\kappa T$ remain constant. In this fluid limit, the plasma parameter g, being of the same order as the discreteness parameters, also vanishes.

A strongly turbulent state of a plasma may be defined as that in which *the energy spectrum of the fluctuations remains finite even in the fluid limit*; the correlations persist in this state with macroscopic intensities. It becomes essential, therefore, for a theory of plasma turbulence to deal directly with a state with strong correlations rather than starting from conventional quiescent description of the plasma state which would ignore the existence of these correlations.

A number of attempts have been made in recent years to formulate a theory of turbulent plasmas (3). The advent of the quasilinear theory (4, 5) of plasma oscillations marked an important step toward the solution of the problems associated with weakly turbulent plasmas. This approach takes explicit account of the feedback action of the growing oscillations upon the single-particle distribution function and treats the wave-wave interaction in a perturbation-theoretical way. Many significant contributions (6, 7, 8) have been made toward the understanding of nonlinear interactions between various collective modes in plasmas.

For a treatment of strong plasma turbulence, Dupree (9) advanced a perturbation theory based on the use of a statistical set of exact particle orbits, instead of the unperturbed orbits conventionally used in a solution of the Vlasov equation. It is then argued (10) that the dominant nonlinear effect of low-frequency instabilities is an incoherent scattering of particle orbits by waves, which causes particle diffusion and appears in the theory as an enhanced viscosity. Weinstock (11) has developed a theory of strong plasma turbulence along a similar line of arguments.

In this paper we wish to review another approach toward the theory of strong plasma turbulence, developed in recent years by the author and his collaborators. After briefly describing the idea of a self-consistent approach, we proceed to discuss the theory of strong plasma turbulence based on a kinetic-theoretical approach and the physical consequences arising from it. Finally, we shall touch on a number of topics related to plasma turbulence.

I. Self-Consistent Approach

As we remarked earlier, a turbulent state of a plasma is characterized by the existence of fluctuations and correlations with macroscopic intensities. A main objective of the theory is then to provide a formalism by which the energy spectrum of turbulence may be determined.

It has been widely recognized that a simple perturbation-theoretic calculation, starting from a quiescent description of the plasma state and regarding the fluctuations as perturbations, cannot lead to an adequate description of plasma turbulence. An example of this difficulty may be found in the theory of critical fluctuations (12) in the plasma according to which the spectral amplitudes of the fluctuations *diverge* as the plasma approaches from the region of stability a critical point corresponding to the onset of a plasma-wave instability.

In order to avoid such difficulties, we subsequently proposed a self-consistent approach to the theory of plasma turbulence (13). The approach consists of three steps.

1. Assume a stationary state of a turbulent plasma, which may be characterized by the existence of finite fluctuations superposed on an ordinary

quiescent state; the energy spectrum $\mathcal{E}_{\mathbf{k}}$ is an unknown quantity to be determined later.

2. Apply a weak perturbation to this turbulent state, and calculate the relevant linear response functions; these response functions now depend on $\mathcal{E}_{\mathbf{k}}$ as well as on other parameters describing quiescent plasmas.

3. Establish a relationship between the fluctuation spectrum and the response functions; the resulting self-consistent equation may then be solved, and the turbulent state determined.

We have applied this approach to a theoretical study of a turbulent state arising from the ion-acoustic wave instability. The results have shown reasonably good agreement with experiments.

It is also of interest to note a close connection between this self-consistent approach and a recent development in the theory of electron gas at metallic densities (14). Here a standard random-phase approximation (RPA) is known to predict physically unacceptable (i.e., negative) values for the electron correlation function at short distances. To treat the strong correlations existing between the electrons at short distances in a self-consistent fashion, Hubbard (15) and Singwi and his coworkers (16) independently applied an approach similar to the above. The linear response function relevant to the problem is the dielectric response function expressed as a functional of the correlation function; the fluctuation-dissipation theorem (17) provides the relationship between the correlation function and the dielectric response function. The results of their calculations have shown a remarkable improvement over a conventional calculation based on the RPA for short-range electron correlations.

In spite of these successes, a number of shortcomings have been noted in the self-consistent approach. First, the calculation of response functions in a turbulent plasma always involves some kind of approximation; in many cases, it is not easy to assess the accuracy of the treatment or the extent of the approximation in a systematic way. Second, there exist no exact theories, such as the fluctuation-dissipation theorem, which would relate the response functions to the fluctuation spectrum of a turbulent plasma; the latter plasma is by definition a system far from thermodynamic equilibrium. The dielectric superposition principle (13, 18) does offer an approximate relationship to be used for this purpose; the accuracy of this principle as applied to a turbulent plasma, however, remains to be critically examined.

In order to obtain an alternative and more systematic approach to the theory of plasma turbulence, we have therefore attempted to develop a kinetic-theoretic approach based on the BBGKY hierarchy of plasma kinetic equations (19, 20). We now turn to a discussion of this topic.

II. Kinetic-Theoretical Approach

With a knowledge of the pair correlation function, or the fluctuation spectrum, the first equation of the BBGKY hierarchy described the behavior of the single-particle distribution function; the correlation function enters in the collision term of the kinetic equation. Turbulence therefore affects the transport processes of the plasma through this equation.

The second equation of the hierarchy determines the pair correlation function. It is in this sense that we regard the second equation as equivalent in physical content to the self-consistent equation for $\mathcal{E}_{\mathbf{k}}$ which we sought to establish in the previous approach. By working directly on the second equation, therefore, we must be able to arrive at the equation for $\mathcal{E}_{\mathbf{k}}$ without going through those intermediate steps.

In order to determine the pair correlation function from the second equation, it is necessary to know the ternary correlation function, as well as the single-particle distribution function. Conversely, we note that, if the latter two functions are known, a formal solution can always be obtained. The problem thus reduces to that of finding the expression for the ternary correlation function which is appropriate to the turbulent situation under consideration. The first two equations of the BBGKY hierarchy will then provide a set of coupled equations for the single-particle distribution function and the pair correlation function.

For a quiescent plasma, there exists a well-known hierarchy among the correlation functions in terms of various orders in the plasma parameter, g. One can use this fact to truncate the BBGKY hierarchy systematically and thus can, in principle, determine the ternary correlation function to an arbitrary degree of accuracy in the plasma parameter (21).

In the case of turbulence, the above procedure is no longer applicable; by our definition of the strongly turbulent state, the pair correlation function must be of the same order in the plasma parameter as the single-particle distribution function. A similar statement may also be true for all the higher-order correlation functions. In fact, for the theory of strong plasma turbulence, we are interested in accurately determining only those parts of the many-particle distribution functions which remain finite in the fluid limit. Therefore we are led to adopt the following guideline for finding an appropriate ternary correlation function: Many-particle distribution functions, as well as the single-particle distribution function, remain finite in the limit $g \rightarrow 0$ and satisfy the BBGKY hierarchy also in this limit.

We have not been able to find a general expression for the many-particle distribution functions which exactly satisfies this criterion. However, an inspection of the third equation of the hierarchy in the fluid limit reveals that the ternary correlation function constructed according to Kirkwood's superposition approximation (22) and the higher-order correlation functions similarly constructed indeed satisfy the criterion to a good degree of approximation. We therefore adopt this approximation; it is the major approximation involved in this kinetic theory of plasma turbulence. The accuracy of the theory can be assessed and improved in a systematic way from the third equation of the hierarchy.

III. Turbulence Spectrum

The theory of plasma turbulence (20) resulting from the kinetic-theoretical approach described above contains an important physical feature; it takes explicit account of the fact that the effective interactions between two particles may be drastically modified because of the presence of strong correlations in a turbulent plasma. Let $V(r)$ denote the interaction potential between two particles at a distance r. When the effects of the particle correlations are taken into account, the effective potential may be written as $\tilde{V}(r) = V(r)[1 + g(r)]$, where $g(r)$ is a suitably normalized pair correlation function.

More accurately, the ratio between the Fourier transforms of $\tilde{V}(r)$ and $V(r)$ is given by

$$t(\mathbf{k}) = 1 + \frac{1}{n} \sum_{\mathbf{q}} \frac{\mathbf{k} \cdot \mathbf{q}}{q^2} S(\mathbf{k} - \mathbf{q}) \tag{1}$$

where $S(\mathbf{k})$ is the structure factor, calculated from the statistical average of the fluctuating density field $n(\mathbf{r}, t)$ as

$$S(\mathbf{k}) = \frac{1}{n} \int d\mathbf{r} \, \langle n(\mathbf{r}' + \mathbf{r}, t) n(\mathbf{r}', t) \rangle \exp(-i\mathbf{k} \cdot \mathbf{r}) \tag{2}$$

This function is proportional to the density-fluctuation spectrum in $\mathbf{k}$ space; it is directly related to the energy spectrum of turbulence (see eq. 11 below). In deriving eq. 1, we have assumed the Coulomb interaction for $V(r)$; general cases are treated in Ref. 14. In eq. 2, $\langle \; \rangle$ denotes a statistical average; the intergration is to be carried out over a unit volume. The appearance of this $\mathbf{k}$-dependent factor, $t(\mathbf{k})$, represents a major effect of strong correlations in a turbulent plasma. We emphasize that this factor describes the modification of effective particle interactions arising from the presence of turbulence.

A stationary solution of the second equation in the BBGKY hierarchy has been obtained in Ref. 20; it has the form

$$S(\mathbf{k}) = \frac{1}{n} \int d\omega \, \frac{S^{(0)}(\mathbf{k}, \omega)}{|\tilde{\epsilon}(\mathbf{k}, \omega)|^2} \tag{3}$$

Here

$$S^{(0)}(\mathbf{k}, \omega) = n \int d\mathbf{v}\, f(\mathbf{v})\, \delta(\omega - \mathbf{k}\cdot\mathbf{v})$$

$$\cdot \left\{ 1 + \frac{1}{n} \sum_{\mathbf{q}} \frac{t(\mathbf{q})}{t(\mathbf{k})} \frac{\mathbf{k}\cdot\mathbf{q}}{q^2} S(\mathbf{k}-\mathbf{q})[S(\mathbf{q}) - 1]\cdot[1 - |\tilde{\epsilon}(\mathbf{k}, \omega)|^2] \right\} \quad (4)$$

$$\tilde{\epsilon}(\mathbf{k}, \omega) = 1 + \frac{\omega_p{}^2 t(\mathbf{k})}{k^2} \int d\mathbf{v} \frac{1}{\omega - \mathbf{k}\cdot\mathbf{v}}\, \mathbf{k}\cdot\frac{\partial f(\mathbf{v})}{\partial \mathbf{v}} \quad (5)$$

$$\omega_p{}^2 = \frac{4\pi n e^2}{m} \quad (6)$$

and $f(\mathbf{v})$ is the velocity distribution function of the particles normalized to unity. Equation 3 thus represents the self-consistent integral equation which should be solved for $S(\mathbf{k})$.

A number of important physical consequences follow directly from this integral equation; some of them are discussed in the following.

A. Stable Turbulent State

As the form of eq. 3 clearly indicates, the zeroes of the function, $\tilde{\epsilon}(\mathbf{k}, \omega)$, on the complex ω plane give rise to a special kind of density-fluctuation excitations, the collective modes; in this sense, $\tilde{\epsilon}(\mathbf{k}, \omega)$ may be regarded as an effective dielectric function of the turbulent plasma, although it fails to satisfy the f-sum rule. Examination of the equation

$$\tilde{\epsilon}(\mathbf{k}, \omega) = 0 \quad (7)$$

tells us about the stability of the collective modes.

Let us first note that eq. 5 depends not only on $f(\mathbf{v})$ but also on $S(\mathbf{k})$ through $t(\mathbf{k})$. The solution of eq. 7, written as

$$\omega = \tilde{\omega}_{\mathbf{k}} + i\tilde{\gamma}_{\mathbf{k}} \quad (8)$$

may now be a functional of both the velocity distribution function and the fluctuation spectrum. This, starting from a conventionally unstable state of a plasma, we can conceive of two basically different mechanisms by which the plasma is eventually stabilized.

One is the mechanism which is well known to the quasilinear theory, that is, change of $f(\mathbf{v})$. A stabilization may be achieved when the shape of $f(\mathbf{v})$ changes in such a way that the solution

$$\omega = \omega_{\mathbf{k}} + i\gamma_{\mathbf{k}} \quad (9)$$

obtained from the conventional dispersion relation,

$$\epsilon(\mathbf{k}, \omega) \equiv 1 + \frac{\omega_p{}^2}{k^2} \int d\mathbf{v} \frac{1}{\omega - \mathbf{k}\cdot\mathbf{v}} \mathbf{k} \cdot \frac{\partial f(\mathbf{v})}{\partial \mathbf{v}} = 0 \tag{10}$$

takes on a negative imaginary value, $\gamma_{\mathbf{k}} < 0$.

If the situation is such that the shape of the velocity distribution is kept unstable by some external means (e.g., application of electric field or injection of beam), the second mechanism becomes more important. Here large-amplitude fluctuations or strong correlations are established in the plasma so that the effective interactions between particles are drastically modified. This is turn modifies the properties of the collective modes as given by eq. 8. When the velocity distribution has a form such that $\gamma_{\mathbf{k}}$ of eq. 9 takes on positive values over a certain range of the $\mathbf{k}$ space, the fluctuation spectrum in a turbulent state may adjust itself so that $\tilde{\gamma}_{\mathbf{k}}$ may still remain negative over the entire $\mathbf{k}$ space. Under these circumstances, a conventionally unstable particle distribution and large-amplitude fluctuation spectrum can coexist in a stable manner, forming a turbulent stationary state. Some examples are considered in Ref. 20.

B. *Wave Kinetic Equation*

The energy $\mathcal{E}_{\mathbf{k}}$ contained in the $\mathbf{k}$ mode of the plasma oscillations is given by

$$\mathcal{E}_{\mathbf{k}} = \frac{2\pi ne^2}{k^2} [\tilde{\omega}_{\mathbf{k}} \tilde{\epsilon}'(\mathbf{k}, \tilde{\omega}_{\mathbf{k}})] S(\mathbf{k}) \tag{11}$$

where

$$\tilde{\epsilon}'(\mathbf{k}, \tilde{\omega}_{\mathbf{k}}) = \left[\frac{\partial \tilde{\epsilon}(\mathbf{k}, \omega)}{\partial \omega}\right]_{\omega = \tilde{\omega}_{\mathbf{k}}} \tag{12}$$

and $S(\mathbf{k})$ is defined as that part of eq. 3 coming from the integration only in the vicinity of the collective mode, $\omega = \tilde{\omega}_{\mathbf{k}}$. In this vicinity, we note also that $|\tilde{\epsilon}(\mathbf{k}, \omega)|^2 \simeq 0$,

$$\frac{1}{|\tilde{\epsilon}(\mathbf{k}, \omega)|^2} \simeq \frac{\pi}{[\tilde{\epsilon}'(\mathbf{k}, \tilde{\omega}_{\mathbf{k}})]^2 \tilde{\gamma}_{\mathbf{k}}} \delta(\omega - \tilde{\omega}_{\mathbf{k}})$$

With the aid of these formulas, as well as eqs. 4 and 11, we may rewrite eq. 3 in the following way:

$$\begin{aligned}\mathcal{E}_{\mathbf{k}} &= \frac{-1}{2\tilde{\gamma}_{\mathbf{k}}} \left[\Theta + \sum_{\mathbf{q}} \tilde{B}(\mathbf{k}, \mathbf{q}) \mathcal{E}_{\mathbf{k}-\mathbf{q}} \mathcal{E}_{\mathbf{q}}\right] \\ &= \frac{\Theta + \sum_{\mathbf{q}} \tilde{B}(\mathbf{k}, \mathbf{q}) \mathcal{E}_{\mathbf{k}-\mathbf{q}} \mathcal{E}_{\mathbf{q}}}{-2\gamma_{\mathbf{k}} + \sum_{\mathbf{q}} \tilde{A}(\mathbf{k}, \mathbf{q}) \mathcal{E}_{\mathbf{q}}}\end{aligned} \tag{13}$$

Here

$$\Theta \equiv \frac{\pi m \omega_p{}^2 \tilde{\omega}_{\mathbf{k}}}{\mathbf{k}^2 \tilde{\epsilon}'(\mathbf{k}, \tilde{\omega}_{\mathbf{k}})} \int d\mathbf{v}\, \delta(\tilde{\omega}_{\mathbf{k}} - \mathbf{k}\cdot\mathbf{v}) f(\mathbf{v}) \tag{14}$$

is the rate of spontaneous emission of the plasma oscillations (23)

$$\sum_{\mathbf{q}} \tilde{A}(\mathbf{k}, \mathbf{q}) \mathcal{E}_{\mathbf{q}} = 2(\gamma_{\mathbf{k}} - \tilde{\gamma}_{\mathbf{k}}) \tag{15}$$

as may be clear from eq. 1 and comparison between eqs. 5 and 10; and

$$\tilde{B}(\mathbf{k}, \mathbf{q}) \equiv \frac{\tilde{\omega}_{\mathbf{k}}}{n^2 e^2 k^2 \tilde{\epsilon}'(\mathbf{k}, \tilde{\omega}_{\mathbf{k}})} \int d\mathbf{v}\, \delta(\tilde{\omega}_{\mathbf{k}} - \mathbf{k}\cdot\mathbf{v}) f(\mathbf{v})$$

$$\cdot \frac{t(\mathbf{q})}{t(\mathbf{k})} \frac{\mathbf{k}\cdot\mathbf{q}(\mathbf{k}-\mathbf{q})^2}{\tilde{\omega}_{\mathbf{q}} \tilde{\omega}_{\mathbf{k}-\mathbf{q}} \tilde{\epsilon}'(\mathbf{q}, \tilde{\omega}_{\mathbf{q}}) \tilde{\epsilon}'(\mathbf{k}-\mathbf{q}, \tilde{\omega}_{\mathbf{k}-\mathbf{q}})} \tag{16}$$

We now construct a wave kinetic equation for $\mathcal{E}_{\mathbf{k}}$ in such a way that its stationary solution should agree with eq. 13. Since for a quiescent plasma we know that (23)

$$\frac{\partial \mathcal{E}_{\mathbf{k}}}{\partial t} = \Theta + 2\gamma_{\mathbf{k}} \mathcal{E}_{\mathbf{k}}$$

the required wave kinetic equation for a turbulent plasma must be

$$\frac{\partial \mathcal{E}_{\mathbf{k}}}{\partial t} = \Theta + 2\gamma_{\mathbf{k}} \mathcal{E}_{\mathbf{k}} - \sum_{\mathbf{q}} \tilde{A}(\mathbf{k}, \mathbf{q}) \mathcal{E}_{\mathbf{k}} \mathcal{E}_{\mathbf{q}} + \sum_{\mathbf{q}} \tilde{B}(\mathbf{k}, \mathbf{q}) \mathcal{E}_{\mathbf{k}-\mathbf{q}} \mathcal{E}_{\mathbf{q}} \tag{17}$$

Thus $\tilde{A}(\mathbf{k}, \mathbf{q})$ and $\tilde{B}(\mathbf{k}, \mathbf{q})$ may be interpreted as the mode-coupling constants, which themselves depend on the spectral density of turbulence.

C. Ordering of Various Quantities with Respect to the Plasma Parameter

Equation 13 makes it possible to order various physical parameters with respect to the powers of the plasma parameter, g. It is easy to note from eqs. 14 and 16 that $\Theta \sim g^1$ and $\tilde{B} \sim g^0$. We may also note from eqs. 5 and 10 that $(\gamma_{\mathbf{k}} \sim \tilde{\gamma}_{\mathbf{k}}) \sim g^0 S(\mathbf{q})/n$; hence, in the light of eqs. 11 and 15, we find that $\tilde{A} \sim g^0$.

Let us now introduce formal definitions of the three states of a plasma: quiescent, weakly turbulent, and strongly turbulent.

A quiescent state is defined as that in which $\gamma_{\mathbf{k}}$ is negative (i.e., conventionally stable) and $|\gamma_{\mathbf{k}}/\omega_{\mathbf{k}}| \sim g^0$. Equation 13 then yields

$$\mathcal{E}_{\mathbf{k}} = -\frac{\Theta}{2\gamma_{\mathbf{k}}} \sim \mathrm{g}^1 \qquad \text{(i.e., thermal fluctuations)} \tag{18}$$

In this state $\tilde{\gamma}_{\mathbf{k}}$ is essentially the same as $\gamma_{\mathbf{k}}$; hence $|\tilde{\gamma}_{\mathbf{k}}/\tilde{\omega}_{\mathbf{k}}| \sim g^0$, as $\tilde{\omega}_{\mathbf{k}}$ always remains of the order g^0.

A weakly turbulent state is defined as a state in the very vicinity of the critical point ($\gamma_{\mathbf{k}} = 0$), so that $|\gamma_{\mathbf{k}}/\omega_{\mathbf{k}}| \leqslant g^{1/2}$. We must now take account of the entire experssion of eq. 13; the equation yields $\mathcal{E}_{\mathbf{k}} \sim g^{1/2}$, some enhancement over the thermal-fluctuation level of eq. 18. Each term on the right-hand side of the wave kinetic eq. 17 is now of the order g^1; the presence of turbulence may be ignored in the evaluation of the mode-coupling constants, $\tilde{A}$ and $\tilde{B}$. Hence, as has been postulated in the quasilinear theory, the effects of mode coupling may be treated in a perturbation-theoretical way for a weakly turbulent plasma. The term $\tilde{\gamma}_{\mathbf{k}}$, which remains negative always, obeys the ordering $|\tilde{\gamma}_{\mathbf{k}}/\tilde{\omega}_{\mathbf{k}}| \sim g^{1/2}$; thus the coherence, or the inverse of the frequency bandwidth of the fluctuation spectrum, is also enhanced substantially in this state. The ordering described in this paragraph is essentially the same as that discussed originally by Frieman and Rutherford (24).

A strongly turbulent state is defined as one in which $\gamma_{\mathbf{k}}$ is positive (i.e., conventionally unstable) and $|\gamma_{\mathbf{k}}/\omega_{\mathbf{k}}| \sim g^0$. Equation 13 then gives $\mathcal{E}_{\mathbf{k}} \sim g^0$; $\sum g\tilde{A}(\mathbf{k},\mathbf{q})\mathcal{E}_{\mathbf{q}}$ should be greater than $2\gamma_{\mathbf{k}}$ in this state. We thus find that Θ is negligible in eq. 13. The thermal fluctuations, or discreteness of the particles, virtually have no effect in determining the state of strong turbulence; fluid description of the plasma appears sufficient. Now $|\tilde{\gamma}_{\mathbf{k}}/\tilde{\omega}_{\mathbf{k}}|$ goes up to g^0; the bandwidth of the frequency spectrum of turbulence, therefore, greatly increases over the corresponding value in a weakly turbulent plasma.

D. *Similarity Law and Universal Spectrum*

When the plasma is strongly turbulent, $S(\mathbf{k}) \gg 1$; hence, in the vicinity of $\omega = \tilde{\omega}_{\mathbf{k}}$, we may simplify eq. 4 as follows:

$$S^{(0)}(\mathbf{k}, \omega) = \sum_{\mathbf{q}} \frac{t(\mathbf{q})}{t(\mathbf{k})} \frac{\mathbf{k}\cdot\mathbf{q}}{q^2} S(\mathbf{k}-\mathbf{q})S(\mathbf{q}) \int d\mathbf{v}\, f(\mathbf{v})\, \delta(\omega - \mathbf{k}\cdot\mathbf{v}) \tag{19}$$

It is then possible to derive a similarity law for strong plasma turbulence from the integral eq. 3.

To begin, let us consider a set of similarity transformations for the following physical parameters:

$$n \to \alpha n, \qquad v \to \beta v, \qquad m \to \mu m, \qquad e \to \eta e, \qquad S \to \zeta S \tag{20}$$

Since the plasma frequency ω_p and the Debye wavenumber k_{D} transform as

$$\omega_p \to \alpha^{1/2}\eta\mu^{-1/2}\omega_p, \qquad k_{\mathrm{D}} \to \alpha^{1/2}\eta\mu^{-1/2}\beta^{-1}k_{\mathrm{D}} \tag{21}$$

the variables, ω and $\mathbf{k}$, must scale in the same way.

For a strongly turbulent plasma, the function $t(\mathbf{k})$ defined by eq. 1 differs appreciably from unity; the effective interactions are drastically modified by the presence of strong correlations. In terms of the similarity transformation, this statement is equivalent to saying that the last term in eq. 1 should scale as 1; we thus find a scaling law:

$$\alpha\eta^6\mu^{-3}\beta^{-6}\zeta^2 = 0 \tag{22}$$

It is not difficult to show that, when eq. 22 is satisfied, $\tilde{\epsilon}(\mathbf{k}, \omega)$ is kept invariant under the similarity transformation because $f(\mathbf{v})$ scales so as to conserve the normalization condition. Finally, we may examine the scaling of the integral equation for strong plasma turbulence, that is, eq. 3 combined with eq. 19; we then find that this integral equation, too, is kept invariant as long as eq. 22 is satisfied. We here observe a kind of self-consistency built into the theory; the equation for strong turbulence and the condition for significant modification of effective interactions by turbulence are indeed governed by the same scaling law.

Equation 22 is, therefore, the only similarity law that we can derive from our theory of strong turbulence. We now rewrite it in a slightly different way so that its physical meaning becomes more obvious. To do so, we note from eq. 11 that the energy spectrum transforms as

$$\mathcal{E}_{\mathbf{k}} \to \mu\beta^2\zeta\mathcal{E}_{\mathbf{k}} \tag{23}$$

We pick up the scaling factors of the energy spectrum, the wavenumber, and the kinetic energy $\frac{1}{2}mv^2$ per particle; and define them, respectively, as

$$E \equiv \mu\beta^2\zeta, \qquad \xi \equiv \alpha^{1/2}\eta\mu^{-1/2}\beta^{-1}, \qquad \theta \equiv \mu\beta^2 \tag{24}$$

Thus eq. 22 reduces to

$$E = \alpha\theta\xi^{-3} \tag{25}$$

From this similarity law, then, we can conclude that the energy spectrum $\mathcal{E}_{\mathbf{k}}$ of turbulence must be expressed in the form

$$\mathcal{E}_{\mathbf{k}} = n\langle\tfrac{1}{2}mv^2\rangle k^{-3}F\left(\frac{k}{k_{\mathrm{D}}}, \cos\chi; g, \{z\}\right) \tag{26}$$

Here $\langle\frac{1}{2}mv^2\rangle$ is an average kinetic energy per particle; F is a nondimensional function of dimensionless variables, k/k_{D} and $\cos\chi$, the direction cosine of $\mathbf{k}$ with respect to an axis of anisotropy of the system (if such exists). Generally, the function F may contain as parameters any dimensionless parameters that can be constructed from the physical parameters describing the plasma. The plasma parameter, g, is certainly one of them. If the drift motion of the particles is involved, the ratio between the drift velocity and the average random velocity, for example, may also be such a parameter; $\{z\}$ in eq. 26 collectively denotes the dimensionless parameters involved.

Previously we noted that the fluid limit suffices for a description of strong turbulence. Hence we may take the limit of $g \to 0$ in eq. 26; defining

$F_0(k/k_D, \cos\chi; \{z\}) \equiv \lim_{g\to 0} F(k/k_D, \cos\chi; g, \{z\})$, we have

$$\mathcal{E}_{\mathbf{k}} = n\langle \tfrac{1}{2}mv^2\rangle k^{-3}F_0\left(\frac{k}{k_D}, \cos\chi; \{z\}\right) \tag{27}$$

Not surprisingly, the scaling law, eq. 26 or eq. 27, conforms to the idea of turbulent heating. As the level of turbulence increases, so does the average kinetic-energy density of the particles. The relationship is almost that of proportionality, except for a possible shift in the k spectrum.

Although eq. 27 is suggestive of the k^{-3} spectrum, it is premature to conclude so, because any power of (k/k_D) can be factored out of the function F_0. If, however, one can define in the k space of the plasma turbulence spectrum something similar to the inertial region in the hydrodynamic turbulence, then in such a domain it is possible to establish from eq. 27 a universal spectrum of strong plasma turbulence, analogous to the Kolmogorov-Obukhov law of hydrodynamics.

Let us first note that the Debye wavenumber k_D gives a measure of wavenumber above which the collective modes are substantially damped; there the fluctuations are dissipated into heat through interaction with particles. Hence we may look for an inertial region of strong plasma turbulence in a long-wavelength domain away from k_D so that the turbulence spectrum may not depend on k_D. If such a region exists, we can then write eq. 27 as

$$\mathcal{E}_{\mathbf{k}} = n\langle \tfrac{1}{2}mv^2\rangle k^{-3}F_1(\cos\chi; \{z\}) \tag{28}$$

where F_1 is another nondimensional function. A universal spectrum proportional to k^{-3} is, therefore, indicated in the inertial region of strong plasma turbulence. With additional assumptions such as isotropy of the spectrum one could further simplify eq. 28.

Needless to say, the validity of the argument in the preceding paragraph depends critically on the existence of the inertial domain in strong plasma turbulence; at the moment this is still an open question. In this connection, it is quite interesting to note that Kadomtsev (3) has obtained, from his specific model of ion-acoustic wave turbulence, a spectral distribution that essentially agrees with eq. 27; F_0 is shown to be proportional to $\ln(k_D/k)$ in his calculation.

Thus far, we have considered the similarity law of plasma turbulence with a three-dimensional spectrum. When a magnetic field is applied to the plasma, the particles move with helical orbits along the magnetic lines of force. If we pass to the limit of an extremely strong magnetic field, the plasma behaves as if it were a one-dimensional gas. One can repeat the entire procedure of similarity transformation as applied to such a one-dimensional plasma. In place of eq. 25, we then have

$$E = \alpha\theta\xi^{-1} \tag{29}$$

Consequently, eq. 27 is replaced by

$$\mathcal{E}_{\mathbf{k}} = n\langle \tfrac{1}{2}mv^2\rangle k^{-1} F_2\left(\frac{k}{k_{\mathrm{D}}}; \{z\}\right) \tag{30}$$

We might then guess that, with an arbitrary strength of magnetic field, the universal spectrum, if it exists at all, would have a k-dependence somewhere between k^{-1} and k^{-3}.

E. Size-Dependent Effect

We have emphasized, on a number of occasions, the importance of taking into account the modification of the effective interactions between two particles caused by the presence of the strong correlations in a turbulent plasma. This modification is described by the function $t(\mathbf{k})$. It must be noted, however, that this function approaches unity as $\mathbf{k} \to 0$; $\tilde{V}(r)$ and $V(r)$ become the same for large values of r. The presence of turbulence can have no effects on the properties of the collective modes in the vicinity of $\mathbf{k} = 0$. Turbulence in those plasma systems, which may inherently sustain unstable collective modes down to $\mathbf{k} = 0$, therefore, requires special consideration in terms of the size effects of a finite plasma.

Physically, the possible existence of such a size-dependent effect may be anticipated from the critical role that the correlation function plays in determining the stability of a turbulent plasma. The finite dimension of the system may impose a boundary condition that $S(\mathbf{k})$ should vanish within a certain (small) domain of $\mathbf{k}$ around $\mathbf{k} = 0$. Thus this condition in turn affects the stability of the system, and hence the level and spectral distribution of turbulence.

The size effect enters into the mathematical picture of the theory because of the increasing difficulty in creating the difference between $\tilde{\epsilon}(\mathbf{k}, \omega)$ and $\epsilon(\mathbf{k}, \omega)$ through $t(\mathbf{k})$ as the wavenumber decreases. When a typical dimension of the plasma is given by L, it is necessary to secure the stability of the collective modes for all values of the wavenumbers down to the order of $1/L$; the most difficult part to do usually lies in the long-wavelength domain, $k \sim 1/L$. As L increases, the amplitudes and the spectral distribution of the turbulence must adjust themselves so that the resulting value of $|t(1/L) - 1|$ is such as to assure a stability of the collective mode at $k = 1/L$.

In the theory of hydrodynamic turbulence, a similar size-dependent effect occurs through the Reynolds number; the resemblance may have more than casual physical significance.

IV. Related Topics

Many interesting phenomena in plasmas are closely related to turbulence. We wish to discuss a number of examples here.

A. *Turbulent Heating and Anomalous Resistivity*

The collision term in the first equation of the BBGKY hierarchy may generally be written as (19)

$$\left[\frac{\partial f(\mathbf{v})}{\partial t}\right]_c = -\omega_p{}^2 \sum_{\mathbf{k}} \frac{\mathbf{k}}{k^2} \cdot \frac{\partial}{\partial \mathbf{v}} \int d\mathbf{v}' \, G_{\mathbf{k}}(\mathbf{v}, \mathbf{v}') \tag{31}$$

where $G_{\mathbf{k}}(\mathbf{v}, \mathbf{v}')$ is the spatial Fourier transform of the pair correlation function; it is related to $S(\mathbf{k})$ through the equation

$$S(\mathbf{k}) = 1 + n \int d\mathbf{v} \int d\mathbf{v}' \, G_{\mathbf{k}}(\mathbf{v}, \mathbf{v}') \tag{32}$$

For a strongly turbulent plasma, we may investigate eq. 31 with the aid of the similarity transformation (eq. 20). Introducing the collision frequency $\nu(\mathbf{v})$ via $[\partial f(\mathbf{v})/\partial t]_c \equiv -\nu(\mathbf{v}) f(\mathbf{v})$, we find

$$\nu(\mathbf{v}) = \frac{e^2 \omega_p{}^2}{m^2 v^5} \langle \mathcal{E}_{\mathbf{k}} \rangle F_3 \left(\frac{v k_D}{\omega_p}, \cos\phi; \{z\} \right) \tag{33}$$

where $\cos\phi$ represents the direction cosine of $\mathbf{v}$ with respect to the axis of anisotropy in turbulence.

It immediately follows from eq. 33 that the rate of turbulent heating or the rate of increase in particle energy, $w = \frac{1}{2} m v^2$, may be written as

$$\frac{dw}{dt} = \frac{m^{1/2} e^2 \omega_p{}^2}{w^{3/2}} \langle \mathcal{E}_{\mathbf{k}} \rangle F_4 \tag{34}$$

An expression equivalent to this one has been obtained by Tsytovich (25); the dimensionless factor F_4 is shown to be equal to $(2)^{-3/2}$ in his calculation.

The collision frequency, eq. 33, also determines the electrical conductivity σ in a strongly turbulent plasma. With the aid of the scaling law, eq. 22, we first find from eq. 33 that the average collision frequency, $\langle \nu \rangle$, is expressed as

$$\langle \nu \rangle = \omega_p F_5(\{z\}) \tag{35}$$

In strong turbulence, the average collision frequency is enhanced by a factor of g^{-1} over the value in a quiescent plasma. The conductivity is then calculated to be

$$\begin{aligned} \sigma &= \frac{\omega_p{}^2}{4\pi \langle \nu \rangle} \\ &= \omega_p F_6(\{x\}) \end{aligned} \tag{36}$$

A similar expression has been obtained by O'Neil (26).

B. *Anomalous Diffusion*

It has been widely speculated that the enhanced fluctuations or turbulence may be the cause of anomalous diffusion phenomena of the Bohm type. Rosenbluth and I (27) have calculated the diffusion coefficient of the electrons across a magnetic field (with a strength B) in terms of the spectral function of the electric-field fluctuations. For a Maxwellian plasma, we find that the diffusion contains an anomalous term $D_\perp^*$ in addition to the usual classical term. It has an expression

$$D_\perp^* = \frac{(2\pi)^{1/2} ne^2 c^2 m_e{}^{1/2}}{B^2(\kappa T_e)^{1/2}} \left(1 + \frac{T_i}{T_i}\right) \ln\left(\frac{m_i}{m_e}\right) \times \left\{\exp(y) K_0(y) - \tfrac{1}{2}\left(1 - \frac{T_i}{T_e}\right)\left[\exp(y) K_1(y) - \frac{1}{y}\right]\right\} \tag{37}$$

where K_0 and K_1 are modified Bessel functions of the second kind,

$$y \equiv \left(1 + \frac{T_e}{T_i}\right) \frac{\omega_e{}^2}{\Omega_e{}^2} \tag{38}$$

ω_e is the electron-plasma frequency, Ω_e is the electron-cyclotron frequency, and the subscripts e and i refer to the electrons and the ions, respectively. The electron fluctuations in the low-frequency domain such that $\omega < |\Omega_e|$ give rise to the anomalous term.

When such fluctuations are enhanced above the thermal level owing to the onset of an instability, the anomalous term will grow accordingly, resulting in an enhanced diffusion. As we have discussed in Sections IV-C and IV-D and have, in particular, demonstrated in eq. 27, the enhancement of fluctuations in strong plasma turbulence is measured by g^{-1}. We also note that, because of the factor $(1 + T_e/T_i)$, which is usually quite large, $y \gg 1$ over a wide and physically important range of the magnetic field. Under these circumstances, we may expand $K_0(y)$ and $K_1(y)$ in eq. 37 in asymptotic series and keep only the leading terms. We thus have

$$\tilde{D}_\perp^* \simeq \frac{1}{32\pi} \frac{c}{eB} \kappa (T_e T_i)^{1/2} \left(1 + \frac{T_i}{T_e}\right)^{3/2} \ln\left(\frac{m_i}{m_e}\right), \qquad y \gg 1 \tag{39}$$

This is essentially equivalent to the value predicted by Bohm.

C. *Radiation from Turbulence*

It has been supposed that plasma turbulence, being produced by the instability of longitudinal collective modes, would not emit electromagnetic radiation, unless nonlinear coupling mechanisms with transverse modes are taken into consideration. However, Starr and I (28) have pointed out the existence of a radiation mechanism of plasma turbulence which stems solely from the anisotropy of the plasma; such an anisotropy may be provided by a drift motion of particles or by an external magnetic field.

Consider the frequency-wave-vector dispersion relationship of the electromagnetic waves

$$\det \left| \boldsymbol{K}(\mathbf{k}, \omega) - \left(\frac{kc}{\omega}\right)^2 \left(\boldsymbol{I} - \frac{\mathbf{kk}}{k^2}\right) \right| = 0 \tag{40}$$

where $\boldsymbol{K}$ is the dielectric tensor, and $\boldsymbol{I}$ is a unit tensor. When the system is isotropic, eq. 40 gives two transverse modes and one longitudinal mode. For an anisotropic system, such a decoupling is not generally possible; instead, one finds two quasitransverse modes and one quasilongitudinal mode. Those modes approach the corresponding pure modes when the parameters characterizing the strength of the anisotropy vanish or when **k** falls in the direction of the anisotropy. The quasilongitudinal mode, although small in magnitude, contains a component of the electromagnetic field perpendicular to **k**; this component thereby contributes to radiation spectrum.

Radiation from a turbulent plasma as predicted by this theory has since been observed by experiment (29). This emission mechanism may play a significant part in explaining some of the astrophysical phenomena (30).

Acknowledgments

I would like to thank Professor D. Pines for many useful discussions of the topics described in this review. This work was supported in part by the U.S. Army Research Office (Durham) under Grant DA HC04-69-C-0007.

References

1. N. N. Bogoliubov, in *Studies in Statistical Mechanics*, Vol. 1 (transl. by E. K. Gora), ed. by J. de Boer and G. E. Uhlenbeck, North-Holland, Amsterdam, 1962, p. 1.
2. N. Rostoker and M. N. Rosenbluth, *Phys. Fluids*, **3**, 1 (1960).
3. B. B. Kadomtsev, *Plasma Turbulence* (transl. by L. C. Ronson and M. G. Rusbridge), Academic Press, New York, 1965.
4. W. E. Drummond and D. Pines, *Nucl. Fusion*, Suppl., **3**, 1049 (1962); *Ann. Phys. (N.Y.)*, **28**, 478 (1964).
5. A. A. Vedenov, E. P. Velikhov, and R. Z. Sagdeev, *Nucl. Fusion*, Suppl., **2**, 465 (1962); A. A. Vedenov, *J. Nucl. Energy*, **C5**, 169 (1963).
6. V. N. Tsytovich, *Usp. Fiz. Nauk*, **90**, 435 (1966); *Nonlinear Effects in Plasma* (transl. by J. S. Wood), ed. by S. M. Hamberger, Plenum, New York, 1970.

7. R. Z. Sagdeev and A. A. Galeev, *Nonlinear Plasma Theory*, rev. and ed. by T. M. O'Neil and D. L. Book, Benjamin, New York, 1969.
8. M. N. Rosenbluth, B. Coppi, and R. N. Sudan, *Proceedings of the Third International Conference on Plasma Physics and Controlled Nuclear Fusion Research*, Novosibirsk, 1968 (IAEA, 1969), p. 771; R. E. Aamodt and M. L. Sloan, *Phys. Fluids*, **11**, 2218 (1968); and many others.
9. T. H. Dupree, *Phys. Fluids*, **9**, 1773 (1966).
10. T. H. Dupree, *Phys. Fluids*, **11**, 2680 (1968).
11. J. Weinstock, *Phys. Fluids*, **12**, 1045 (1969); J. Weinstock and R. H. Williams, *Phys. Fluids*, **14**, 1472 (1971).
12. S. Ichimaru, D. Pines, and N. Rostoker, *Phys. Rev. Lett.*, **8**, 231 (1962); S. Ichimaru, *Ann. Phys. (N.Y.)*, **20**, 78 (1962).
13. S. Ichimaru and T. Nakano, *Phys. Lett.*, **25A**, 163 (1967); *Phys. Rev.*, **165**, 231 (1968).
14. S. Ichimaru, *Phys. Rev.*, **A2**, 494 (1970).
15. J. Hubbard, *Phys. Lett.*, **25A**, 709 (1967).
16. K. S. Singwi, M. P. Tosi, R. H. Land, and A. S. Sjölander, *Phys. Rev.*, **176**, 589 (1968); *Solid State Commun.*, **7**, 1503 (1969); *Phys. Rev.*, **B1**, 1044 (1971).
17. H. B. Callen and T. A. Welton, *Phys. Rev.*, **83**, 34 (1951); L. D. Landau and E. M. Lifshitz, *Statistical Physics*, Pergamon, London, 1958, Chap. XII.
18. S. Ichimaru, in *1968 Tokyo Summer Lectures in Theoretical Physics: Statistical Physics of Charged Particle Systems*, ed. by R. Kubo and T. Kihara, Benjamin, New York, 1969, p. 69.
19. S. Ichimaru, *Phys. Rev.*, **174**, 289, 300 (1968).
20. S. Ichimaru, *Phys. Fluids*, **13**, 1560 (1970).
21. T. O'Neil and N. Rostoker, *Phys. Fluids*, **8**, 1109 (1965).
22. S. A. Rice and P. Gray, *The Statistical Mechanics of Simple Liquids*, Interscience, New York, 1965, p. 74.
23. D. Pines and J. R. Schrieffer, *Phys. Rev.*, **125**, 804 (1962).
24. E. A. Frieman and P. H. Rutherford, *Ann. Phys. (N.Y.)*, **28**, 134 (1964).
25. V. N. Tsytovich, *Usp. Fiz. Nauk*, **89**, 89 (1966).
26. T. M. O'Neil, *Phys. Rev. Lett.*, **25**, 995 (1970).
27. S. Ichimaru and M. N. Rosenbluth, *Proceedings of the International Conference on Plasma Physics and Controlled Nuclear Fusion Research*, Madison, Wis., 1971, Paper No. CN-28/F-2.
28. S. Ichimaru and S. H. Starr, *Phys. Rev.*, **A2**, 821 (1970).
29. L. D. Bollinger and H. Böhmer, *Phys. Rev. Lett.*, **26**, 535 (1971).
30. S. Ichimaru, *Nature*, **226**, 731 (1970).

Numerical Methods in Problems of Low-Temperature Plasma

A. A. SAMARSKI

Institute of Applied Mathematics, Academy of Sciences of the USSR, Moscow, USSR

I. Numerical Experiments

In connection with the appearance of high-speed computers and the development of numerical methods numerical experiment (which is often called "mathematical simulation") begins to acquire great importance along with traditional theoretical and experimental methods of research. Formerly, a physicist was compelled to confine himself, in the theoretical treatment of a complicated phenomenon, to a qualitative dimensional analysis, studying limiting cases described by equations that could be solved analytically (the degree of physical approximation is defined to a great extent by the means of analytical investigation at a scientist's disposal), but now the situation has radically changed.

A sufficiently complete approximation (model) covering the main effects is now considered. Some mathematical problem corresponds to the model. A computational algorithm is constructed for the problem, so that its numerical solution can be obtained on the computer.

By varying different parameters of the problem, we can carry out a detailed study of the physical process in the framework of the accepted model, discover the main laws of the process, estimate the influence of different factors, and so on. Then we can consider a new physical model and take into account other effects, which were not included in the first approximation; also, we can evaluate the limits of application of the first approximation.

A numerical experiment can replace successfully some complicated, long, and expensive physical experiments. The creation of large complexes of programs for the numerical treatment of complicated physical processes has become a reality, and it is in some sense equivalent to the creation of experimental plants.

It should be noted that planning physical experiments is impossible without mathematical processing of supposed experimental data by computers.

Certainly, a numerical experiment does not replace other ways of study completely, but only supplements them. A successful numerical analysis of a problem is impossible without a clear physical and mathematical formulation, without knowledge of physical parameters (and the degree of their reliability), without using all the traditional methods for preliminary study of the problem. Experimental data, dimensional analysis, self-similarity solutions, exact solutions in particular cases, and asymptotic estimates (i.e., all sources that give information on the problem and on the qualitative aspect of the phenomenon in question) must supplement the numerical experiment and often precede it. Moreover, exact or self-similarity solutions are necessary for the verification of numerical methods.

II. Magnetohydrodynamic Problems

The physics of low-temperature plasma is one of the branches to which numerical examination has been applied. The problem of plasma moving electric and magnetic fields is closely related to the problem of the magnetohydrodynamic generation of energy, to the creation of plasma engines and powerful sources of radiation, to plasma chemistry, and to a series of other important technical problems. We shall use the MHD approximation, taking into account heat conductivity, viscosity, finite conductivity, radiation transfer, phase transitions, and heterogeneity of medium to describe processes in a low-temperature plasma. The corresponding system of differential and integro-differential equations is nonlinear and can, in general, be solved only numerically. To carry out calculations of some concrete problem it is necessary to know such characteristics of a substance as the equation of state, coefficients of conductivity, and heat conductivity over a wide range of temperature and density. Methods of computing these characteristics were reported by

N. N. Kalitkin. Let us note only the essentially nonlinear character of conductivity as a function of temperature and density.

At present the numerical simulation of low-temperature plasmas is proceeding in two directions.

1. The solving of one-dimensional nonstationary problems, taking into account the whole complex of physical processes and making it possible to evaluate the influence of different factors and to reveal the main physical laws.

2. The solving of two-dimensional problems for simpler physical models to clear up the question of stability, of the influence of boundary effects, and so on.

The main method of numerical solution in magnetohydrodynamics is a finite difference method. Although the difference method depends on the nature of a problem, there are general principles for constructing difference methods (schemes) that are strictly valid for linear problems and are confirmed by numerical computations in nonlinear cases.

III. Finite Difference Schemes

In recent years the essential theory of finite difference methods for solving the problems of mathematical physics has been developed. The state of the theory of these difference schemes will be outlined briefly here; more detailed information is given in my recent book, which also contains an extensive bibliography.

Let us consider two classes of problems.

1. Nonstationary problems (heat conductivity, diffusion, vibration), leading to second-order parabolic and hyperbolic equations:

$$\frac{\partial u}{\partial t} = Lu + f(x, t), \qquad \frac{\partial^2 u}{\partial t^2} = Lu + f(x, t) \tag{1}$$

where Lu is the elliptic operator, for example,

$$Lu = \operatorname{div}\,(k \operatorname{grad} u)$$

$$Lu = \sum_{\alpha, \beta=1}^{p} \frac{\partial}{\partial x_\alpha}\left[K_{\alpha\beta}(x, t)\frac{\partial u}{\partial x_\beta}\right], \qquad x = (x_1 \ldots x_p)$$

Here p is dimension and $(K_{\alpha\beta})$ is a positively defined matrix.

2. Stationary problems, leading to elliptic equations:

$$Lu = -f(x) \tag{2}$$

where L is one of the above-stated operators.

As is known, the main idea of the finite difference method is the following. We introduce the grids $\omega_h = \{x_{(i)} \in G\}$ and $\omega_\tau = \{t_j = j\tau, j = 0, 1, \ldots\}$, with steps $h_1 \ldots h_p$ in the directions $x_1 \ldots x_p$ and with step τ in the t direction in the domains G and $[0, T]$, $x \in G$ and $t \in [0, T]$. The derivatives in differential equations are replaced by difference ratios on grid points. As a result, we obtain a system of algebraic (difference) equations, the order of which is equal to the number of grid points, instead of a differential equation. A series of such systems for all kinds of grids is named a scheme.

Constructing difference schemes constitutes a difficult problem, and many theoretical questions arise in the process. The analysis of some of them forms the object of the theory of difference schemes. We shall mention only a few of the requirements here.

1. The scheme must approximate the solution of an initial problem with prescribed accuracy.

2. The scheme must be stable, that is, the solution of the system of difference equations must depend continuously on initial data, on the right parts of equations, and on coefficients; moreover, this dependence must be uniform in relation to grid steps.

3. The difference scheme must be economic, that is, it must require a minimum of arithmetical and logical operations to solve the system of difference equations; in other words, the computer time must be as short as possible. Note that the criterion of a minimum number of arithmetical operations is not absolute. We usually speak about a minimum number of operations in respect to the number of grid points; for instance, under the requirement of economy for nonstationary problems we understand that the number of operations to determine the solution by transition from the moment $t = t_j$ to the moment $t = t_{j+1}$ (from the j level to the $j + 1$ level) is in proportion to the number of grid points ω_h.

The nature of computer problem solving requires that one and the same computational algorithm (program) be used for the widest classes of problems and not just for a particular one (i.e., a general-purpose algorithm).

There are many problems leading to equations with rapidly changing and even discontinuous coefficients (e.g., heat conductivity or diffusion in a nonhomogeneous medium). Homogeneous difference schemes are developed to solve such problems ("smoothed calculating schemes"). These schemes do not provide an explicit isolation of points of discontinuities in the coefficients but permit one to carry out the calculations throughout the domain using one and the same formula (1). It is known that the methods of smoothed calculation for shock waves are used in gas dynamics and are based on the introduction of quasiviscosity, having the property of "smearing" shock waves.

Note that the homogeneous difference schemes generally used in computational practice are naturally derived from the conservation law for elementary grid cells by the integral-interpolation method. Such schemes have an important property of conservation and express the conservation laws on a grid. The condition of homogeneity imposes strict limitations on the scheme structure. Conservation of the homogeneous scheme is a necessary condition for convergence in a class of discontinuous coefficients (2). The main a priori characteristic of such a scheme is an approximation error, defined as a discrepancy arising when we substitute the solution u of a differential equation into a difference equation instead of the difference solution y. The most important result obtained in recent years is that approximation error must be estimated, not in a point or uniform norm, but in a special integral or negative norm [1]. The assertion has been made that, if a scheme is stable and approximates some differential equations, then it converges—that is, the solution y of a difference problem converges to the solution of a differential equation (when the grid steps tend to zero). Hence the study of the accuracy of the difference scheme is reduced to a study of approximation and stability. The main problem is the stability of difference schemes.

IV. Stability of Difference Schemes

At present, necessary and sufficient conditions for stability have been obtained for a wide class of difference schemes corresponding to the non-stationary problems of mathematical physics. They are very readily applied.

Consider, for example, the two-level difference scheme

$$B\frac{y^{j+1} - y^j}{\tau} + Ay^j = \varphi^j, \qquad j = 0, 1, 2, \ldots \tag{3}$$

where $y^j = y(t_j)$ and $\varphi^j = \varphi(t_j)$ are functions defined on grid ω_h, A and B are linear difference operators acting on y as a function of point $x_{(i)} \in \omega_h$. If (,) is a scalar product in the space of functions $\mathring{\Omega}$ defined on the grid ω_h and vanishing on its boundary, and A is a self-adjoint positive operator, the sufficient condition for stability of scheme 3 is

$$B \geqslant 0.5\tau A$$

or

$$(By, y) \geqslant 0.5\tau(Ay, y)$$

for all $y \in \mathring{\Omega}$. If we have a class of stable schemes (e.g., two- or three-level schemes), we can look for such schemes in the classes that satisfy supplementary conditions of prescribed accuracy and economy. A method of constructing stable schemes with the prescribed quality or "regularization method" (see Ref. 1, chaps. VI, VII) has been developed for parabolic and hyperbolic systems.

V. Convergence of Iteration Schemes

Any iterational method for solving the system

$$Au = f$$

arising, for example, from difference approximation of the Poisson equation, is reduced to solving the nonstationary problem for stabilization. It is clear, as we know, for instance, that the solution of the heat equation $\partial u/\partial t = \Delta u + f(x)$ with stationary (time-independent) right part and boundary conditions tends (for any initial data) to the solution of the Poisson equation $\Delta u + f(x) = 0$ as $t \to \infty$.

Any one-step iteration method for solving equation $Au = f$ may be written in the form of a two-level scheme:

$$B_k \frac{y_{k+1} - y_k}{\tau_{k+1}} + Ay_k = f, \qquad k = 0, 1, 2, \ldots \tag{4}$$

where B_k is some linear operator and k is the number of iterations.

The proof of iteration convergence is reduced to study of the stability of the two-level scheme (eq. 4), so that the theory of iteration methods is part of the general theory of the stability of difference schemes.

A great number of effective iteration methods are available for solving stationary problems of mathematical physics. We present some results for the operator equation of the first kind, $Au = f$, supposing that A is a linear operator defined on a linear finite-dimensional space with scalar product (,) and norm $\|y\| = \sqrt{(y, y)}$.

1. Let $A = A_1 + A_2 + \cdots + A_p$, where $A_1, A_2, \ldots, A_p$ are self-adjoint positive and commutative operators, $A_\alpha A_\beta = A_\beta A_\alpha$, $\alpha \neq \beta$. Then the implicit alternating direction method (1) gives a solution of the equation with accuracy $\epsilon > 0$ for a number of iterations, $n_0(\epsilon) = 0[\ln(1/\eta) \ln(1/\epsilon)]$, on choosing iteration parameters $\tau_1, \tau_2, \ldots, \tau_{k_0}$ in a special way. Here $\eta = \delta/\Delta$; δ is the lower and Δ is the upper bound of the operator spectrum $A_1, \ldots, A_p$. (We suppose for simplicity that these bounds are equal for all $A_1, A_2, \ldots, A_p$.) If $p = 2$, choosing parameters by Jordan's method is optimal. If $p > 2$, the cyclic set of parameters, $\tau_1, \tau_2, \ldots, \tau_{k_0}$, is applied. For the Poisson equation this result means that $n_0(\epsilon) = 0[\ln(1/h), \ln(1/\xi)]$, where h is a grid step under the condition that the domain of changing variables $x = (x_1, x_2, \ldots, x_p)$ is a rectangle, as the difference operators $A_\alpha y = -y_{\bar{x}_\alpha x_\alpha}$ and $A_\beta y = -y_{\bar{x}_\beta x_\beta}$ are commutative only in this case.

Now methods with the same rate of convergence for the difference Dirichlet problem are obtained in the case of the Poisson equation in curvilinear coordinates $[(r, \varphi), (r, \theta)$, and $(r, z)]$.

2. If the commutative conditions for A_α and A_β are not satisfied, one can obtain only iteration methods of the type of eq. 4 with the iteration number

$$n_0(\epsilon) = 0\left(\frac{1}{\sqrt[4]{\eta}} \ln \frac{1}{\epsilon}\right)$$

These schemes provide a method for solving difference elliptic problems in a domain of intricate configuration with the iteration number

$$n_0(\epsilon) = 0[(1/\sqrt{h}) \ln (1/\epsilon)].$$

The idea of alternating directions, on the one hand, and a stable family of Chebishev parameters $\{\tau_k\}$, on the other hand, is used (1, chap. VIII, §2.1). In recent years direct methods for solving difference elliptic equations have quite often been used (in the main by American authors). The best of these methods can prove to be more profitable than iteration methods when one and the same Poisson equation with different right parts is solved many times. But the domain must be rectangular.

VI. Multidimensional Problems

At present, difference methods of solving parabolic systems and equations (linear and quasilinear) are the most thoroughly studied and verified in practice. I shall point out, first of all, the one-dimensional quasilinear heat equation:

$$c(x, t, u)\frac{\partial u}{\partial t} = \left[k(x, t, u)\frac{\partial u}{\partial x}\right] + f\left(x, t, u, \frac{\partial u}{\partial x}\right)$$

with boundary conditions of the first, second, and third kinds, and also with conditions of the concentrated heat conductivity type. The method of smoothed computation is adapted to the problem of phase transition (Stephan's problem), by a smearing of enthalpy. This method is applicable when there is not one but several phase transition fronts and when a problem involves more than one dimension.

For multidimensional problems many economic schemes are suggested. All are based, in fact, on one and the same algorithmic idea of alternating directions: the solution of a multidimensional problem is found as a sequence of solutions to one-dimensional algebraic problems. The method of additive approximation, which is based on a new concept of a scheme (an additive one) as a system of p intermediate schemes, has proved to be the most general and effective. Each scheme does not approximate the initial problem, but the sum

$$\psi = \psi_1 + \psi_2 + \cdots + \psi_p$$

of approximation errors ψ_α approaches zero when the grid steps tend to zero. Hence the approximation error of the additive scheme is defined as the sum

$$\psi = \psi_1 + \psi_2 + \cdots + \psi_p$$

The main subject is the constructing of economic additive schemes. To do this one can use the following heuristic approach.

In a rectangle $(0 \leqslant x_1 \leqslant l_1, 0 \leqslant x_2 \leqslant l_2)$ we have the heat equation

$$\frac{\partial u}{\partial t} = (L_1 + L_2)u, \qquad L_\alpha u = \frac{\partial^2 u}{\partial x_\alpha{}^2}; \qquad u|_\Gamma = 0, \qquad u|_{t=0} = u_0(x) \tag{5}$$

where $x = (x_1, x_2)$, $\alpha = 1, 2$, and Γ is the boundary of the rectangle. Suppose that we need to find $u(x, t)$ when $t = t_0 > 0$.

First of all, we solve the one-dimensional problem (in the x_1 direction):

$$\frac{\partial v_{(1)}}{\partial t} = L_1 v_{(1)}, \qquad 0 < t < t_0, \qquad v_{(1)}|_\Gamma = 0, \qquad v_{(1)}(x, 0) = u_0(x) \tag{6}$$

and find $v_{(1)}(x, t_0)$; after that, we solve a one-dimensional heat equation in the x_2 direction:

$$\frac{\partial v_{(2)}}{\partial t} = L_2 v_{(2)}, \qquad 0 < t < t_0, \qquad v_{(2)}|_\Gamma = 0 \tag{7}$$

with initial data

$$v_{(2)}(x, 0) = u(x, 0) \tag{8}$$

Then the following assertion holds:

$$v_{(2)}(x, t_0) = u(x, t_0)$$

This theorem is true for the case in which

$$L_\alpha u = \frac{\partial}{\partial x_\alpha}\left[k_\alpha(x_2, t)\frac{\partial u}{\partial x_\alpha}\right], \qquad \alpha = 1, 2$$

The above method for solving eq. 5 by means of successive solutions of one-dimensional problems (eqs. 6 and 7) $(L_1 \to L_2)$ permits a simple physical interpretation, the following model of the heat conductivity process. In the first stage the heat conductivity is "switched off" in the x_2 direction (i.e., we introduce adiabatic diaphragms in the x_2 direction)—that is, the problem (eq. 6) of heat conductivity along x_1 is considered; then, if $t = t_0$, we obtain the temperature distribution $v_{(1)}(x, t_0)$. Then we start from $t = 0$ and take $v_{(1)}(x, t_0)$ for the initial temperature; we "switch off" heat conductivity in the x_1 direction and consider heat conductivity along x_2 (the problem of eq. 7). At the moment of time when $t = t_0$ we obtain the temperature $v_{(2)}(x, t_0)$, which coincides with the true value, $u(x, t_0)$.

In accordance with this model the real physical heat conductivity process is prolonged and takes a time $2t_0$, not t_0, as actually occurs for the two-dimensional heat conductivity process.

Every one-dimensional problem (eqs. 6 and 7) can be solved by some suitable method (analytical, difference, a combination of these, etc.).

Consider another example of the reduction of a multidimensional problem to a chain of one-dimensional ones. This is the Cauchy problem for the transfer equation (cf. Ref. 2):

$$\frac{\partial u}{\partial t} = (L_1 + L_2)u, \qquad L_1 u = \frac{\partial u}{\partial x_1}, \qquad L_2 u = \frac{\partial u}{\partial x_2}, \qquad -\infty < x < \infty$$

$$u(x, 0) = \mu(x) = \mu(x_1, x_2)$$

It has solution $u(x, t) = \mu(x_1 - t, x_2 - t)$. The process of transfer can be realized in two steps, the first along x_1 within the time $t = t_0$:

$$\frac{\partial v_{(1)}}{\partial t} = L_1 v_{(1)}, \qquad v_{(1)}(x, 0) = \mu(x), \qquad 0 \leqslant t \leqslant t_0$$

and the second along x_2 within the same period of time:

$$\frac{\partial v_{(2)}}{\partial t} = L_2 v_{(2)}, \qquad v_2(x, 0) = v_{(1)}(x, t_0)$$

It is easy to see that

$$v_{(1)}(x, t) = \mu(x_1 - t, x_2),$$

$$v_{(2)}(x, t) = v_{(1)}(x_1, x_2 - t, t_0) = \mu(x_1 - t, x_2 - t)$$

and, consequently,

$$v_{(2)}(x, t_0) = u(x, t_0)$$

In the general case, when L_1 and L_2 have variable coefficients and the domain of the variable, $x = (x_1, x_2)$, has an intricate configuration, equality 8 does not hold. In this case we act in the following way. We introduce grid $\omega_\tau = \{t_j = j\tau\}$ with step τ on the segment $0 \leqslant t \leqslant t_0$ and solve the problems of eqs. 6 and 7 in succession on each segment $t_j \leqslant t \leqslant t_{j+1}$.

Then, instead of eq. 8,

$$v_{(2)}(x, t_0) = u(x, t_0) + 0(\tau) \tag{9}$$

For simplicity we have considered the case of homogeneous equations and boundary conditions.

Estimate 9 is also valid in the case of nonhomogeneous equations and boundary conditions. Here $f_1 + f_2 = f$, where f_1 and f_2 are the right-hand parts of eqs. 6 and 7 and f is the right-hand part of eq. 5.

Let us indicate one more method, widely utilized in practice, of dividing (factorizing) a multidimensional problem into one-dimensional ones solved in succession. We take an equation

$$Pu = \frac{\partial u}{\partial t} - \sum_{\alpha=1}^{p} L_\alpha u - f(x, t) = 0, \qquad x \in G, \qquad 0 \leqslant t \leqslant t_0$$

$$L_\alpha u = \frac{\partial}{\partial x_\alpha}\left[k_\alpha(x, t)\frac{\partial u}{\partial x_\alpha}\right], \qquad k_\alpha(x, t) > 0 \tag{10}$$

Here $x = (x_1, x_2, \ldots, x_p)$ is a point of the p-dimensional domain G ($p = 2, 3, \ldots$). The boundary conditions are prescribed on a boundary Γ of the domain

$$u(x, t) = \mu(t), \qquad x \in \Gamma \tag{11}$$

and for $t = 0$ the initial data

$$u(x, 0) = u_0(x) \tag{12}$$

are prescribed.

We denote

$$P_\alpha u = \frac{1}{p}\frac{\partial u}{\partial t} - L_\alpha u - f_\alpha$$

where f_α, $\alpha = 1, 2, \ldots, p$, are chosen so that

$$\sum_{\alpha=1}^{p} f_\alpha = f$$

Then it is evident that $Pu = \sum_{\alpha=1}^{p} P_\alpha u$. We introduce again grid $\omega_\tau = \{t_j = j\tau\}$ on the segment $0 \leqslant t \leqslant t_0$; in addition, we divide each interval (t_j, t_{j+1}) into p equal parts:

$$\Delta_\alpha = \left\{ t_{j+(\alpha-1)/p} \leqslant t \leqslant t_{j+(\alpha/p)} = t_j + \frac{\tau\alpha}{p} = \left(j + \frac{\alpha}{p}\right)\tau \right\}, \qquad \alpha = 1, \ldots, p$$

At intervals Δ_α we solve the equation

$$P_\alpha v_{(\alpha)} = 0, \qquad \frac{1}{p}\frac{\partial v_{(\alpha)}}{\partial t} = L_\alpha v_{(\alpha)} + f_\alpha(x, t) \tag{13}$$

with initial data $v_{(\alpha)}[x, t_{j+(\alpha-1/p)}] = v_{(\alpha-1)}[x, t_{j+(\alpha-1/p)}]$, $\alpha = 1, 2, \ldots, p$, and boundary conditions $v_{(\alpha)} = 0$ on corresponding parts Γ_α of boundary Γ of domain G(Γ_α consists of the points of intersection of straight lines parallel to the 0, x_2 axis with boundary Γ).

Let $\alpha = 1$ and $v_{(1)}(x, t_j) = v(\alpha, t_j)$; then find $v_{(1)}[x, t_{j+(1/p)}]$ and so on until $v(x, t_{j+1}) = v_{(p)}(x, t_{j+1})$ is determined. It seems that in this case an estimate $v(\alpha, t_j) = u(x, t_j) + 0(\tau)$ is valid for all $j = 1, 2, \ldots$ if $v(x, 0) = u_0(x)$.

Another physical model corresponds to this way of reducing the problem of eqs. 10 to 12 to the chain of problems in eq. 13: the heat conductivity coefficient decreases p times in each stage, and therefore the duration of the chain of one-dimensional heat processes is the same as that of the multidimensional process. Note that in the case where $k_2 = 1, f = 0$, $\mu = 0$, and G is p-dimensional parallelepiped, the two models coincide and

$$v(x, t_{j+1}) = u(x, t_{j+1})$$

Effective and economic schemes are available for solving each of the one-dimensional problems in eqs. 6, 7, and 13. That is why we obtain an economic algorithm for solving the two-dimensional problem of eq. 5, replacing, for example, eqs. 6 and 7 by corresponding schemes. Such a local one-dimensional method has proved to be acceptable for solving heat equations in domains of intricate configuration. No serious difficulties arise in transferring from the two-dimensional equation to three-dimensional and quasilinear heat equations.

For the hyperbolic equation

$$\frac{\partial^2 u}{\partial t^2} = \sum_{\alpha=1}^{p} L_\alpha u, \qquad L_\alpha u = \frac{\partial}{\partial x_2}\left[k_\alpha(x, t)\frac{\partial u}{\partial x_2}\right]$$

$$u(x, 0) = u_0(x), \qquad u_t(x, 0) = \bar{u}_0(x), \qquad u|_\Gamma = \mu(x, t)$$

a local one-dimensional model can be constructed (D. G. Gordeziany and A. A. Samarski):

$$\frac{\partial^2 v_{(\alpha)}}{\partial t^2} = L_\alpha v_{(\alpha)}, \qquad t_j \leqslant t \leqslant t_{j+1}, \qquad \alpha = 1, 2, \ldots, p$$

All the equations are solved with one and the same initial data:

$$v_{(\alpha)}{}^j = v^j, \qquad \left.\frac{\partial v_{(\alpha)}}{\partial t}\right|_{t=t_j} = \left(\frac{\partial v}{\partial t}\right)_{t=t_j}$$

and boundary conditions, $v|_\Gamma = \mu$. Using $v_{(\alpha)}{}^{j+1}$, we find

$$v^{j+1} = \frac{1}{p}\sum_{\alpha=1}^{p} v_{(\alpha)}{}^{j+1}$$

The estimate $v^{j+1} - u^{j+1} = 0(\tau^2)$ takes place for all $j = 0, 1, \ldots$. If we replace each equation with the number α by the three-level scheme with weights, we obtain an economic local one-dimensional scheme.

VII. Plasma Applications–Principles

It is absolutely evident that multidimensional problems, such as the problem of plasma physics, can be solved practically only on grids that are not very small (about 100 points in each direction is, as a matter of fact, the minimum number in the case of three-dimensional problems, even for future computers, especially if we take into account the processes of radiation transfer). Also, it is necessary to bear in mind the need to solve a great number of variants in numerical experiments; therefore the grid must not be very small.

The problem is to construct schemes that possess optimal properties on real grids not only when the grid steps tend to zero. When we solve such a problem, it is insufficient to study the asymptotic properties of the scheme and its accuracy. Moreover, for many nonlinear problems not even an investigation of this type is feasible. It is necessary to use arguments of qualitative character as well: the difference scheme must correctly simulate the main properties of differential equations in the space of the grid functions. One qualitative requirement is the demand of conservativity of difference schemes.

To be concrete we shall speak about the problems of continuum mechanics and magnetohydrodynamics which are described by partial differential equations. Any difference scheme gives a mathematical description of a discrete medium model. It is natural to demand that the discrete model correctly represent the main properties of the continuum, such as the laws of conservation (momentum, mass, energy, etc.). Difference schemes expressing the laws of conservation on grid are called conservative ones.

It was mentioned above that conservativity is necessary for the convergence of homogeneous schemes in the case of eqs. 1 and 2 with discontinuous coefficients K. Therefore is it necessary for gas dynamics and MHD problems.

Summarizing, we may mention three ways of constructing difference schemes of the prescribed quality.

1. The integral-interpolation method of constructing conservative schemes.
2. The method of regularization in a class of stable schemes.
3. The method of additive approximation.

These methods are applicable not only for linear but also for nonlinear problems.

VIII. Plasma Applications–The High-Current Pulsed Discharge

It is a wrong idea that the principles of constructing difference schemes mentioned above represent a completed system of requirements. The problems confronting computational mathematics become ever more complicated, and sometimes the available algorithms cannot be used effectively for their solution. That is why we inevitably have to introduce new requirements and formulate new principles leading, sometimes, to the construction of already known classes of difference schemes.

This idea can be illustrated by the principle of total conservation, which was mentioned in connection with the solution of the MHD problem involving high-current radiating impulse discharges in low-temperature dense plasma. A heavy battery of capacitors is discharged through a column of plasma being formed as a result of the electric explosion of a thin metal wire. The intricate MHD flow arising is characterized by a sharp change of discharge parameters in time and space (quick multiple pinching of plasma columns, arising from a narrow high-temperature zone–the T-layer, etc.). Former numerical methods proved to be ineffective for the accurate solution of this problem.

For simplicity we shall explain the matter by using the example of the gas dynamics system of equations. This system is an expression of three conservation laws: mass, impulse, and energy, and in Lagrangian mass coordinates for the one-dimensional nonstationary case has the form

$$\frac{\partial v}{\partial t} = -\frac{\partial p}{\partial x}, \qquad \frac{\partial r}{\partial t} = v, \qquad \frac{\partial}{\partial t}\left(\frac{1}{\rho}\right) = \frac{\partial v}{\partial x} \tag{14}$$

$$\frac{\partial}{\partial t}(\epsilon + 0.5v^2) = -\frac{\partial}{\partial x}(pv) \tag{15}$$

where t is time, r is the Euler coordinate, ρ is density, $x(dx = \rho\, dr)$ is the Lagrangian coordinate, p is pressure, ϵ is internal energy, and v is the longitudinal component of velocity.

The equation of energy (eq. 15), expressing the law of conservation of energy, may be rewritten in nondivergent form as

$$\frac{\partial \epsilon}{\partial t} = -p\frac{\partial v}{\partial x} \tag{16}$$

or

$$\frac{\partial \epsilon}{\partial t} = -p\frac{\partial}{\partial t}\left(\frac{1}{\rho}\right) \tag{17}$$

eqs. 16 and 17 having a simple physical meaning; namely, expressions 16 and 17 indicate that the change of internal energy in a system is determined by the work of compression. It should be noted that eqs. 15 to 17 are equivalent in the sense that they reduce to each other with the help of the rest of eqs. 14. When we solve the system by the method of finite differences, the system of differential equations is approximated by some difference scheme, and the continuous medium is replaced by some discrete model. The energy difference equation can be constructed on the base of any equivalent forms, eqs. 15 to 17.

It may be shown that the indicated property of equivalence in difference form does not take place in a general case. If in a scheme a nondivergent difference equation of the type of 16 or 17 (e.g., the well-known "cross" scheme) is used, the conservation law of total energy will be broken. Calculating high-current discharge by means of such schemes, we obtained the following physically absurd result: at the end of the process the energy quitting the system as light surpassed the initial energy stock in the battery of capacitors.

When we use arbitrary conservative schemes, the law of conservation of energy is valid, of course. But in this case the difference analogs of eqs. 16 and 17 will not be valid. As a result, the internal energy balance will prove to be destroyed. Such a defect is no less dangerous than violation of the conservation law, especially in problems where there are functions sensitively dependent on temperature, such as electrical and heat conductivity coefficients. For example, when we do the calculation of the problem of the high-current discharge, the above-indicated defect of the conservative schemes leads to a decrease of temperature of the separate plasma mass in the stage of compression in the presence of Joule heating. The presence of energy imbalance in the scheme can be interpreted as the presence of some energy sources of a purely difference nature, connected with "disconcordance" of separate difference equations in the scheme. The imbalance depends upon the character of solution: they are small for smooth functions, but the imbalance terms can reach a considerable magnitude, comparable with the total system energy, for solutions changing rapidly in time and space.

The energy imbalance in a scheme is also conditioned by the time step of the difference grid τ; for absolutely implicit schemes it has an order of $0(\tau)$. In principle, we can remove the defects described by infinite reduction of the step τ. But, if we formulate the problem of sufficiently accurate computation of processes on finite grids changing rapidly in time and space, we must look for new approaches to constructing difference schemes. Therefore the principle of total conservation was advanced, which in the gas dynamics system states that the difference scheme must be a scheme for which the property of equivalence of different forms of the energy equation valid in the differential form will be valid in the difference form.

In the general case this means that in a difference scheme, besides the main conservation laws, some additional relations must be employed. Naturally, the completely conservative schemes represent the narrowing of a class of common conservative schemes. Such schemes have been constructed both for gas dynamics and for the MHD equations. The above-mentioned problem of the high-current discharge was solved, in particular, with their help.

IX. Plasma Applications—Two-Dimensional Problems

We consider some characteristic two-dimensional problems arising in low-temperature plasma dynamics. First of all we consider the problem of finding the distribution of electric fields and currents in a plasma with anisotropic conductivity (taking into account the Hall currents) and with small values of the magnetic Reynolds number. In this case we have Maxwell equations for the electric field intensity **E** and current density **j**.

$$\text{rot}\,\mathbf{E} = 0, \qquad \text{div}\,\mathbf{j} = 0$$

and Ohm's law:

$$\mathbf{j} + \mathbf{j} \times \boldsymbol{\beta} = \sigma\left(\mathbf{E} + \frac{1}{c}\,\mathbf{v} \times \mathbf{H}\right)$$

where σ is conductivity, $\boldsymbol{\beta} = e\mathbf{H}/mev$ is the Hall parameter, $\mathbf{H}$ is the magnetic field intensity, ν is the collision frequency of the current carrier, and the other symbols are generally accepted. We restrict ourselves to the examination of the two-dimensional case, in which $j_z = 0$ and all functions depend only on x and y. Let $\mathbf{H} = (0, 0, H)$. Introduce the vector potential $\boldsymbol{\psi} = (0, 0, \psi)$, and assume that $\mathbf{j} = \text{rot}\ \boldsymbol{\psi}$, so that

$$j_x = \frac{\partial \psi}{\partial y}, \qquad j_y = -\frac{\partial \psi}{\partial x}$$

As a result we obtain for ψ the non-self-adjoint elliptic equation

$$\frac{\partial}{\partial x}\left(K\frac{\partial \psi}{\partial x}\right) + \frac{\partial}{\partial y}\left(K\frac{\partial \psi}{\partial y}\right) + \frac{\partial}{\partial x}\left(r\frac{\partial \psi}{\partial y}\right) - \frac{\partial}{\partial y}\left(r\frac{\partial \psi}{\partial x}\right) = f \tag{18}$$

where $K = 1/\sigma$, $r = \beta/\sigma$, and

$$f = \text{div}\ \mathbf{v}H = \frac{\partial(v_x H)}{\partial x} + \frac{\partial}{\partial y}(v_y H)$$

Most often we find boundary conditions of two types.

(*a*) If a part of the boundary is an ideal sectioned electrode through which j_n is prescribed, $\psi = \psi(s)$ is a prescribed function of arch length s; the condition $\psi = \text{const}$ means "nonpassing" of current through an ideal dielectric.

(*b*) If a part of the boundary is an ideal conducting electrode, the tangential component of electric field $E_s = 0$; it is equivalent to the vanishing of the directional derivative:

$$k\frac{\partial \psi}{\partial n} - r\frac{\partial \psi}{\partial s} = 0$$

In the general case the first kind of condition is prescribed on part of the boundary, and the condition of directional derivative on the rest. To obtain a homogeneous difference scheme to the second order of accuracy a variation-difference method was applied: the bilinear form $Q[\psi, \varphi]$ for eq. 18 was considered; this was replaced by a difference analog, $Q_h[\psi, \varphi]$. The difference equations were obtained from the minimum condition for $Q_h[\psi, \varphi]$. To solve the system of difference equations, iterational alternating direction methods were used.

X. Plasma Applications–The Ionization Instability

The next problem is closely related to the phenomenon of the ionization instability theoretically predicted by E. P. Velihov. The purpose of the numerical experiment is to study the influence of the nonlinear electronic heat conductivity process on the evolution of the ionization instability, and to investigate the topological structure of the fully developed instability and the influence of boundary conditions (the phenomenon of current contraction between finite electrodes, etc.).

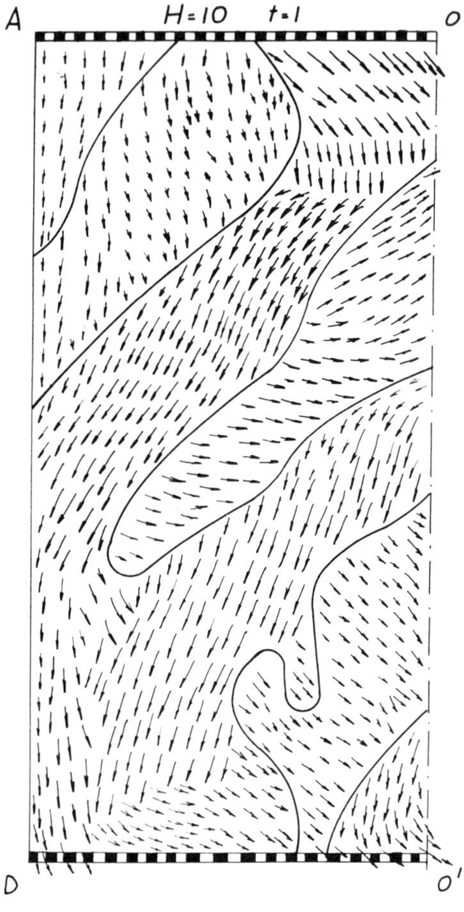

Fig. 1. Development of the "ionization instability," current flow and electron density.

The known model of low-temperature equilibrium plasma with a negligibly small temperature for the heavy component was used. We shall not write down a system of nonlinear equations; we note only that the above-mentioned equation for ψ is one of the equations of the system. The second equation is an energy balance equation for electrons, taking into account nonlinear electronic heat conductivity. On the whole, a nonstationary quasilinear two-dimensional problem is obtained. To solve it, a local one-dimensional method was used; and in every interval of time an elliptic problem was solved to determine ψ by an iteration method. The region occupied by plasma is a rectangle, *ABCD.* The horizontal segments *AB* and *CD* are ideal sectioned electrodes. ($j_y = -1$) or ideal conducting electrodes ($E_x = 0$) electrodes.

Fig. 2. Development of the "ionization instability," current flow and electron density.

Total current through the electrodes is given. On the vertical dielectrics *AD* and *CB* we put $j_x = 0$. For the whole boundary of the domain *ABCD* temperature T was prescribed. The initial homogeneous plasma state was perturbed by changing the concentration n (or the temperature T) by a value Δn (or ΔT). The characteristic properties of the evolution of the disturbances inserted into the initially homogeneous stationary background are shown in Figures 1 to 4.

Some consequences of the analysis of computations can be noted. First, the accepted physical model reflects the main properties of the dynamics of a phenomenon known from experiment: the formation of layers and the stabilization of the quasistationary non-one-dimensional structure of electron concentration distribution and current density at the stage of fully-developed instability. It is found that the development of instability passes successively through two qualitatively different stages.

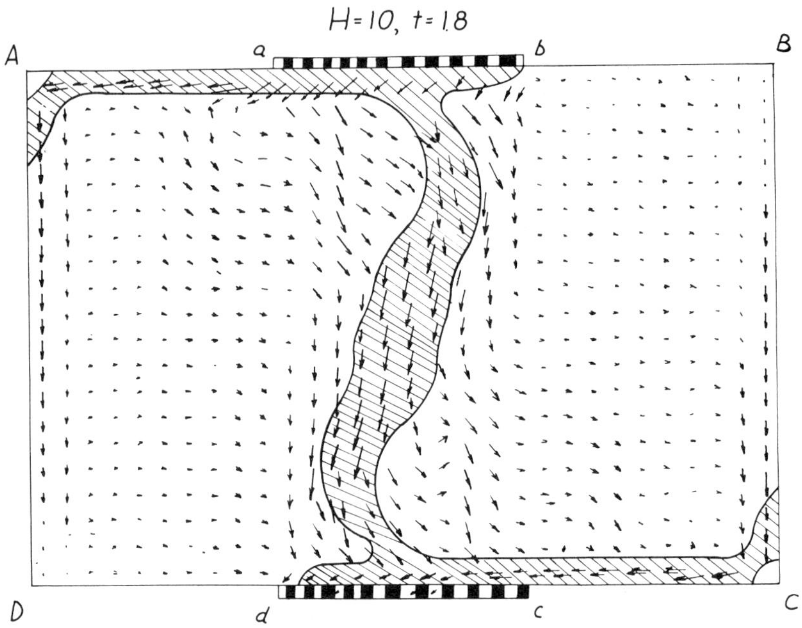

Fig. 3. Development of the "ionization instability," formation of a current channel between sectional electrodes.

In the first stage layers with the higher electron concentration are formed (Fig. 1). The interaction of the nonlinear processes of Joule heating and energy losses of electrons by collisions with heavy particles lies at the root of this formation. The formation of layers is accompanied by a decrease in effective plasma conductivity σ_{eff} and its effective Hall parameter, β_{eff}.

The beginning of the second stage is characterized by the appearance of breakdowns and by a flowover of current along "cofferdams" between layers. Thus the topology of "current channels" proves to be typical of the stage of developed instability (Fig. 3). This circumstance prevents further concentration of current in layers and consequently prevents a decrease in σ_{eff} and β_{eff}.

The variation of boundary conditions in computations showed that the dynamics of layer formation depends essentially on this factor. Hence the "current channel" between ideal sectioned electrodes are directed mainly from

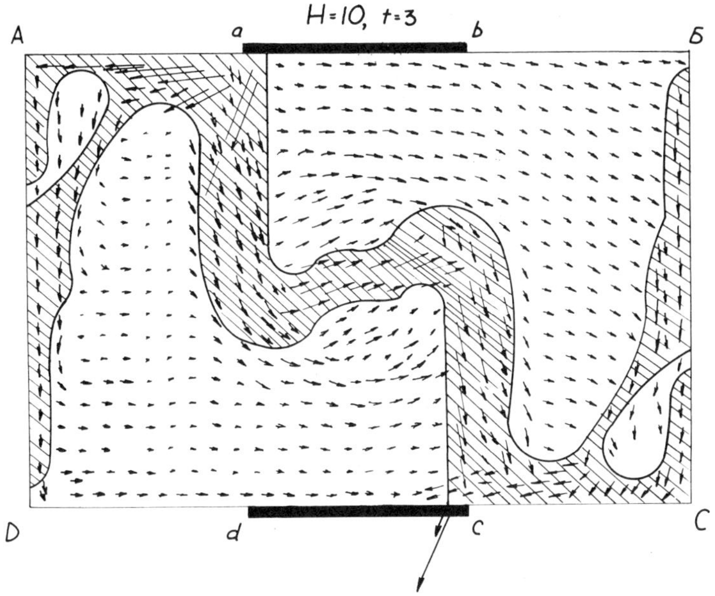

Fig. 4. Development of the "ionization instability," localization of current flow between continuous electrodes.

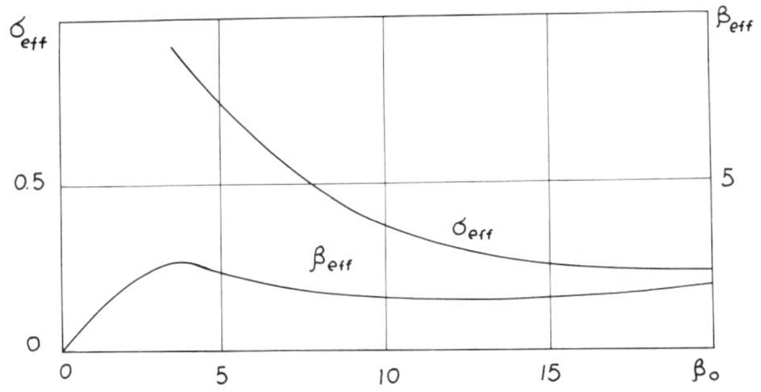

Fig. 5. The "ionization instability," effective conductivity σ_{eff} and Hall coefficient, β_{eff}.

one electrode to another (Fig. 3). In the case of continuous electrodes, layers settle down parallel to electrodes (Fig. 4). The computations were carried out for different values of Hall's parameter, β_0. This made it possible to obtain the dependence of effective plasma parameters σ_{eff} and β_{eff} on β_0 in the stage of developed instability (Fig. 5). The dependence obtained by computation satisfactorily conforms to the experimental result.

XI. Plasma Applications–MHD Channel Flow

In the next problem we study two-dimensional effects of MHD plasma current in a plane channel of constant width. At the entry to the channel a homogeneous flux with parameters ρ_1 (density), p_1 (pressure, and u_1 and v_1 (velocity components) is prescribed. The gas has a finite conductivity σ_1. The horizontal walls are ideal conducting electrodes connected to an external load. At the moment $t = 0$ an external magnetic field $\mathbf{H}(0, 0, H)$ appears in some domain. We suppose that $\mathrm{Re}_m \ll 1$, and H_z is a given function of x:

$$H_z = \begin{cases} e^{\alpha x}, & x < 0 \\ 1, & 0 < x \leqslant L \\ e^{-\alpha(x-L)}, & x > L \end{cases}$$

As a result, the stabilized current is described by stationary nonlinear equations of magnetohydrodynamics in the (x, y) plane. The indicated system of MHD equations for the plasma and the equation for electric current potential ψ discussed above were solved simultaneously by the method of stabilization. A nonstationary system of gas dynamics was solved, as a matter of fact, and in every interval of time the elliptic non-self-adjoint boundary value problem for ψ_0 was solved. The main dimensionless parameter of the problem in such a formulation is the parameter of MHD interaction S:

$$S = \frac{\sigma_0 H_0{}^2 d}{c^2 \rho_0 u_0}$$

Unlike a number of papers in which analytical methods with the approximation $S \ll 1$ are used, the application of numerical methods makes it possible to investigate the influence of edge effects and the Hall effect at $S \sim 1$.

The result of numerous computations showed the following:

1. When a flux comes into a rapidly-increasing magnetic field, breaking takes place in a compression wave localized in the zone $0 < x < \delta$, which is narrow in comparison with channel width d (Fig. 6).

When the flux comes into a slowly increasing field ($d = 3 \rightarrow 5$), the zone of the compression wave spreads up along the flux, and a dilute zone appears near the walls (Fig. 7).

When S increases, the degree of braking grows, and beginning with some $S = S_{cr}$ stationary current becomes impossible. As the computations show,

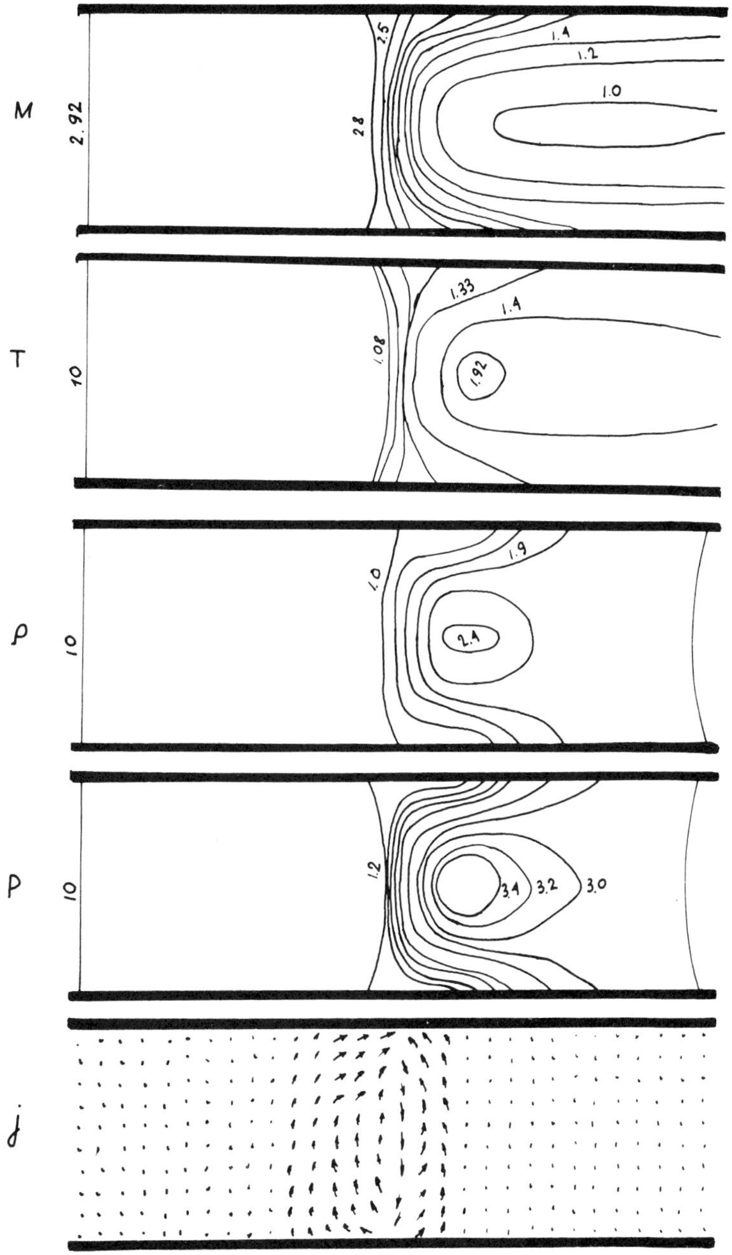

Fig. 6. Channel flow. Current distribution for a rapidly changing magnetic field.

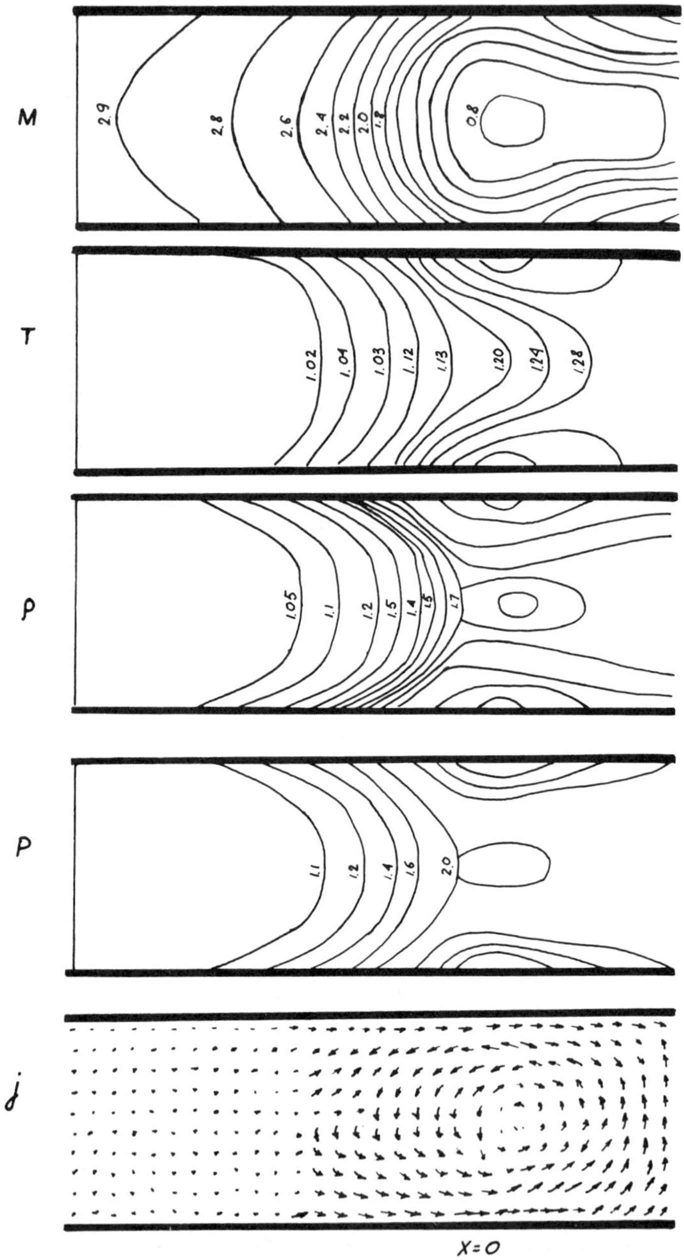

Fig. 7. Channel flow. Current distribution for a slowly changing magnetic field.

Fig. 8. Channel flow. Influence of the Hall effect.

the reformation of the current takes place as a result of moving up along the flux of the shock wave arising in the compression zone (Fig. 8).

2. The exit of a supersonic flux from the magnetic field for $S \sim 1$ is also accompanied by braking and heating under the interaction of flux with current vortex. However, this interaction has a qualitatively different character, than in the above-investigated case of "entry," namely, the current in the zone of "exit" is always smooth, and the crossing of the sonic speed line proceeds continuously. The critical value $S = S_{cr}$ exists; exceeding it leads to bifurcation of the current vortex, and one fraction spreads up along the flux, leaving behind the subsonic current.

3. It is characteristic that the smooth change of magnetic field decreases the braking of flux but does not improve its homogeneity in section.

4. The influence of the Hall effect on the current in the electrode zone of the channel becomes substantional, starting from $\Omega \sim 1$. In this case (Fig. 8) the flux deviates to one cathode and may be "reflected" from it under the action of an increasing transverse gradient of pressure. Moreover, this process may be repeated along the channel length. As a result, the space parameter distribution and the distribution along electrodes have a nonmonotonic character and lead to the existence of two-dimensional local energy release.

References

1. A. A. Samarski, Introduction to the Theory of Difference Schemes, *Progress*, 1971 (Russian).
2. V. Ya. Goldin, G. V. Danilova, and N. N. Kalitkin, The Numerical Integration of Multidimensional Transfer Equation, *Zh. Vychisl. Mat. Mat. Fiz., Suppl.*, **4** 190–193 (1966) (Russian).
3. Yu. P. Popov and A. A. Samarski, The Completely Conservative Difference Schemes, *Zh. Vychisl. Mat. Mat. Fiz.*, **9**, No. 4 (1969) (Russian).
4. Yu. P. Popov and A. A. Samarski, The Completely Conservative Difference Schemes for Magnetohydrodynamics Equations, *Zh. Vychisl. Mat. Mat. Fiz.*, **10**, No. 4 (1970) (Russian).
5. L. M. Degtiarev, A. A. Samarski, and A. P. Favorski, The Numerical Solution of Internal Stationary Electrodynamics Problems, *Zh. Vychisl. Mat. Mat. Fiz.*, **10**, No. 6 (1970) (Russian).
6. E. P. Velihov, L. M. Degtiarev, A. A. Samarski, and A. P. Favorski, The Computation of Ionization Instability in Low-Temperature Magnetized Plasma, *Dokl. Akad. Nauk SSSR*, **184**, No. 3 (1969) (Russian).
7. E. P. Velihov, L. M. Degtiarev, and A. P. Favorski, The Numerical Treatment of Ionization Instability, *Report on Fifth International Symposium on MHD Electrical Power Generation*, Munich, 1971, Vol. III, p. 307.
8. A. V. Gubarev, L. M. Degtiarev, A. A. Samarski, and A. P. Favorski, The Current of Supersonic Flux of Conducting Gas in Inhomogeneous Magnetic Field, *Dokl. Akad. Nauk SSSR*, **192**, No. 3 (1970).

9. A. V. Gubarev, L. M. Degtiarev, A. A. Samarski, and A. P. Favorski, Some Two-Dimensional Effects of Supersonic Current of Conducting Gas in MHD Channels, *Report on Fifth International Symposium on MHD Electrical Power Generation*, Munich, 1971, Vol. II, p. 159.

Computational and Experimental Plasma Physics for Theoreticians

BURTON D. FRIED

University of California, Los Angeles, California, USA

Introduction

Since the central problems in modern plasma physics are nonlinear, a completely analytic treatment, from first principles, is difficult and complex at best; frequently it is impossible. As in the theory of elementary particles, guidance from experiment is indispensable. Moreover, it is often necessary to supplement the purely analytic approach with computational methods. At present, much effort is devoted to computer "simulation" of a plasma, that is, the numerical solution of the initial value problem for the motion of N charged particles (with N of the order of 1000 to 1,000,000) in the presence of external and self-consistent electromagnetic fields. Such simulations are of most value when the computational results are treated on the same basis as data from a physical experiment, to be explained as quantitatively as possible on theoretical grounds, an approach that has been used with considerable success by Professor John Dawson and his group. Unfortunately, this approach is not suitable for steady-state, "boundary value" type problems, and it requires extensive computational facilities, in terms of both running time and memory size, particularly if the phenomena under study involve widely disparate time scales, as in the case of problems concerning ionic oscillations or diffusion, when realistic values for the electron-ion mass ratio are used. Moreover, most theoreticians lack not only the massive computational facilities (and computer time!) required for such work but also the necessary computer expertise, since the detailed knowledge of programming methods and other specialized computational techniques needed is comparable in magnitude to the technical proficiency required for laboratory experiments. Therefore we shall not discuss conventional simulation problems here.

Large electronic computers are also useful to the theoretician as a direct aid to analysis, *provided that* the computer can be used to do "experimental mathematics." Typically, we want to experiment with the physical approximations used in formulating a problem, with the kind of mathematical algorithm used to solve it, and with the numerical methods used to implement the algorithm. If iterative techniques are employed, we need to know quickly what convergence properties they have. In this mode of operation, the need is for an interactive computer system, with quick graphical response, rather than efficiency for production runs. If the theoretician learns to do the programming himself, as many now do, he may easily be diverted from the physics itself to an unacceptable extent. On the other hand, using a professional programmer introduces an extra element that tends to dilute the desired intimate interaction between theoretician and computer. We shall review presently a unique "on-line" approach to computing, due originally to Professor Glen Culler, which makes it comfortable and convenient for a theoretician who is not familiar with conventional programming (and not anxious to become familiar with it) to do interactive computing and the experimental mathematics which this makes possible. In this way it is possible to take advantage of the complementarity that exists between analytic and computational methods. Analytic techniques are most feasible when there exists some small or large parameter in the problem, but they encounter difficulty when all quantities are of comparable order of magnitude. Conversely, numerical methods are awkward in the former case but work well in the latter. (In addition, of course, computation sometimes is necessary when the problem is very complicated.)

Turning now to the experimental side, we note that the large fusion devices are typically very complex, with so many phenomena occurring simultaneously that their usefulness to the theoretician is quite limited, as is his ability to explain the observations quantitatively, let alone make reliable predictions. Naturally, such experiments are necessary in order to reach the regimes appropriate to controlled fusion, but it is important to supplement them with *interactive* experiments, that is, relatively small, flexible, easily diagnosed configurations in which a single particular effect or phenomenon of interest can be studied without interference from other processes. Although this can be accomplished in various ways, we will review here some new, large-volume, very quiescent plasma configurations, developed by Professor Kenneth MacKenzie and his group, which are of particular interest to theoreticians because they allow a direct test of theoretical ideas. Like the "on-line" computational facilities described in the next section, these experimental devices are "on-line" in that the theoretician can manipulate the controls, observe the diagnostics on the oscilloscope, and thus acquire a direct feeling for the dynamics of the nonlinear processes.

Our theme, then, is how a theoretician can actively participate in both computational and experimental plasma physics, avoiding the tiresome book-keeping associated with conventional programming, on the one hand, and the technical complexities of "hardware," on the other. We shall first describe the currently operating, on-line mathematical facilities, then discuss the experimental devices, and finally show some illustrative problems.

I. On-Line Mathematical Systems (OLMS)

Although it is impossible to describe in words the most important property of on-line systems, that is, their highly *interactive* character, we shall sketch here the principal features to the extent required for an understanding of the examples given in Section III. (A more complete account of an early version of such systems can be found in *On-line Computing*, edited by W. Karplus, McGraw-Hill, New York, 1967, Chap. VI and Appendix; Russian Edition, Izdatelstvo Mir, 1969.) The console consists of a standard laboratory storage oscilloscope and two typewriter keyboards, designated as the *operator* and *operand* keyboards. Each key push of the operator keyboard initiates some subroutine in the computer, while each key of the operand keyboard corresponds to a storage location. The computer output either graphical or alphanumeric, appears on the oscilloscope when requested by the user.

The most important operators of the system are those that refer to functions of a real or complex variable. An arbitrary function can be represented in the computer by an ordered list of numbers, real or complex. Accordingly, a portion of the computer memory is set aside as a pair of vector registers or accumulators, called the x and y registers. Pushing any key of the operator keyboard causes some transformation of these registers. In the REAL mode (which is in effect whenever the operator key labeled REAL has been pushed), only the y register is transformed; in the complex mode (that is, after the COMPLEX key has been pushed), the x and y registers are coupled together as a complex $x + iy$ register. The STORE operator transfers the contents of the vector register to a storage location indicated by any one of the 26 keys of the operand keyboard. (These keys are labeled from A to Z.) Conversely, the LOAD operator transfers a vector from any storage location into the vector register. This is best made clear by some elementary examples.

Pushing the dimension key (DIM), followed by any integer, N, up to 747, sets the dimension of the vector registers at N. The identity key (ID) generates the identity function ($y = x$) in the sense that it fills the x and y registers with N numbers between -1 and 1. Thus the sequence of key pushes† DIM 125 ID ⊕ 1 ⊙ 1.8 STORE T, for example, transforms the contents of the y register to a vector consisting of 125 equally spaced numbers in the range from 0 to 3.6 and then stores those values in a storage location which the user need think of as simply the key T on the operand keyboard. Following this with the key pushes SQ SIN ⊙ .6 transforms the y register to a vector whose components are 0.6 sin (t^2), for 125 equidistant values of t ranging from 0 to 3.6. STORE F

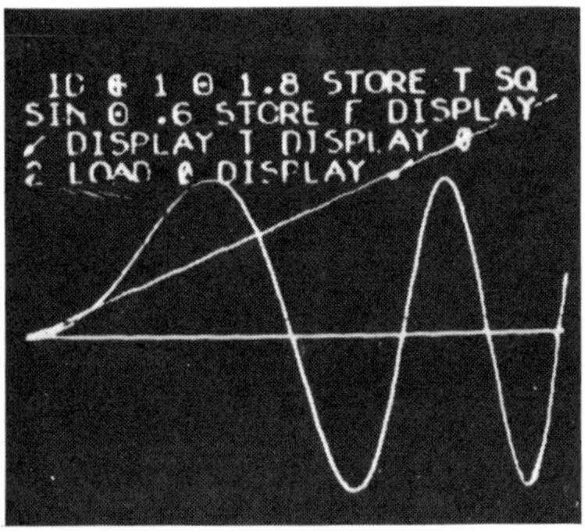

Fig. 1. The curves $T(0 \leqslant T \leqslant 3.6)$ and $F = 0.6 \sin (T^2)$.

transfers that vector to storage location F (also, of course, leaving it in the y register). Pushing DISPLAY ↙ (where ↙ is the carriage-return key on the operand keyboard) displays on the oscilloscope a plot of the y resister versus the x register, that is, the oscillatory curve in Fig. 1, representing .6 sin t^2. Similarly, DISPLAY T first transfers the vector stored in T to the y register, and then plots the y register versus the x register, giving the straight line shown in Fig. 1. (The convention is made that the limits of the oscilloscope display are ±1; i.e., we really see the mantissas of the vector being displayed. When we want to see its binary scale, we push DISPLAY 0, the 0th element of the vector being defined as its binary scale. For our T vector, the binary scale is 2, while for F it is 0.) To ascertain the numerical value of the kth component of the y register we push DISPLAY k (where k is any integer typed on the numerical

† DIM ID ⊕ ⊙ and STORE are operator keys; the numerical and alphabetic symbols are on the operand keyboard.

keys of the operand keyboard); successive pushes display the $(k + 1)$, $(k + 2)$, etc., components. (The horizontal axis in Fig. 1 results from LOAD 0, which generates a constant vector, of value 0, in the y register, followed by DISPLAY ↙.)

To this *very* brief description of the *style* of operation of the on-line system, we add three important remarks, of which the third is the most crucial.

1. The basic operations of the system include, in addition to the simple ones illustrated above, a set from which all of the fundamental processes of classical analysis can conveniently be composed. For example, DIFF computes

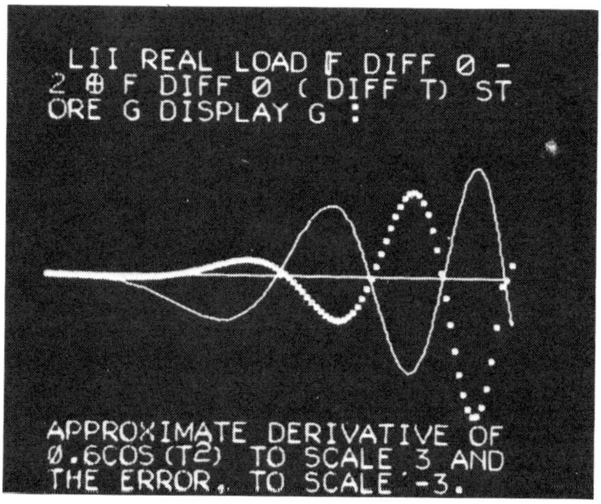

Fig. 2. The derivative of $F = 0.6 \sin (T^2)$, as computed with the key pushes shown at the top of the figure, that is, $\Delta[F - \frac{1}{2}\Delta F]/\Delta T$. The difference between that and the curve $1.2T \cos (T^2)$ is also shown. The error is about 1 part in 2^6.

the first differences of the y register, that is, it transforms the y register accord- to $y_i \to y_{i+1} - y_i$, $1 \leqslant i \leqslant N$, while SUM computes the running sum, that is, effects the transformation $y_i \to \sum_{j+1}^{i}$. With these, the various numerical approximations to differentiation and integration, definite and indefinite, can be readily constructed. As a simple example, using

$$\frac{\Delta(F - \frac{1}{2}\Delta F)}{\Delta T}$$

as an approximation for the first derivative, the key pushes

LOAD F DIFF ⊘ −2 ⊕ F DIFF ⊘ (DIFF T) STORE G DISPLAY G

where F and T contain the curves shown in Fig. 1, produce the curve shown in Fig. 2, its binary scale being 3. The dotted curve is the error, that is, the difference between G and the vector $1.2t \cos (t^2)$; its scale is −4. (The left and right parentheses used in the above example are "punctuation" keys, located on the operand keyboard. The dotted display, which is always available as an option, shows the N points actually used for computations. The normal line display differs only in having straight line segments connecting the data points.)

Other operators on the operator keyboard include the elementary functions sin, cos, log, exp, arc tan; maximum and absolute value; a Dirac-Kronecker delta function which transforms the y register to one that is 1 wherever there is an exact zero or a change of sign and is 0 everywhere else; a linear interpolation operator; and so on. In the real mode, these operate on the y register; in the complex mode they operate, in the correct, complex-variable fashion, on the combined $x + iy$ register. (As may be inferred from the differentiation example above, an operator, like DIFF, operates on the vector register unless it is followed by an operand key like T, in which case it acts on the indicated operand. Experience has shown this feature to be very convenient.)

2. Although vectors are the principal objects of concern, it is sometimes necessary to deal with scalars, on the one hand, or (two-dimensional) arrays of vectors, on the other. Both facilities are conveniently provided by introducing the concept of "levels," which can be thought of as analogous to "shifts" or "cases" on a typewriter. If the level I key on the operator keyboard is pushed, all of the operations that have meaning for scalars (i.e., excluding things like DIFF and SUM) transform a scalar register, and the operand keyboard keys address storage locations for scalars. If the level II key is pushed, then all operators refer to vectors, as in the above examples. Similarly, level III refers to arrays of vectors. Interlevel operations are allowed (e.g., substituting a scalar from level I for the kth component of a level II vector, and conversely), and all levels may be used in either real or complex mode.

3. A simple procedure is provided for composing *any* sequence of key pushes, on either keyboard, into a new subroutine or "console program," meaning simply a program generated at the console. Six "user levels" on the operator keyboard are available for storing such console programs; that is, one such program (or list of key pushes) can be assigned, by the user, to any one of 26 keys on the operator keyboard, on any one of the six user levels. To execute such a program, we simply push, for example, USER LI ⊕, which will initiate whatever console program was previously stored under the ⊕ key on user level I. *The critical pyramidal feature, which makes this console a computer in the von Neumann sense, rather than a fancy desk calculator, follows from the fact that any console program can include among its key pushes not only the basic operations of levels I, II, and III but also the key pushes which call any other, previously constructed console program*, for instance, USER LI +.

Although all of the facilities described here are quite simple, the ease with which they can be composed and combined and the highly interactive, graphical character of this type of on-line system give it sufficient power to make it useful for the difficult problems typical of our field. Of course, this cannot really be demonstrated, since one cannot intelligently discuss complicated problems that are still unsolved, and all solved problems appear trivial a posteriori. The examples given in Section III, however, provide very simple illustrations of some of the applications of such systems to plasma physics.

II. Experimental Configurations

Experiments suitable for quantitative theoretical interpretation are best performed on large-volume, steady-state (or very slowly decaying), quiescent plasmas. It is desirable for an external magnetic field to be an optional feature, and hence such a field should not be required for plasma confinement. Professor Kenneth MacKenzie has developed a number of such plasma devices, but we concentrate here on the "double-plasma" or DP machine. The plasma is generated in each of two, electrically isolated chambers by the ionizing electron current from a number of hot, negatively biased filaments. The chambers are separated by a negatively biased grid, whose principal function is to isolate the electrons in the two chambers and hence permit the chambers to be at different potentials. (In the absence of the grid, electrons would quickly "short out" such a potential difference, confining it to sheaths at the chamber walls.) If one chamber (the "driver" chamber) is biased positively with respect to the other (the "target") chamber, a steady, broad beam of ions will flow through the target chamber. If, with or without such a bias, an alternating potential difference is applied, ion-acoustic waves will propagate into the target chamber. If an increasing potential difference is applied (a voltage "ramp"), a ramp-shaped ion-acoustic wave will propagate into the target chamber and steepen into a collisionless ion-acoustic shock, with a trailing wave structure. By means of such devices, it has been possible to study in detail the properties of these shocks, as well as the nonlinear properties of ion-acoustic waves, which have large amplitude initially or which show spatial growth due to the ion streaming produced by a constant bias between the chambers.

An array of very fine wires (diameter 0.02 mm), immersed in the plasma and biased positively, acts as a kind of "Maxwell demon," preferentially removing slow electrons, which strike the wire, while the higher-energy electrons, conserving angular momentum, miss the wire. Varying the bias on these wires permits the electron temperature to be varied continuously over an order of magnitude up to a maximum of about 5 eV. The containment of electrons is greatly improved by a very simple confinement configuration, an array of small permanent magnets, covering the walls of the chamber, with their axes normal to the chamber walls. This array of dipoles reflects the ions very effectively but leaves the interior of the chambers free of magnetic field. Of course, an external B field can be added when desired. Present machines have dimensions of the order of 1 meter in diameter and 2 or more meters in length, so that many of the experimental phenomena can quite reasonably be treated with one-dimensional theory, a feature of inestimable value to the theoretician. It seems entirely feasible to build similar devices on a much larger scale.

The examples discussed in Section III give some indication of the versatility of these machines and their usefulness for experiments of theoretical interest. However, as with the on-line mathematical systems, it is not really possible to convey the *interactive* aspect of these devices. Since the large-volume, steady-state character facilitates the diagnostics, it is easy to observe, in on-line fashion on an oscilloscope, the variations in structure with distance, electron temperature, composition, density, and other parameters and thus obtain far more clues to the nonlinear phenomena involved than are contained in conventional presentations of data. In addition, new experimental features suggested by theory can readily be implemented.

III. Simple Examples

A. *Cross-Field Instabilities*

Linear instabilities and the associated dispersion relations underlie many current problems in plasma physics. We illustrate this with the cross-field instability which occurs when there is relative streaming of electrons and ions across a magnetic field, due to differential drifts or other circumstances. Although this instability may be very important in collisionless shocks and other anomalous transport configurations, it is difficult to identify amidst the other phenomena occurring with it. A simple experiment in which it can be studied directly has been carried out by R. J. Taylor and Peter Barrett, using a DP machine to produce an ion beam, streaming with velocity $\mathbf{V}$ across a magnetic field, with sufficient energy so that the curvature of the ion orbits can be neglected. This corresponds to the situations of principal interest, in which the electrons are strongly magnetized whereas the ions are not. The instability is observed to occur at low frequencies ($\omega \ll \omega_{ce}$) and to have the character of a modified ion-acoustic wave. The principal difference from the normal ($B = 0$) case is that when $kr_{ce} \ll 1$ and $k_{\parallel}/k \ll 1$, the growth rate can be enhanced by a factor $k/k_{\parallel}$. This is easy to understand. Since the electron thermal motion is along B, its projection along the wave vector, $\mathbf{k}$, gives an effective distribution function with a temperature diminished by the ratio $k_{\parallel}/k$. As a consequence, even for small beam velocities, of the order of the ion-acoustic speed, c_s, the phase velocity of the waves, which propagate at a velocity of the order of c_s in the ion frame, can coincide with the location of maximum slope of the effective electron distribution function. From this we see also that the threshold for beam velocity is not very sensitive to T_i/T_e, remaining of the order of c_s as the temperature ratio varies from 0 to 1. The dispersion relation, as derived by Arefev, can be written in the form

$$\kappa \equiv \frac{1 - \Gamma_0 + k^2/k_D{}^2}{\Gamma_0} = \tfrac{1}{2}\left[Z'\left(\frac{\omega}{k_\| a_e}\right) + \left(\frac{T_e}{T_i \Gamma_0}\right) Z'\left(\frac{\bar{\omega}}{k a_i}\right)\right] \tag{1}$$

where

$$\Gamma_0(b) = \exp(-b) I_0(b),\ b = (k_\perp r_{ce})^2/2,\ \bar{\omega} = \omega - \mathbf{k} \cdot \mathbf{V}$$

and the other notation is standard. Although this is a relatively simple dispersion relation, it is typical in that it involves transcendental functions, which have a simple form only for large or small arguments, and a considerable number of independent parameters. Thus our analysis of its properties can serve as a model for more complicated cases as well, and we shall discuss it in some detail.

For given values of the parameters, it is easy to solve this equation for the complex roots in the ω plane, using simple consequences of Cauchy's theorem. Consider any analytic function, $F(U)$; a closed curve C in the U plane which encloses no sigularities of F; and the image, $C' = F(C)$. If C' encloses the origin in the F plane once, then $F(U)$ has one root inside C, and the value of the enclosed root is given by

$$U_0 = (2\pi i)^{-1} \oint_C \frac{dU\, UF'(U)}{F} = (2\pi i)^{-1} \oint_{C'} \frac{dFU}{F} \tag{2}$$

Thus we simply construct a console program which, for a given curve, C, in the U plane, computes and displays the curve $F(C)$, where

$$F(U) = +1/2Z'\left[s\left(U + \frac{V}{c_s}\right)\right] + \frac{1}{\Gamma_0 U^2} - \kappa$$

$$U = \frac{\bar{\omega}}{kc_s}, \qquad s = \frac{k}{k_z}\left(\frac{m}{2M}\right)^{1/2} \tag{3}$$

(For simplicity, we are considering the case of small T_i/T_e, where the ion Z' function can be approximated by its analytic argument, but the general procedure would be the same for arbitrary T_i/T_e.) In addition, we need a console program to construct $Z'(C)$ and another to compute the integral in eq. 2. We consider these in turn.

1. The Plasma Dispersion Function

Since we are looking for unstable roots, C will lie in the upper half-plane, and we find that it is quite adequate to compute

$$Z'(T) \equiv \pi^{-1/2} \int dw \exp(-w^2)(w - T)^{-2} \tag{4}$$

by choosing for the path of the w integration a line below, and parallel to, the real axis running from $(-3 - i)$ to $(3 - i)$. A very coarse grid along that line (typically 10 intervals) suffices for most purposes; when necessary, of course, we can use a finer grid. Thus we want to calculate

$$Z'(T) = \frac{6\pi^{-1/2}}{N} \sum_{J=1}^{N} \exp\left(-w_j^2\right)\left(w_j - T\right)^{-2}$$

$$w_j = \frac{6_j}{N} - 3 + i \tag{5}$$

This is most conveniently done by making two console programs, as shown in Fig. 3: one, stored under USER II ARG, which adds one more term to the

Fig. 3. Console programs for $Z'(T)$.

sum (eq. 5) and a second, stored under USER II ATAN, which does the initial preparation and then repeats USER II ARG N times; we usually choose N to be 10, but can easily increase it if necessary. (A simple test is to examine the difference between the results with $N = 10$ and $N = 20$, for example.)

2. Root-Finding Program

This simply involves the console program shown in Fig. 4, where the contour integration is done using the trapezoidal rule,

$$\int G\, dF = \Sigma(G + 0.5\,\Delta G) \cdot \Delta F \tag{6}$$

The resulting value of the root is displayed in numerical form on the oscilloscope and also left as a constant vector in II W, ready to be used as the center for a new contour in a (rapidly converging) iteration procedure.

3. Computation of $F(U)$

The console program which computes $F(U)$ is shown in Fig. 5. Note that it uses the console program for Z', described above, calling it by simply "pushing" USER II ATAN.

With these simple console programs, we can quickly find the roots of eq. 3 for given values of k, k_D, s, V, c_s, and b by using the Cauchy formula,

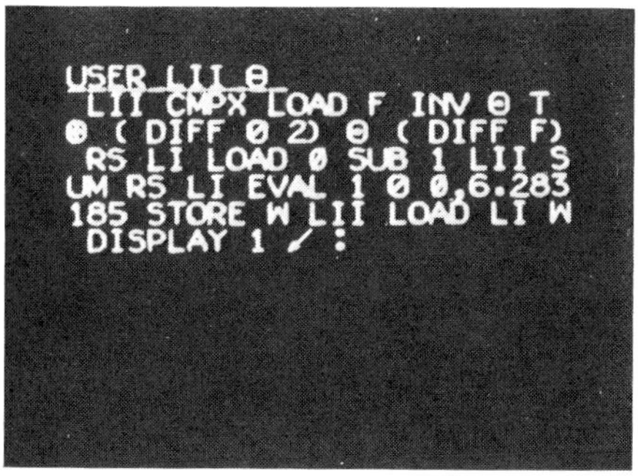

Fig. 4. Console program for finding roots in the complex plane.

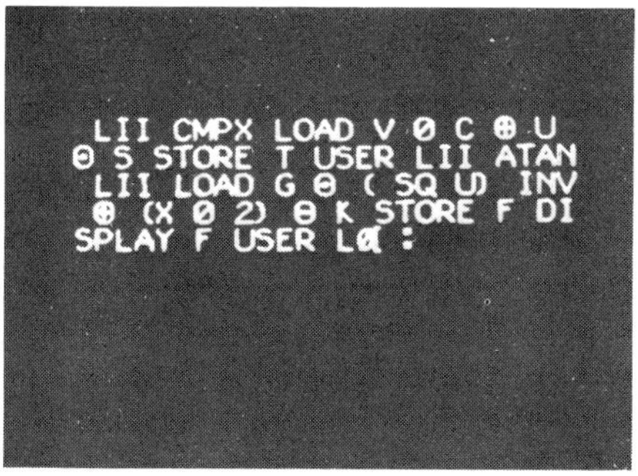

Fig. 5. Console program for computing the dispersion relation, $F(U)$, eq. 3.

eq. 2. [A simple console program, not shown here, computes the modified Bessel function $I_0(b)$, so we simply assume here that Γ_0 has a known value.] We first experiment with various choices for the contour C until we find one whose image encircles the origin once, for example, the unit square shown in Fig. 6. Running the Cauchy console program (Fig. 4), we find a first approximation to the unstable root, as shown in Fig. 6. Using this as the center of a small circle gives us a new C, as shown in Fig. 7, whose image, C', is smaller and which provides a better approximation to the root. In the unusual case where extreme accuracy is required, we could repeat this iteration process, increase the number of points used for the computation of Z' (presently 10), increase the dimension of the level II vectors (presently 101), and so on.

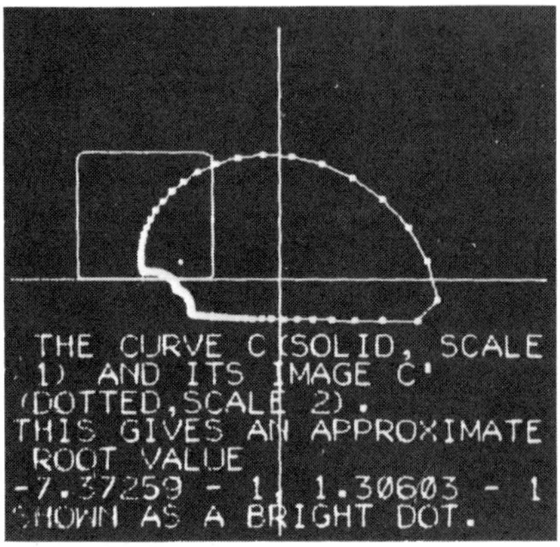

Fig. 6. First approximation to the root of $F(u)$, obtained by using the square contour, C, and eq. 2. The image, $C' = F(C)$, and the resultant root value are shown.

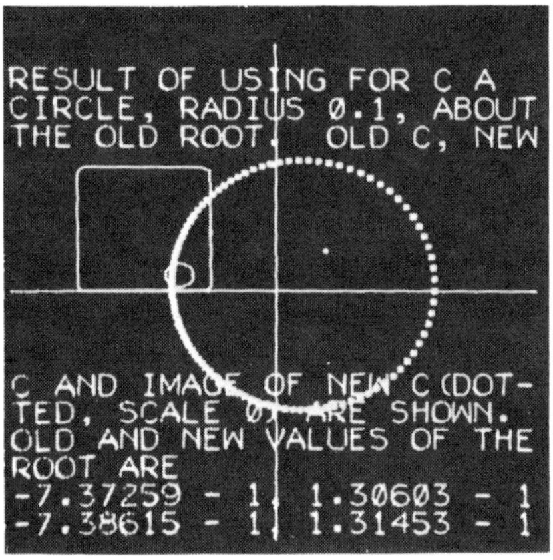

Fig. 7. Second approximation to the root of $F(u)$, using for C a small circle about the root found before (Fig. 6). The correction is very small, indicating the rapid convergence of this iterative procedure.

Although this approach is a general one that can always be used, even when ion Landau damping is included, another alternative is available for this problem, namely, to write eq. 3 in a form well suited for iteration in the usual case (far from threshold) where V/c_s is larger than U, which is typically of the orders of 1:

$$U = (\Gamma_0)^{-1/2} \left\{ \kappa - \tfrac{1}{2} Z' \left[s \left(U + \frac{V}{c_s} \right) \right] \right\}^{-1/2} \tag{7}$$

This has the advantage that we can treat any one of the parameters as a variable, and hence obtain directly the growth rate, Im ω or Im k, as a function of b or V or k, etc. In Fig. 8 we show the first few iterations of eq. 7 for values of s, V, and c_s appropriate to the experimental data. The plot shows Im U versus Re U (i.e., the locus of the roots of eq. 3 in the complex U plane), After the third

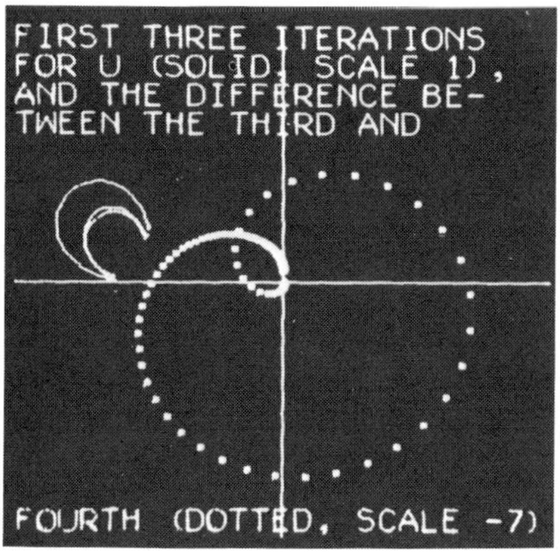

Fig. 8. Iterative solution of eq. 7. Note that after three iterations we have convergence to 1 part in 2^8, since the binary scale of the U curves is +1 while that of the difference, $U_3 - U_4$, is -7.

iteration, the curves can no longer be distinguished on the oscilloscope; hence it is more useful to compute and display the difference of successive iterations. As we see from Fig. 8, convergence to 1 part in 2^8 has been achieved after three iterations with b as a variable parameter.

For comparison with experiment, we plot the relative growth rate, Im k/Re k, as a function of $1/b$ (which is proportional to B^2), with $k_\parallel$ chosen to be π/L, where L is the plasma dimension along the magnetic field (about 40 cm). Figure 9 gives these curves with L chosen as 30, 40, and 50 cm; the experimental points of Barrett and Taylor for an ion beam energy of 80 V, frequency 1 Mc, electron temperature of the order of 2 eV, and density $3 \cdot 10^8$ cm^{-3} are also shown. From this it appears that $k/k_\parallel$ varies with B. Allowing $s = (k/k_\parallel)(m/2M)^{1/2}$ to vary linearly with $1/b$ from 0.2 to 0.09 gives

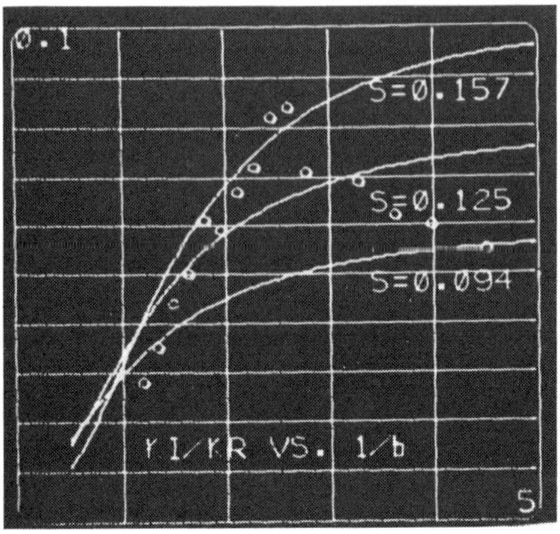

Fig. 9. Comparison of experimental and theoretical growth rates, k_I/k_R versus 6^{-1}. Assuming plasma lengths of 30, 40, and 50 cm gives s values of 0.094, 0.125, and 0.157. No single theoretical curve shows good agreement with the experimental data.

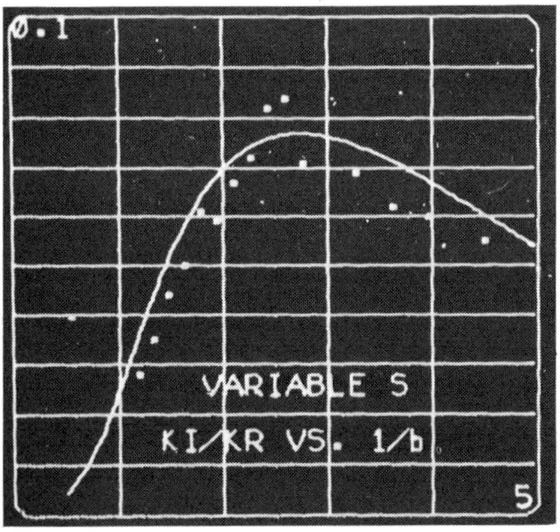

Fig. 10. Comparison of experimental and theoretical points when $s = (k/k_{\parallel})\,(m/2M)^{1/2}$ is allowed to vary with b as shown in Fig. 11.

an excellent fit to the experimental points, as shown in Fig. 10. It is also of interest to choose s so as to maximize the growth rate at each value of b, for given V and c_s. This is most conveniently done by using the iterative formulation of eq. 7, choosing some range of values for s, and observing which one maximizes Im U. The resultant values of s, together with the linear variation that leads to the good fit of Fig. 10, are shown in Fig. 11. The optimized values of s lead to

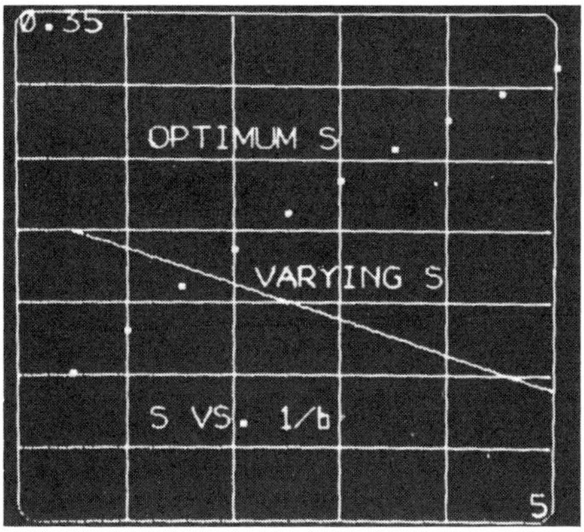

Fig. 11. Values of s that maximize growth rate (shown as dots). Allowing s to vary linearly with $1/b$, as shown by the solid curve, gives close agreement with experiment, as shown in Fig. 10.

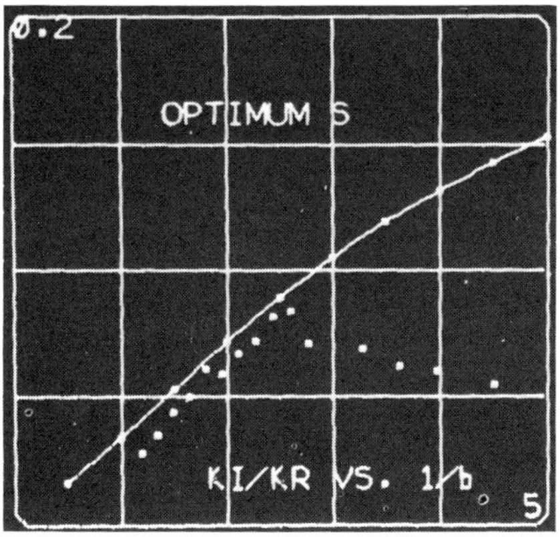

Fig. 12. Comparison of experimental growth rates with the maximum values obtainable by varying s at each value of b.

growth rates comparable to the experimental values for small B, as shown in Fig. 12, but the agreement fails for larger B (i.e., smaller b).

B. Collisionless Shock Waves

As a second example, of a quite different character, we consider the propagation of collisionless electrostatic shock waves. These can conveniently be generated with a DP machine by applying a linear ramp voltage form to the driver chamber, the target chamber being held at ground. The resulting ion-acoustic wave steepens into a shock, with trailing wave structure. Because of the flexibility and accessibility of the DP machines, it is relatively easy to study the properties of this shock in some detail. Using the "Maxwell demon" heating technique, we can continuously vary T_e/T_i, observing, among other things, the changes in the number of reflected particles and the consequent transition from a laminar shock to a turbulent one. The same effect is produced by adding a few per cent of a light ion species (for example, helium in argon), as suggested by Professor Charles Kennel and Academician Roald Sagdeev, thus enhancing the population of reflected particles without changing any other properties. Moving the receiving probe shows the development of the structure in space, while use of an energy-analyzing probe provides data on the distribution function at various points, showing in detail the properties of the reflected particles.

The simplest theoretical model of such shocks is provided by a time-dependent version of the original fluid formulation due to Moiseev and Sagdeev. Although we do not have the space to describe this calculation in detail, one feature of it illustrates a different aspect of the mathematical on-line approach. To avoid the complexities associated with the sheath around the negatively biased grid, located at $X = 0$, we assume the potential and the electric field to be continuous there, but allow the electron density to have a discontinuity. Thus, if $V(t)$ is the potential of the driver chamber, a Boltzmann (isothermal) ansatz for the electrons takes the form (all potentials in units of T_e)

$$n_e(x, t) = \begin{cases} n_0 \exp\,[\phi(x, t) - V(t)], & x < 0 \\ n_0 \exp\,[\phi(x, t)] & x > 0 \end{cases} \tag{8}$$

The problem is then to solve the time-dependent fluid equations for the ion density and mean velocity, plus Poisson's equation (with $k_D{}^{-1}$ as the unit of length):

$$\frac{\partial^2 \phi}{\partial x^2} = n_e - n_i \equiv F(x, t) \equiv \exp\,[\phi - V\theta(-x)] - n_i \tag{9}$$

It is straightforward to put the fluid equations in finite difference form with respect to the time variable and to step them forward in time. At each time step, however, we must solve Poisson's equation with two-point boundary conditions

$$\phi = \begin{cases} V(t), & X = -L \\ 0, & X = L \end{cases} \tag{10}$$

where $x = \pm L$ at the ends of the DP machine.

A straightforward iteration approach would be to write eq. 9 (suppressing the t dependence) as an integral equation:

$$\phi(x) = \int_{-L}^{x} dx'\,(x - x')F(x') + a(x + L) + b \tag{11}$$

with a and b chosen to satisfy eq. 10:

$$b = V, \qquad a = \frac{-\left[V + \int_{-L}^{L} dx'\,(L - x')F(x')\right]}{2L} \tag{12}$$

The console program corresponding to eqs. 11 and 12 is, of course, quite simple. Using it, we quickly find that the iteration process diverges badly if L is large compared to the Debye length (the case of interest), as shown in Fig. 13 for $L = 3$.

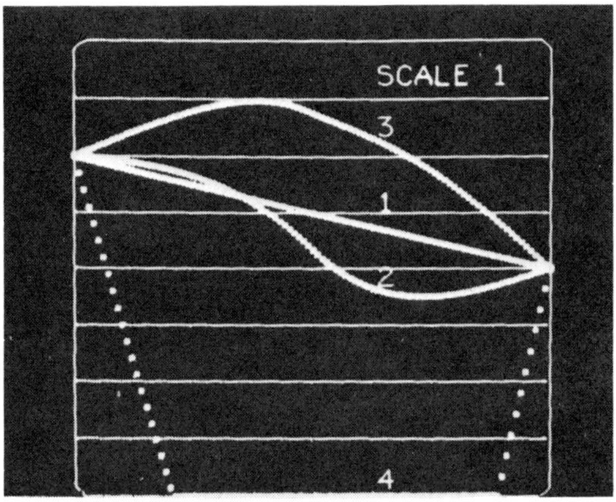

Fig. 13. Result of solving eqs. 11 and 12 by iteration. Using the straight line, labeled 1, as a first guess gives the curve labeled 2. Using this function on the right-hand side of eqs. 11 and 12 gives the curve labeled 3. This, in turn, leads to the curve labeled 4, etc. (All curves have binary scale 1.)

A better alternative is to write eq. 9 in the form

$$\phi'' - \phi = G(x) \equiv e^{\phi} - \phi - n_i, \quad x > 0 \tag{13}$$

which leads to the integral equation form

$$\phi(x) = \int_{-L}^{X} \sinh(x - x')G(x') + A_{+} \sinh(x - L) \tag{14}$$

plus a similar expression for $x < 0$, the two constants $A_{\pm}$ being determined by the continuity of ϕ and ϕ' at $x = 0$. (The result is an integral equation since G is a function of ϕ and $A_{\pm}$ involve integrals of ϕ.) Although this involves somewhat more effort than using eqs. 11 and 12, the excellent convergence properties,

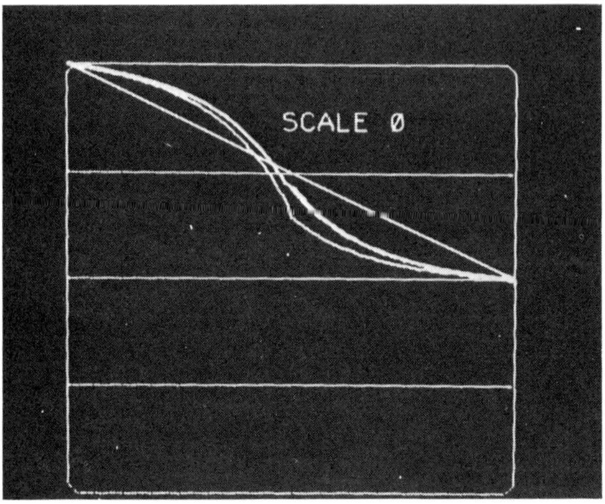

Figure 14. Iterative solution of eqs. 13 and 14, plus the corresponding equations for negative x. As in Fig. 13, the first guess is the straight line, but now the process is rapidly convergent. Starting with the straight line, one iteration results in the lowest of the other curves. It, in turn, gives the next highest curve, and one more iteration brings us to a curve that no longer changes visibly with further iterations. (All curves have binary scale 0.)

illustrated in Fig. 14, make it well worth while, since this portion of the problem (i.e., the solution of Poisson's equation) now becomes a trivial matter.

C. Nonlinear Ion-Acoustic Waves

As the next example, we consider a problem involving both mode coupling and wave-particle interaction. As already noted, a positive bias applied to the driver chamber of a DP machine causes a steady flow of ions through the target chamber. An alternating-current signal added to the bias gives rise to an ion-acoustic wave propagating in the driver chamber, and the ion beam can cause this wave to be unstable. An experiment in which the signal consists of *two* frequencies, ω_1 and ω_2, was performed by Robert J. Taylor and H. Ikezi, who found that in addition to the primary waves, ω_1 and ω_2, a number of secondary waves, with frequencies $n\omega_1 + m\omega_2$ (n, m integers) also grow for some distance, saturate, and then decay. The usual mode coupling treatments must be modified to take into account the fact that the growth rates of the waves are not very small compared to the frequencies. An analysis by P. Martin leads to equations of the usual mode coupling form, but with an additional term resulting from the linear growth.

For a small number of waves (three to seven), it is easy to integrate these equations by iterating the associated integral equations. (We do not give the details here, the general approach being similar to that used for the Poisson equation in the previous example.) The surprising result is that the waves show

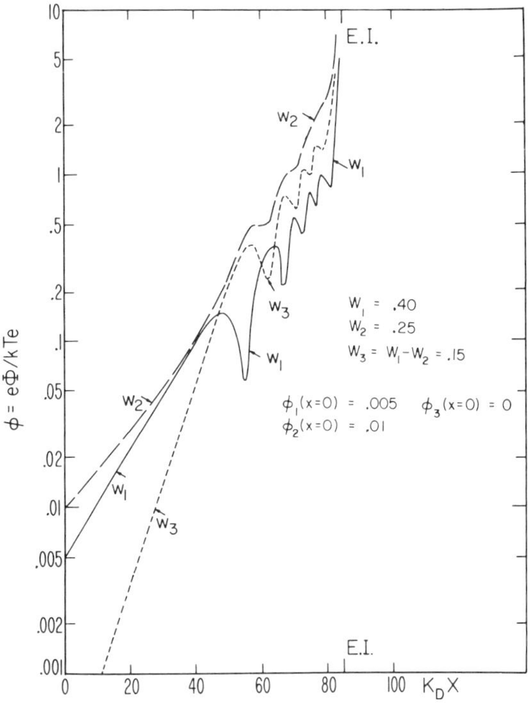

Fig. 15. Explosive growth of positive-energy waves.

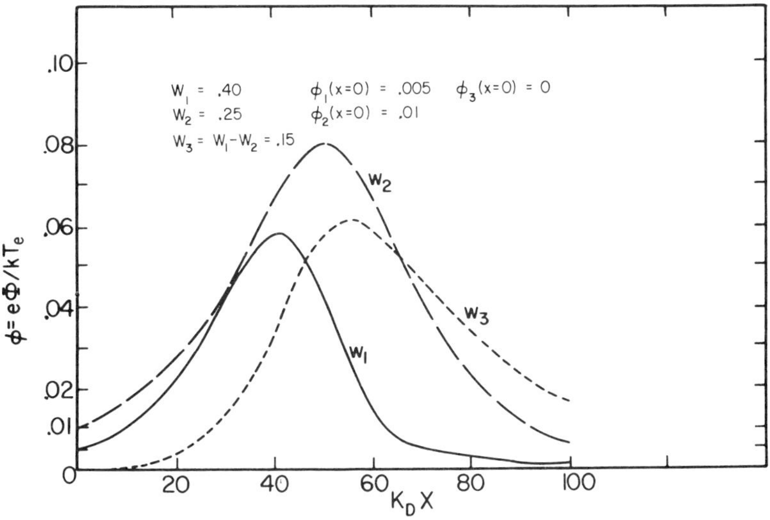

Fig. 16. Growth of unstable ion-acoustic waves when both mode coupling and wave-particle interactions are included.

an explosive growth, that is, all waves become singular at finite values of x, as shown in Fig. 15. This occurs even though all of the waves have positive energy, being simply a consequence of the linear growth combined with the mode coupling effects. (For clarity, the original oscilloscope pictures have been redrawn.) Inclusion of wave-particle effects, that is, the modification of the beam distribution function by the waves, changes the linear term in the mode coupling equations, adding a damping term proportional to $\int d\,x\,E_\omega{}^2$, where E_ω is the wave amplitude. As would be expected, this removes the explosive instability, as shown in Fig. 16, these curves being in semiquantitative agreement with the experimental results.

IV. Concluding Remarks

The main purpose of this review has been to show how both on-line computing and on-line experiments can be of value to theoreticians. What is involved here is basically a question of *style*, and that is a very difficult thing to convey. It is best done by examples, but even the very simple ones chosen as illustrations here require more explanation than is really appropriate for a general presentation. I hope that from what has been said here you can extrapolate from these particular, somewhat elementary examples and use your own imagination to see how these kinds of facilities can be employed also for more complex problems. Finally, I must emphasize that, valuable as on-line facilities may be, they should, of course, be used only where appropriate. For example, a large hydrodynamic problem, for which successful and efficient integration techniques and numerical methods are known, should never be done on-line. The only sensible procedure is to use the conventional, batch computing approach. Similarly, many vital questions in the physics of controlled fusion will be answered only by large, complex experiments, no matter how inaccessible these may be to theoreticians. I have tried only to call to your attention the facts that modern developments make it possible for us, as theoreticians, to get significent help in our attempts to understand this difficult and fascinating field and that this assistance, from both computation and experiment, can be obtained without having to become expert in the technology of either computing or experimental physics. Certainly, no one can deny that in this field we need all the help we can possibly get!

Acknowledgment

The work reported here has been partially supported by the U.S. Atomic Energy Commission, the Office of Naval Research, and the National Science Foundation.

Numerical Simulation of the Kinetic Processes in Plasma

Y. N. DNESTROVSKII AND D. P. KOSTOMAROV

Lomonosov State University, Moscow, USSR

The physical problems of plasma kinetic processes can be divided into two classes.

1. Problems in which binary collisions play the secondary role and the fundamental interactions are due to a self-consistent electric field. The Vlasov equation gives a good description of these processes.
2. Problems in which binary collisions determine the process. In this case the Landau equation for the homogeneous plasma is usually the basic equation.

I. The Simulation of Collisionless Plasma

To simulate these problems (1) four different methods are now used.

1. The macroscopic particle method.
2. Expansion of the distribution function on a system of basis functions.
3. The "water-bag" (lines of discontinuity) method.
4. The finite difference method.

The particle method was the first one used for simulation and still remains the most appropriate. Recently some two- and three-dimensional problems were solved by this method, which is very economical in computer time for a great number of problems. For these reasons we shall consider this method in more detail.

A. The Macroscopic Particle Method

To ascertain the advantages and difficulties of the macroscopic particle method, let us consider its evolution, using the example of one-dimensional problems.

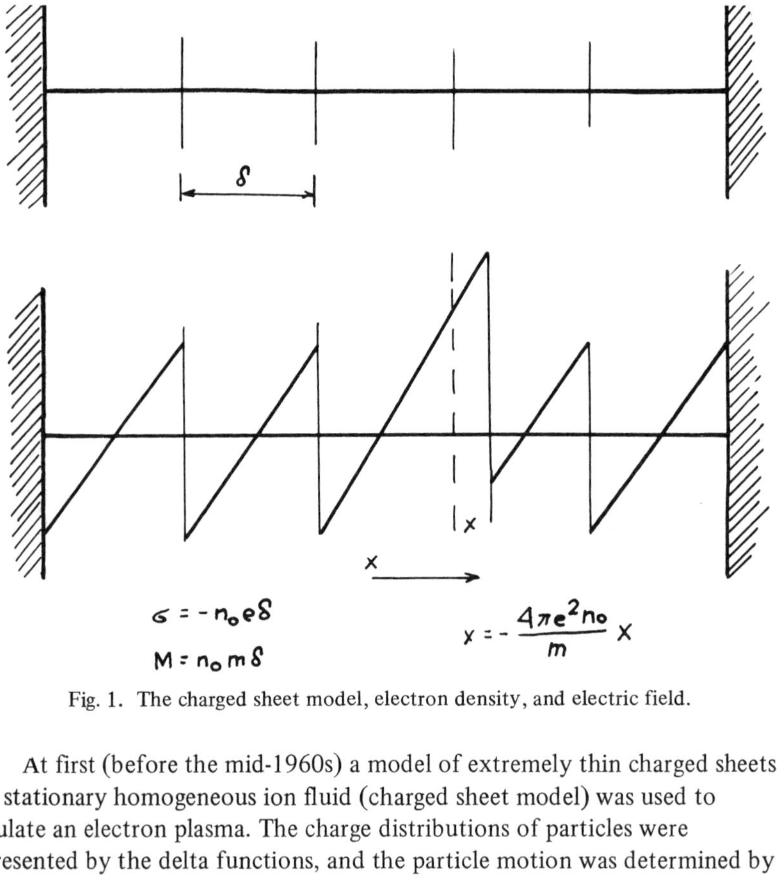

Fig. 1. The charged sheet model, electron density, and electric field.

At first (before the mid-1960s) a model of extremely thin charged sheets in a stationary homogeneous ion fluid (charged sheet model) was used to simulate an electron plasma. The charge distributions of particles were represented by the delta functions, and the particle motion was determined by the self-consistent electric field E. In such a problem the E field is described by piecewise continuous functions, having the discontinuities at points where the particles are placed (Fig. 1) (1).

When a particle passes through another one, the force acting on it jumps. At the beginning of the 1960s the particle equations of motion were numerically integrated by some difference scheme. But this procedure leads to nonconservation of the full plasma energy because of the fast growth of numerical errors. This difficulty can be bypassed by using the fact that the equations of motion in the linear electric field (between particle "collisions") may be integrated analytically. For this reason the calculations are usually organized in the following way. After setting the initial conditions, the analytical solutions of the equations of motion are obtained. Then that pair of particles is found which meets each other first. At the moment of collision, all particle coordinates and velocities are calculated and then the search is repeated. Thus the advancement in time proceeds with a maximal step coinciding with the time interval between collisions. Calculations have shown that the integrals of motion are very well conserved.

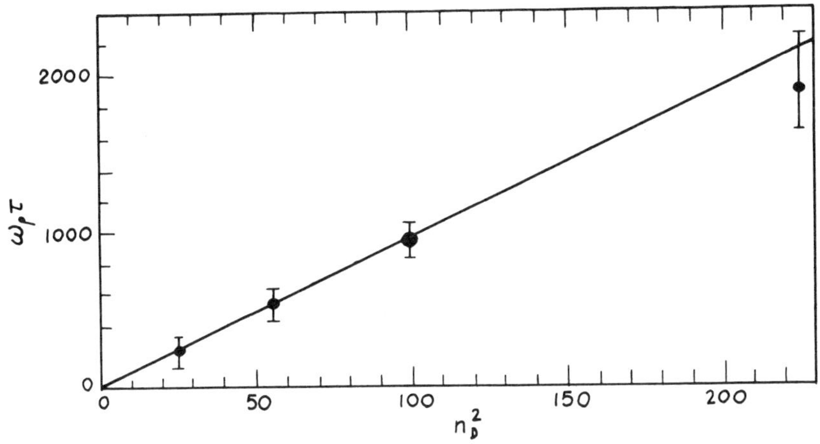

Fig. 2. Relaxation time τ as a function of the number n_D of particles (sheets) per Debye length.

For description of the kinetic processes the number of particles in the Debye length, λ_D, must be large enough. Figure 1 for the E field shows that the sheet model leads to nonphysical fluctuations about the mean value, that is to say, noise has appeared. If the number of particles diminishes, the amplitude of the noise increases. The influence of noise is analogous to the effect of real collisions and leads to relaxation in the system. For the determination of the relaxation velocity the following problem for the homogeneous (in x space) initial distribution:

$$f(x, v, 0) = \left\{ \begin{array}{ll} \text{const}, & |v| \leqslant v_0 \\ 0 & |v| > v_0 \end{array} \right\} = \varphi(v) \tag{1}$$

has been solved. The distribution of eq. 1 is a solution of the Vlasov equation, but the "noise" will relax it to the Maxwellian distribution. In Fig. 2 the dependence of the relaxation time on the number of particles in the Debye length (n)

is shown (1). Here ω_p is the plasma frequency. This diagram shows that it is necessary to have about 10 particles of Debye length for good approximation of the physical process in 100 plasma periods.

For noise reduction some "smoothing" methods are used.

1. Particle in Cell (PIC) Method

With this method the full length of the plasma in x space is divided into cells. The complete charge in each cell equals the sum of particle charges placed in this cell. The field in the cell center is found by solving the difference Poisson equation, the right-hand part containing the cell charges. The field at the points where particles are placed is determined by the interpolation process. In such a problem, the field $E(x)$ is the continuous piecewise smooth function and the noise has a rather diminished amplitude.

2. The Particle-Cloud Method

In this method the charge density of particles is described by the nonlocal rapidly-decreasing function of the space coordinates [e.g., "Π-type" functions or Gaussian functions, $\exp(-x^2/a^2)$].

Very often methods 1 and 2 are applied together. In this case the procedure is called the "cloud in cell" (CIC) method.

These smoothing methods successfully reduce the noise in the system. This is very well shown in Fig. 3 (1), where the dependences of the square amplitudes of noise harmonics versus mode numbers are shown for $a = 0$ (dotted line) and $a = 2\lambda_D$ (a is the "radius" of the particle). Nevertheless new questions have appeared in connection with the admissible size of cell l and particle radius a. It is found that for distinct problems the influence of these parameters can be quite different. In any case l and a must not exceed appreciably λ_D and the minimal wavelength for the process considered. For example, let us consider the beam-plasma problem. For its solution 10^3 particles are used (1) (200 particles for the beam and 800 for the plasma) with $L = 50\lambda_D$ (L is the full length of the plasma). In Fig. 4 the dependences of the relative errors of the instability increments for different modes at radius a are shown. For the fourth mode at $a = 1$ the increment has decreased by a factor of 2, though the number of beam particles on the wavelength equal 50.

The PIC and CIC methods described above have been applied to the solution of two- and three-dimensional problems. In the two-dimensional case, the region in x, y space is divided into rectangular cells and the fields are calculated in the centers of these cells. In the CIC method the charge of each particle is homogeneously distributed, usually on some square (dotted lines in Fig. 5). The hardest problem here is the rapid solution of the Poisson equation. The selection of the optimal method is of high importance, as the problem must

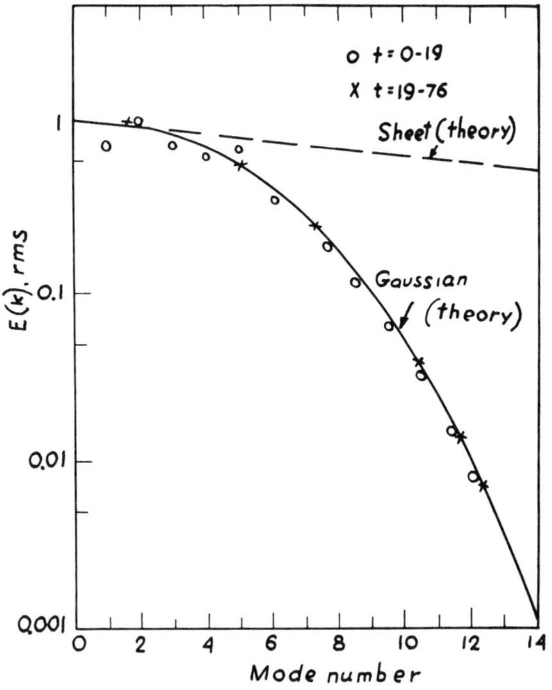

Fig. 3. Noise in a plasma model showing advantages of the "cloud in cell" method.

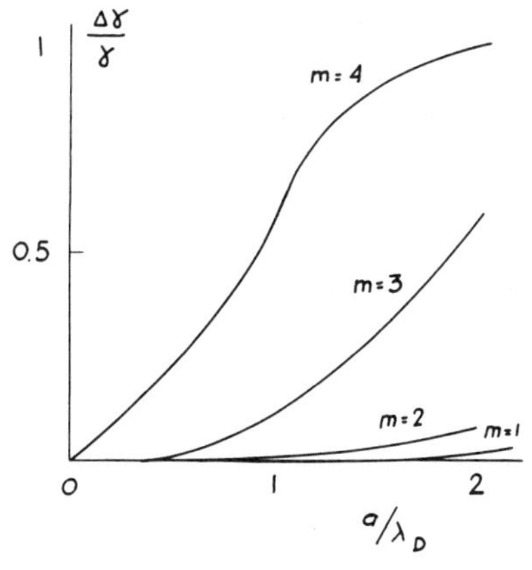

Fig. 4. Relative error in growth rate as a function of mode number and "cloud" size.

be solved for each step in time. If the range of x, y is a rectangle and the boundary conditions are simple enough (e.g., homogeneous or periodic), the fastest methods are the direct ones using the difference Fourier expansion. In the general case it is necessary to apply iteration methods. To accelerate the calculations the upper successive relaxation method is applied. The number of needed iterations is greatly reduced if linear extrapolation from the two preceding steps in time is used as the zeroth approximation. Numerical experiment has indicated that for errors of about 0.1% 10 to 20 iterations are enough. In this case the time required for the field calculations equals 15% of the full computer time.

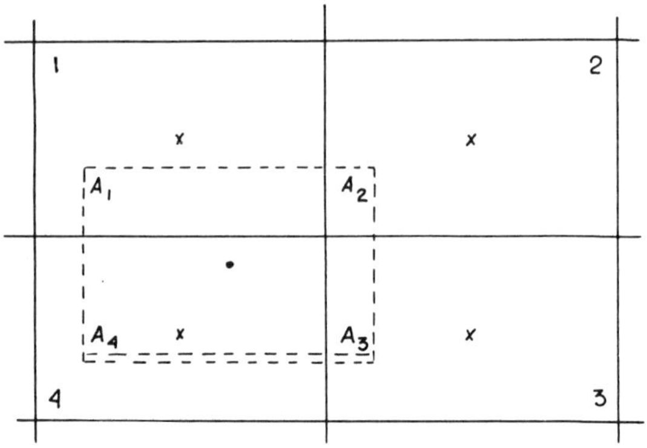

Fig. 5. Cell method in two dimensions.

As an example we shall describe results of the solution of the two-beam instability problem in one-, two-, and three-dimensional models with $L = 100\lambda_D$ (Fig. 6) (1). In the one-dimensional case $20 \cdot 10^3$ particles were used and 64 cells; in the two-dimensional, $80 \cdot 10^3$ particles and 64 x 64 meshes; in the three-dimensional, $332 \cdot 10^3$ particles and 32 x 32 x 32 meshes. In the one-dimensional case, after the increase in the unstable oscillations the stationary BGK waves formed with large vortices in the x, v space. In multidimensional models the vortices rapidly disappear, the oscillation energy monotonically decreases to a level slightly exceeding the initial level of noise in the system and determined by the discreteness of the models.

Let us consider now more briefly the other methods of collisionless plasma simulation. In the one-dimensional case the Vlasov equation has the form

$$\frac{\partial f}{\partial t} + v\frac{\partial f}{\partial x} + \frac{e}{m}E\frac{\partial f}{\partial v} = 0 \tag{2}$$

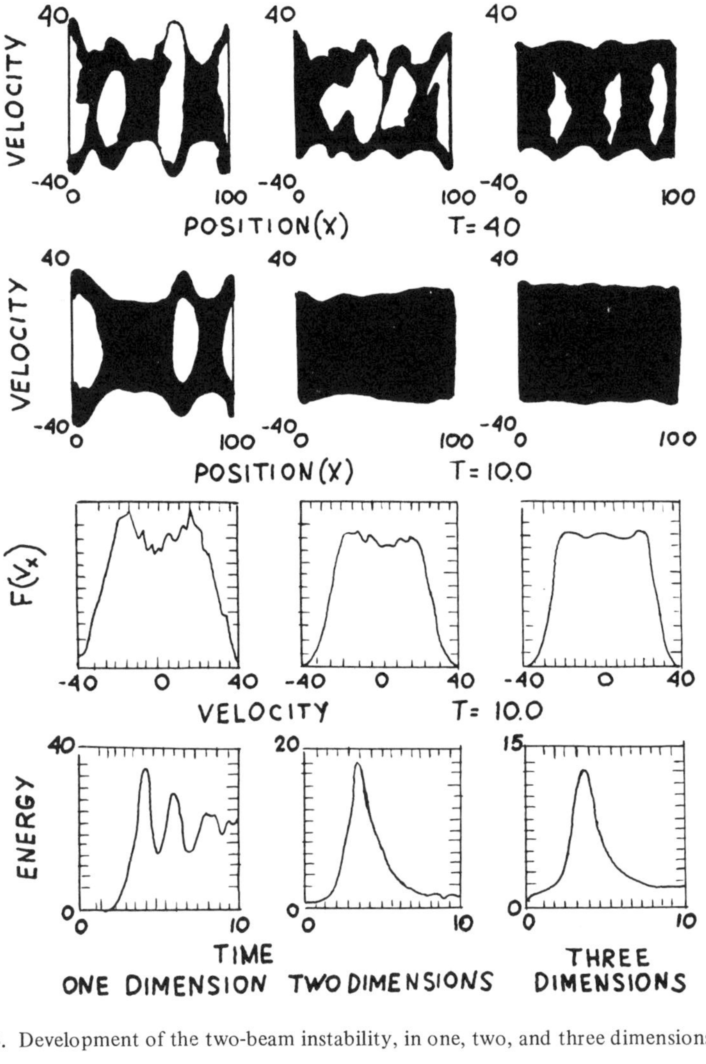

Fig. 6. Development of the two-beam instability, in one, two, and three dimensions.

The very simple case with $E = 0$ and $f = f(x - vt)$ permits us to appreciate the difficulties appearing during the solution of eq. 2. In this case

$$\frac{\partial f}{\partial v} = -t\frac{\partial f}{\partial x} \tag{2a}$$

and the velocity gradients increase without limit in time. This difficulty is the stumbling block for many numerical methods.

B. The Expansion of the Distribution Function in a System of Basis Functions

Usually either Fourier expansion:

$$f(x, v, t) = \frac{1}{2\pi} \sum_n \int_{-\infty}^{\infty} F_n(y, t) \exp\left[i(nk_0 x - vy)\right] dy \tag{3}$$

or Fourier-Hermite expansion:

$$f(x, v, t) = \sum_n \sum_m \exp\left(-\frac{v^2}{2}\right) h_m(v) \exp(ink_0 x) z_{mn}(t) \tag{4}$$

is used.

In the first case the functions $F_n(y, t)$ must satisfy the following system of partial differential equations in

$$\frac{\partial F_n}{\partial t} + nk_0 \frac{\partial F_n}{\partial y} + iy \sum_q E_q(t) F_{n-q} = 0 \tag{3a}$$

In the second case $Z_{mn}(t)$ satisfy the system of ordinary differential equations:

$$\dot{Z}_{mn} + ink_0(\sqrt{m}\, Z_{m-1,n} + \sqrt{m+1}\, Z_{m+1,n}) + \sqrt{m} \sum_q E_{n-q} Z_{m-1,q} = 0 \tag{4a}$$

Here $E_n(t)$ are the Fourier components of the electric field System 3a is rather unpleasant, and very few authors have used expansion 3.

Appreciation of the errors that appeared due to truncation of system 4a in x-space harmonics $(0 \leqslant n \leqslant N)$ is usually not very difficult. The fact is that, if the principal harmonic of perturbation E_n has an amplitude of order ϵ, the secondary harmonics $E_{n\pm q}$ have the order ϵ^q. For this reason the number of x-space harmonics is usually not so high. The problem of truncation of system 4a in velocity space $(0 \leqslant m \leqslant M)$ is more delicate. In this case we have no small parameter due to the difficulty pointed out in eq. 2a. Moreover the terms enclosed in parentheses in eq. 4a compensate for one another in calculations, and if the functions Z_{mn} for $m > M$ are ignored the term Z_{Mn} is strongly excited and then so are the other expansion coefficients. This means in practice that if special measures are not taken, reasonable preciseness in calculations can be obtained only if $\omega_p t \lesssim \sqrt{M}/k_0 Ln$. For this reason usually $M \gg N$ (typical values are $N = 10$, $M = 10^3$).

These difficulties can be overcome by the introduction of the artificial damping in system 4a, which diminishes the highest modes in m: $(\dot{Z})_{coll} = -\nu_c m Z_{mn}$. Experiments have shown that with $\nu_c \sim M^{-1}$ the calculations can be continued to $t \sim M$. For $t > M$ the "collisions" introduced change the physical process.

By means of the method described very many problems have been solved, among them the following:

1. Nonlinear Landau damping.
2. Two-beam instability.
3. Beam-plasma instability.
4. Plasma echo.
5. Landau damping in nonhomogeneous stationary states.

For the last problem the dependence of the Landau decrement on the amplitude of the nonhomogeneous steady state is shown in Fig. 7 (1).

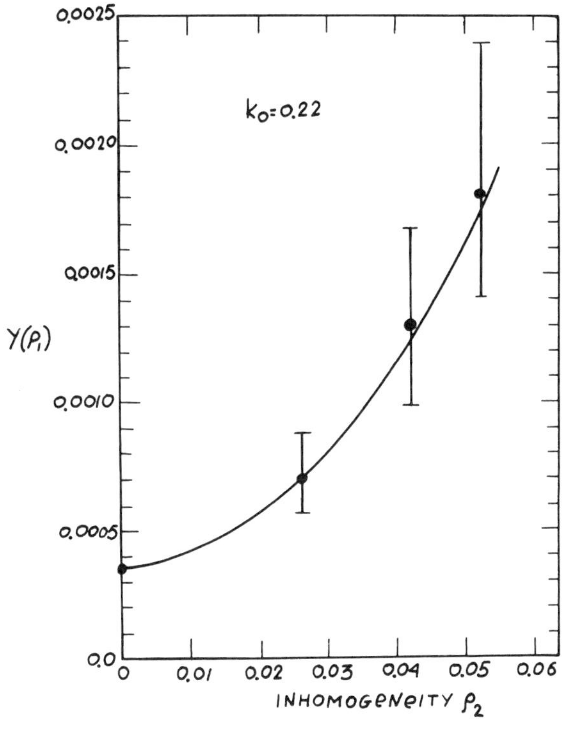

Fig. 7. Influence of inhomogeneity on Landau damping.

We have no information at present about the solution of multidimensional problems by the expansion method.

C. The "Water-Bag" (Lines of Discontinuity) Method

The Vlasov equation can be written in the form $df/dt = 0$ (i.e., $f = \text{const}$ along the Lagrange trajectories). If $f(x, v, 0) = \varphi(x, v)$ is a piecewise constant function, it will keep this character for $t > 0$. For its complete determination it is enough to find only the trajectories passing through the lines of discontinuity C_i of the function $\varphi(x, v)$. The evolution of the lines C_i in time determines the solution of the problem. The number of these lines being small, the number of trajectories required may be many times smaller than the number of particles in the macroscopic particle method. As a result computational time will be shorter.

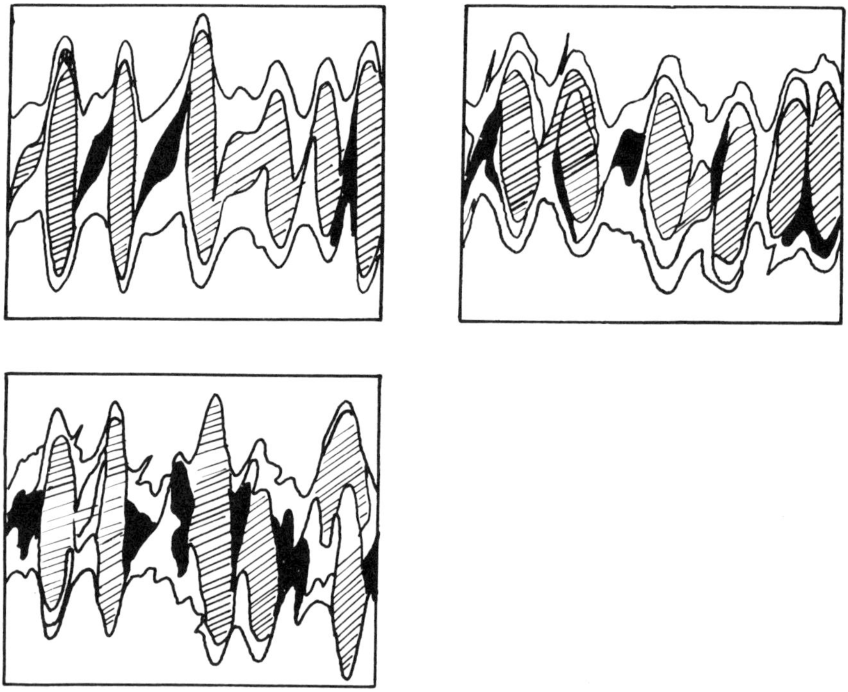

Fig. 8. The two-beam instability by the "Water-Bag Method."

The use of this method is subject to complications of two types.

1. For piecewise constant distributions the resonance particle can be absent. Therefore some kinetic effects can vanish (Landau damping, etc). This difficulty can be overcome by the application of a compound method whereby resonance particles are described by the macroscopic particle method and the rest of the plasma by the water-bag model.

2. The increase of the derivatives with limits v (eq. 2a) leads to very strong twisting of the lines C_i in the plane x, v. Hence one should interrupt computations and change the topology of these lines by hand.

In Fig. 8 (1) the results of calculations for the two-beam problem are shown. The initial distribution was

$$\varphi(x, v) = C \begin{cases} 1, & \begin{cases} v_0 - \epsilon \leqslant v \leqslant v_0 + \epsilon \\ -v_0 - \epsilon \leqslant v \leqslant -v_0 + \epsilon \end{cases} \\ 0 & \text{elsewhere} \end{cases}$$

and had four discontinuities on the horizontal lines v = const in the phase plane x, v. Figure 8 demonstrates that in the nonlinear stage of the process the lines of discontinuity are very strongly twisted. Closed regions in which particles are absent ("holes") form a stable system of vortices in the x, v plane.

D. *Finite Difference Method*

For the simulation of plasma processes fixed meshes in phase space $\mathbf{x}, \mathbf{v}$ are used. Vlasov and Maxwell (Poisson) equations are approximated by the difference equations. A great number of suitable difference schemes are known, though their numerical stability is not always guaranteed. At present there are programs to solve both one- and two-dimensional problems. One of them (Lawrence Radiation Laboratory) uses a mesh with $120 \cdot 10^3$ cells in space (τ, z, v_r, v_z). Very few articles on the finite difference method are now available, and its merits and disadvantages in comparison with other methods have been insufficiently studied.

II. Kinetic Processes in Plasma with Coulomb Collisions

Let us now consider problems of the second type, for which Coulomb collisions are of essential importance.

The Landau collision integral which describes the variation of the distribution function for particles of species, α due to Coulomb collisions with particles of species, β can be written in the form:

$$L_{\alpha\beta}[f_\alpha] = \frac{16\pi^2 e_\alpha{}^2 e_\beta{}^2 L}{m_\alpha{}^2} \frac{\partial}{\partial v_k} \left(\frac{m_\alpha}{m_\beta} \frac{\partial \varphi_\beta}{\partial v_k} f_\alpha - \frac{\partial^2 \psi_\beta}{\partial v_k\, \partial v_e} \frac{\partial f_\alpha}{\partial v_e} \right) \tag{5}$$

Here φ_β and ψ_β are potential functions determined by the distribution function for particles of species β

$$\varphi_\beta(t, \mathbf{v}) = -\frac{1}{4\pi} \int f_\beta(t, \mathbf{v}') \frac{d\mathbf{v}'}{|\mathbf{v}' - \mathbf{v}|}, \qquad \psi_\beta(t, \mathbf{v}) = -\frac{1}{8\pi} \int f_\beta(t, \mathbf{v}') |\mathbf{v}' - \mathbf{v}|\, d\mathbf{v}' \tag{6}$$

According to these formulas, the function φ_β and ψ_β satisfy the equations:

$$\Delta \varphi_\beta = f_\beta, \qquad \Delta \psi_\beta = \varphi_\beta \tag{7}$$

When the time variation of the distribution functions are determined by Coulomb collisions, they satisfy the Landau kinetic equation (Fokker-Planck equation):

$$\frac{\partial f_\alpha}{\partial t} = \sum_\beta L_{\alpha\beta}[f_\alpha] + Q_\alpha \tag{8}$$

where Q_α are sources. These can include terms of two kinds: those that are known functions of time and those that depend on the distribution functions of particles or, in the simplest case, on their densities.

Equations 8 are rather complicated nonlinear integrodifferential equations, and we cannot hope to obtain an analytical solution. It is well known, however, that kinetic processes determined by Coulomb collisions play an important role in many plasma phenomena. Hence computational methods acquire great significance. In the course of the development and application of these methods it is necessary to take into account the following peculiarities of the problems in question:

1. The problems are solved in an unlimited region in velocity space.
2. The representative time scales for electron-electron, ion-ion, and ion-electron interactions due to Coulomb collisions are in the ratio $1:\sqrt{m_i/m_e}:m_i/m_e$.
3. The representative thermal ion and electron velocities are in the ratio $1:\sqrt{m_i/m_e}$.
4. The distribution functions (each in its own scale) are sharply decreasing functions of v, changing by several orders of magnitude with a severalfold increase of velocity.

The ion and electron distribution functions in a magnetized plasma usually possess azimuthal symmetry around the direction of the magnetic field. For problems of this kind the Landau equation has only two independent variables in velocity space, that is, the problems are two-dimensional. Sometimes the problem is reduced to a simpler one-dimensional problem in which the particle distributions are assumed to be isotropic. Therefore we shall start with these problems, the peculiarities of which are displayed when numerical methods are applied.

In the isotropic case the distribution function and the potentials φ and ψ determined by it depend only on $|v|$. Hence after integration of expressions 6 over the angles, they can be written as one-dimensional, not three-dimensional, integrals which have no singularity. The kinetic equation reduced to the variable v becomes rather simple.

Several problems of this kind have been solved by numerical methods. For example, the evolution from a non-Maxwellian initial distribution to the Maxwellian one for a single species of particles, irrespective of their interaction with particles of other species, has been considered, and the energy exchange between ions and electrons with different temperatures has been studied (6).

We will describe a problem of this kind, and on the basis of this example will discuss the questions in which we are interested. This problem is connected with an investigation of the peculiarities of the velocity distribution of fast superthermal ions in a nonisothermal plasma with different temperatures for ions and electrons (10). Under these conditions the particle distribution functions are non-Maxwellian. The corresponding difference is stronger for ions than for electrons, this difference being especially great for superthermal ions. Calculations connected with an investigation of the magnitude and nature of this difference have been carried out.

If the fact that the Maxwellization time for electrons is considerably less than the characteristic time for ion-electron interaction is taken into consideration, the deviation of the electron distribution from the Maxwellian one can be neglected. Hence our problem consisted in the solution of one kinetic equation for ions where electron coefficients were taken to correspond to the Maxwellian distribution. The electron temperature, $T_e(t)$, was determined from the condition of total energy conservation:

$$T_e(t) + T_i(t) = T_e(0) + T_i(0) \tag{9}$$

Such an approach eliminated difficulties connected with the essential difference between the time scales of electron and ion processes and considerably reduced the calculations. As we shall see further, this approach can be applied also to the solution of more complicated nonisotropic problems.

To find a rather subtle effect that was interesting for us on the basis of a fast decrease of the distribution function in the superthermal region, the solution was found in the form

$$f_i(t, v) = f_0(t, v)u(t, v) \tag{10}$$

Here $u(t, v)$ is a new and unknown function, and $f_0(t, v)$ is a Maxwellian ion distribution in which the temperature $T_0(t)$ was found from the differential equation that determined the energy exchange between Maxwellian ions and electrons:

$$\frac{dT_0}{dt} = \frac{8\sqrt{2\pi}\, e^4 \sqrt{m_e} L}{3m_i} \frac{T_e(t) - T_0(t)}{T_e^{3/2}(t)} \tag{11}$$

The introduction of temperature $T_0(t)$ was connected with the fact that the true ion temperature, $T_i(t)$, was unknown. This was determined only in the process of solution through the distribution function, $f_i(t, v)$ determined by the problem.

The integrodifferential equation for the function $u(t, v)$ was integrated by the finite difference method. The three-point implicit scheme usually applied to the solution of parabolic differential equations was used. The results showed that the difference between the ion distribution function and the Maxwellian one is small. For example, the distribution function changes by more than 40 orders of magnitude on a variation of v from 0 to $10\Omega_{T_i}$. At the same time its values differ from the corresponding Maxwellian function when $T_e(0)/T_i(0) = 3$ by the factor which changes from 1.1 to 0.1 (Fig. 9).

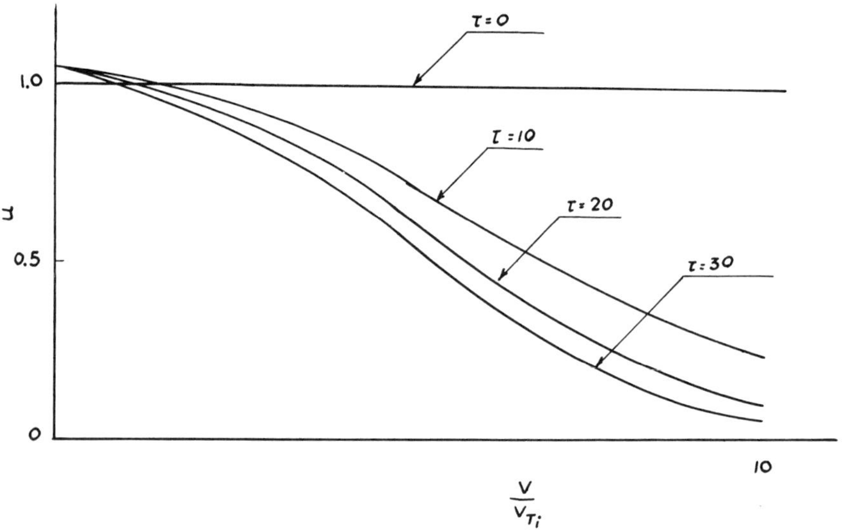

Fig. 9. Relaxation toward a Maxwellian; $n(v)$ versus v as a function of time.

Let us proceed now to a discussion of problems using the Landau equation in the case in which the velocity distribution functions are non-isotropic. But we suppose that the distributions are azimuthally symmetric, that is, depend only on two (not three) variables in the velocity space. We take for these variables the spherical coordinates v and θ.

In the two-dimensional case, as well as the one-dimensional case, problems involving the relaxation of a nonequilibrian initial distribution to Maxwellian one have been solved. But the most important, most discussed, and at the same time most difficult problem is, of course, that of Coulomb losses in mirror machines. It is to the discussion of this problem that we shall now pay attention.

From the mathematical point of view the problem lies in the solution of the coupled equations, eq. 8, for the ion and electron distribution functions. These equations should be solved in the following regions in the velocity space:

$$\sin^2 \theta > \frac{1}{R}\left(1 + \frac{2e_\alpha \phi}{mv^2}\right) \tag{12}$$

Here R is a mirror ratio, and ϕ is an ambipolar potential. If as the result of the higher loss rate for electrons than for ions the plasma is charged positively, eq. 12 defines regions placed outside a hyperboloid of one sheet for ions and a hyperboloid of two sheets for electrons. On the boundary of these regions the corresponding distribution functions should vanish:

$$f_\alpha |_{S_\alpha} = 0 \tag{13}$$

When $\phi = 0$, both regions are transformed into the region outside a cone whose angle θ_0 is determined by the mirror ratio.

Since spatial dependence is not considered in this task, we have no Poisson equation for the determination of the potential θ_0. Therefore some condition for its determination should be formulated supplemental to the problem. Usually the potential is determined from the condition of quasineutrality of plasma, and it regulates in a certain way the loss rate of every component. However, a simpler formulation of the task is possible when the potential is considered to be given—when, for example, it is known from the experimental data.

This problem is very complicated from the point of view of its solution by computer. Therefore we shall discuss methods for its solution with different simplifying suppositions.

The problem is considerably simplified when only a single species (i.e., ions) is considered, and the influence of electrons and the ambipolar potential ϕ is neglected. In this case the ion kinetic equation written in spherical coordinates can be solved by the finite difference method. To achieve this aim ordinary difference schemes can be applied which have been developed for the solution of parabolic differential equations, for example, the alternating-direction implicit (ADI) method. The potential ψ_i determining the coefficients is calculated from the distribution function found in some step by means of numerical integration formulas which, as the result of previous integration through the azimuthal angle, contain the complete elliptic integrals. The problem is slightly simplified by the fact that the coefficients change more slowly than the distribution function and need not be recalculated in each time step.

However, many authors did not go through this valid but rather complicated procedure and instead used a method based on the approximate separation of variables in the equation. The separation of variables is achieved by supposition that the potential ψ_i does not depend on the angle θ. The basis for such a supposition is the smoothing by the integral formula 6 for this function. The solution of the equation simplified in this way is found as

$$f_i(t, v, \theta) = u(t, v)\, \Theta(\theta) \tag{14}$$

it being supposed that in the square term $f^2 \Theta^2(\theta) = \gamma\Theta(\theta)$. Hence one obtained an eigenvalue problem as Legendre's equation for the angle function and a nonlinear integrodifferential equation for the function $u(t, v)$. This equation is analogous to the isotropic equation discussed above, and the solution of such an equation by computer presents no difficulty. Since the distribution function of eq. 14 should be positive, it is necessary to take the first eigenfunction which has a definite sign.

The problem of calculating the parameter γ and normalizing the average potential $\psi_0(t, v)$ which we substitute for the true potential $\psi(t, v, \theta)$ plays an essential role in this model. Comparison of the results of different authors shows that these factors influence rather considerably the results of calculations. These questions were analyzed in the paper of Ben Daniel and Allis (7).

Calculations using the separated variables method are much simpler than those based on the solution of the two-dimensional problem. Results have been produced by different authors in large quantity. Comparison of the results of such calculations with the results of the direct solution of the two-dimensional problem indicates that the true normalization produces an error of about 10 to 20% in the determination of $n\tau$. However, the separated variables method allows consideration only of cases in which the initial distribution function and the sources depend on the angle θ as $\Theta_1(\theta)$. The solutions to problems with another dependence–in particular, with dependence of the delta-function type, corresponding to an injection perpendicular to the magnetic field–cannot be obtained by such a method, and a direct solution of the two-dimensional equation is required (Fig. 10).

A single-species mode does not take into account two physical effects essential for the processes in question, that is, the cooling of ions by electrons and additional ion losses through mirrors owing to the ambipolar potential. However, the total problem mentioned above is too complicated for calculation by computer, because of the different ion and electron time scales and the fact that the collision integral for different species of particles is more complicated than one for a single species of particles. Under these conditions attempts to find a compromise have been undertaken in two directions. One of them is based on further modification of the separated variables method and on substitution of two-dimensional equations by one-dimensional ones. Another is based on the attempt to include the role of electrons in the process without solving their kinetic equation. Let us discuss briefly both of these approaches.

The forced separation of variables in the Landau kinetic equation keeping collisions with particles of the other species does not differ considerably from the one analyzed above. However, with regard to the particles of the second species and to the possible presence of the ambipolar potential, not only the equation but also the domain of the variables resists the separation of variables. In this case we should find loss hyperboloids (eq. 12) of one and two sheets instead of a loss cone. To overcome this new conflict between the wish to separate variables and the possibility of realizing this, Fowler and Rankin (8) made an additional assumption that $\Theta(\theta)$ depended parametrically on v. Such a dependence arises because of the fact that the domain of the angle θ changes with v. As a result the separation constant λ_1, which is determined as the first lowest eigenvalue of the $\Theta(\theta)$ function problem and is included in the $u(t, v)$ function equation, appears to be dependent on velocity v and, the ambipolar potential ϕ.

The second approach to the simplification of the problem is based on considering the electron role without solving the electron equation. Practically speaking, we discussed the idea of such an approach when we discussed the one-dimensional isotropic problem. In the given case the electron distribution

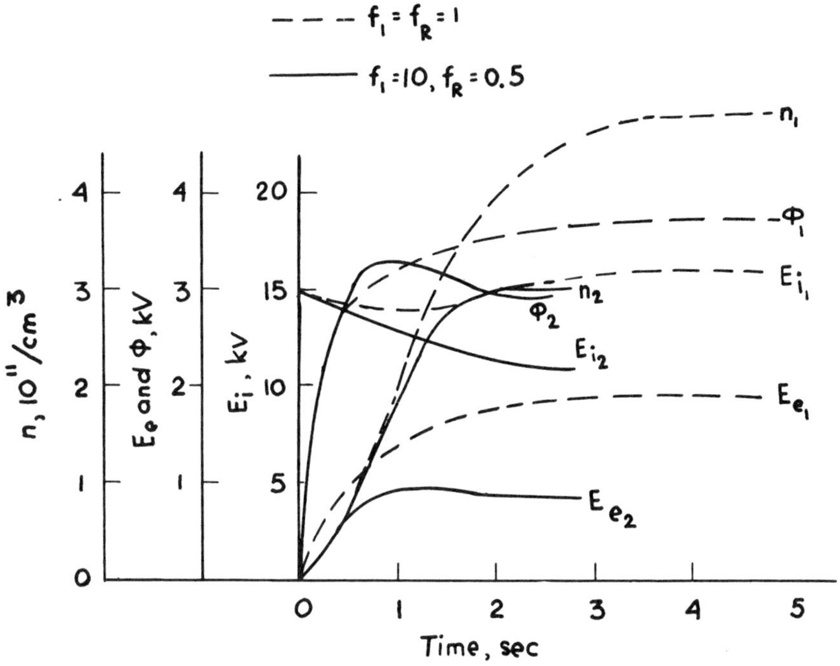

Fig. 10. Dynamics of mirror losses.

function is assumed to be Maxwellian cut by zero in the loss region, and the electron influence on the ions is considered by calculating coefficients with the aid of this distribution function. As for density n_e and temperature T_e, characterizing the electron distribution function, ordinary differential equations determining their time evolution are written (9) from conservation considerations.

It is necessary to note that the ambipolar potential makes the solution of the problem much more complicated, irrespective of the calculation method. It is determined by the method of selection, demands multiple recalculations, and reduces the stability of the calculation procedure.

References

1. Methods in Computational Physics, *Plasma Phys.*, 9 (1970).
2. O. Buneman, *Phys. Rev.*, **115**, 503 (1959).
3. J. Dawson, *Nucl. Fusion*, Suppl., Part 3, p. 1033 (1962).
4. T. P. Armstrong, *Phys. Fluids*, **10**, 1269 (1967).
5. H. L. Berk and K. V. Roberts, *Phys. Fluids*, **10**, 1595 (1967).
6. J. Killeen, W. Heckrotte, and G. Boer, *Nucl. Fusion*, Suppl., Part 1, p. 183 (1962).
7. D. J. Ben Daniel and W. P. Allis, *Plasma Phys*, 4, 31 (1962).
8. T. K. Fowler and M. Rankin, *Plasma Phys.*, 8, 121 (1966).
9. K. D. Marx, *Phys. Fluids*, **13**, 1355 (1970).
10. Y. N. Dnestrovskii and D. P. Kostomarov, *Nucl. Fusion*, **11**, 141 (1971).
11. V. Enalskii and V. Imshennick, *Appl. Math. Techn. Phys.*, No. 1, 1965 (Russian).
12. I. M. Gelfand and others, *J. Comput. Math. Math. Phys.*, No. 2, 1967 (Russian).

Pulsars and Plasma Physics

D. TER HAAR

Magdalen College, Oxford, England

We discuss, on the one hand, how the properties of the interstellar plasma will affect the pulsar signals and, on the other hand, how the properties of the circumstellar plasma in the pulsar magnetosphere may lead to the pulsar emission.

Introduction

Before discussing the relation between plasma physics and pulsar phenomena, we must briefly discuss the properties of pulsar radiation as observed on Earth which are relevant to this discussion. After that, we shall briefly consider what aspects of plasma physics are relevant to pulsar phenomena, and finally we shall discuss the interplay of plasma physics and pulsars.

We have elsewhere (9) given a survey of pulsar data, and we shall mention here only the properties we need for the further argument. Similarly, we shall restrict ourselves here to only part of plasma astrophysics; we have given elsewhere (10) a more extensive review and at the same time provided further references.

I. Pulsar Properties

Pulsars are weak, sporadic, periodic radio emitters (of the 58 pulsars known at the moment of writing only one, the Crab pulsar PSR 0532, emits in the visible range; the X-ray pulsars are as yet not sufficiently well studied for us to say with certainty that they are the same kind of object as the radio pulsars) which were discovered by the Cambridge radioastronomy group. Their periods lie between 0.033 and 4 sec. Their intensity fluctuates strongly. One can divide the fluctuations into three categories: fast fluctuations with a time scale of seconds, fluctuations with a time scale of minutes or hours, and slow fluctuations with a time scale of months. The fast fluctuations show up in that the intensity fluctuates from pulse to pulse. The slow fluctuations are revealed in that the signal sometimes disappears for several weeks; in fact, the discovery of pulsars was delayed because of these fluctuations: Miss Bell had observed the pulsations of PSR 1919 in August 1967, but the intensity of this pulsar did not again become observable until November 1967, when the existence of pulsars was confirmed. As we shall see in Section III, density fluctuations in the interstellar plasma may be responsible for the fluctuations on the medium time scale, but the fast and slow fluctuations are probably intrinsic to the source.

When one integrates a large number of pulses, one obtains for each pulsar a characteristic pulse shape or pulse window. The name pulse window indicates that subpulses will appear only during the part of the period which is covered by the window. The individual pulses consist of subpulses; and while the pulse window characteristically takes up about 4% of the period, that is, typically lasts about 40 msec, the individual subpulses last only a few milliseconds. The pulse profiles, together with the behavior of the subpulses, make it possible to distinguish three classes of pulsar (13, 24): S-type pulsars, which have simple pulse profiles and short periods, and may have large or small degrees of polarization; C-type pulsars, which have complex pulse profiles and long periods, and generally show a large degree of linear polarization; and D-type pulsars, which are weakly polarized and have subpulses that drift across the pulse window.

The subpulses are often very highly polarized, usually linearly. The integrated pulse profiles often show a large degree of linear polarization, but while individual subpulses may have a large degree of circular polarization, the integrated profile only occasionally reveals a net circular polarization. The polarization angle in many cases shows a steady increase or decrease through the pulse window.

If one studies pulsars at different frequencies, one finds that (generally speaking) the profile remains the same, although the separation between different components may vary with frequency (17), and there is also a general broadening of the profile with decreasing frequency, especially for pulsars that are relatively far removed from us. There is also a delay in arrival time, which depends on the frequency and is accurately described by the equation

$$t_1 - t_2 = C\left(\frac{1}{{\omega_1}^2} - \frac{1}{{\omega_2}^2}\right) \tag{1}$$

where t_1 and t_2 are the arrival times of the pulses at frequencies ω_1 and ω_2, while C is a constant.

The spectrum of the pulsar emission generally shows a maximum at a frequency of about 10 GHz, dropping off both at lower and at higher frequencies. The Crab pulsar shows a second maximum at optical frequencies. From the observed flux, the inferred distances (see Section III), and the inferred size of the emitting region one can estimate the brightness temperature T_{br} of the emitting region. One finds that for the radio emission T_{br} is at least 10^{21} °K but may reach a value as high as 10^{31} °K; on the other hand, for the optical emission of PSR 0532, as well as for the X-ray emission of PSR 0532 and the X-ray pulsars Sco X-1, Cyg X-1, and Cen X-3, T_{br} is at most 10^9 °K. This indicates that, while the emission at optical and X-ray frequencies may be thermal emission from relativistic particles, the radio emission must be enhanced in some way: it is usually called coherent emission, the term being used in a rather loose sense to indicate that the emission from a system of N radiating particles exceeds Nu, where u is the emission intensity of a single particle.

Before discussing various aspects of plasma physics, we shall briefly mention some of the conclusions which can be derived from pulsar data about the object that emits the pulses. First of all, a period of about 1 sec indicates an object with a mean density of at least 10^8 g/cm^3. If two stars of equal mass were orbiting around each other in such a way that they would touch (a very unphysical situation), the period τ of the orbital motion would be given by the equation

$$\tau \approx 2.10^4 \bar{\rho}^{-1/2} \text{ sec} \tag{2}$$

where $\bar{\rho}$ is the mean density in grams per cubic centimeter. Similar formulas hold for the periods of the vibrational (pulsational) modes of stars. We are thus led to objects which must be either white dwarfs or neutron stars. The maintaining of a constant period can be due to either orbiting or pulsating or rotating. The first alternative can be ruled out, as it would lead to changes in τ which are several orders of magnitude larger than the observed ones. Also, white dwarf pulsations are ruled out because their periods are too long, and neutron star pulsations because their periods are too short. As white dwarfs would be unstable if they were rotating with periods of the order of 1 sec, by elimination one is led to the conclusion that the central object of a pulsar must be a rotating neutron star. Although neutron stars were studied before the discovery of pulsars, since that discovery this study has been intensified. A typical neutron star has a mass of the order of one solar mass, a radius of about 10 km, and a moment of inertia of 10^{44} to 10^{45} g/cm^2. The dimensions of about 10^6 cm are in agreement with subpulse durations of a few milliseconds.

As a nonmagnetic neutron star would not produce pulsed radiation (there would be no preference for any particular phase in the rotational period), and the same would be true of a magnetic dipole field aligned with the rotational axis, one is led to the oblique rotator model for the pulsars, that is, a neutron star with a magnetic dipole field in which the magnetic axis and the rotational axis make an angle with one another. The magnitude of the dipole field can be estimated from the slowing down of the rotational period. If one equates the loss of rotational energy to the loss of magnetic dipole radiation by a rotating magnetic dipole, one comes characteristically to dipole fields which at the magnetic pole and at the surface reach values of the order of 10^{12} G. The total loss of rotational energy is of the order of 10^{31} ergs/sec. Although this value varies between about 10^{29} and 10^{38} ergs/sec, the maximum field strength varies only between 10^{11} and 4.10^{12} G, but this may be due to observational limitations.

Although originally it was thought that because of the strong gravitational fields at the neutron star surface neutron stars would have no atmosphere, further consideration showed that, for field strengths of the order of those likely to be found for pulsars, electromagnetic forces far exceed gravitational ones. It is likely, therefore, that pulsars are surrounded by magnetospheres with charge densities of the order of 10^{10} cm^{-3} near the neutron star surface. For an investigation of pulsar emission mechanisms we must thus study plasmas with a varying charge density of the order of 10^7 to 10^{10} cm^{-3} and a varying magnetic field of the order of 10^3 to 10^{12} G. The temperature in this plasma at the moment is anybody's guess, but the ambient temperature is often assumed to be about 10^6 °K. Charged particles will be accelerated by the rotating magnetic dipole field, and as a result we must reckon with the possibility that there may be beams of highly relativistic particles in the circumstellar plasma.

II. Plasma Properties

A plasma will produce many effects which can be of relevance to pulsar physics, and we shall briefly consider here a few of them. First of all, we mention the acceleration of particles in a plasma. This occurs essentially through the scattering by fluctuating electric fields. A typical example is the Fermi mechanism for cosmic ray acceleration.

Second, we must consider the interaction of a plasma with electromagnetic radiation. Two aspects are of interest to us here: (*a*) the emission of radiation by the particles in the plasma, and (*b*) the propagation of the radiation through the plasma.

Of the various emission processes we want to mention two in particular: the magneto-bremsstrahlung (synchrotron or cyclotron radiation), which occurs at a frequency ω_H, given by the equation

$$\omega_H = \frac{eH}{\gamma mc} \sim \frac{10^7 H}{\gamma} \text{ Hz} \tag{3}$$

where H is the magnetic field strength in gauss, and γ is the relativistic factor:

$$\gamma = \frac{E}{mc^2} = \left(1 - \frac{v^2}{c^2}\right)^{-1/2} \tag{4}$$

(v = particle velocity, E = particle energy), while e, m, and c are the electron charge, the electron mass, and the velocity of light, respectively; and the plasma oscillations, which occur at the plasma or Langmuir frequency ω_p, given by the equation

$$\omega_p = \left(\frac{4\pi n_e e^2}{m}\right)^{1/2} \sim 10^5\sqrt{n_e}\ \text{Hz} \tag{5}$$

where n_e is the number of electrons per cubic centimeter.

As far as the propagation of electromagnetic radiation through a plasma is concerned, this may be a complicated problem in the case of an energetic plasma where strong plasma turbulence and nonlinear effects have to be taken into account. Although this case is probably important for the magnetospheric phenomena in the immediate neighborhood of the pulsar, we shall confine ourselves at this moment to the case of a dilute plasma and the case in which the influence of the magnetic field may be neglected.

From the Maxwell equations,

$$\operatorname{div} H = 0, \qquad \operatorname{div} E = 0, \qquad \operatorname{curl} E = -\frac{1}{c}\frac{\partial H}{\partial t}, \qquad \operatorname{curl} H = \frac{4\pi j}{c} + \frac{1}{c}\frac{\partial E}{\partial t} \tag{6}$$

the equation for the current density (neglecting ion motion)

$$j = n_e e v \tag{7}$$

and the equation of motion

$$m\dot{v} = eE \tag{8}$$

we easily obtain

$$-\operatorname{div}\operatorname{grad} E = \frac{4\pi n_e e^2}{m} E + \frac{\partial^2 E}{\partial t^2} \tag{9}$$

or, for a plane wave with frequency ω and wave vector $\mathbf{k}$.

$$\omega^2 = \omega_p{}^2 + k^2 c^2 \tag{10}$$

III. Plasma Physics and Pulsars

We must now see how the results of Sections I and II can be brought together. Let us first of all look at some consequences of eq. 10. From this equation it follows that the group velocity v_{gr} of a signal through the dilute plasma will be given by the equation

$$v_{gr} = \frac{d\omega}{dk} \approx c\left(1 - \frac{\omega_p{}^2}{2\omega^2}\right) \tag{11}$$

provided that $\omega_p \ll \omega$, as will be the case for radio signals ($\omega \gtrsim 10^7$ Hz) and $n_e \lesssim 100$ cm^{-3}.

If a broad-band signal passes through a medium of varying density, the components at frequencies ω_1 and ω_2 will arrive at different times t_1 and t_2, where

$$t_1 - t_2 = \left(\frac{1}{\omega_1{}^2} - \frac{1}{\omega_2{}^2}\right)\frac{2\pi e^2}{mc}\int n_e\, dl \tag{12}$$

where the intergration is along the line of sight from the source to the observer. Equation 12 is the same as eq. 1, and we see that from the difference in arrival time of different components we can deduce the dispersion measure, $\int n_e\, dl$, and thus, if we know the average electron density along the line of sight, the distance of the source. In fact, the electron density is different in H I and H II regions, so that the average electron density along the line of sight will differ for different pulsars (21). Once the distances are known, one can reach conclusions about the pulsar distribution in the neighborhood of the Sun and, for instance, conclude that the pulsars are typical high-velocity objects. Conversely, in the few cases where the pulsar distances are known from other evidence, either because the pulsar is related to a known supernova remnant such as the Crab and Vela pulsars, or because 21-cm absorption and emission measurements can pin-point the distance (e.g., PSR 0328 and PSR 2015), one can check whether the assumed average electron densities are reasonable.

Apart from delaying the signal, the plasma also will rotate the angle of polarization of a plane-polarized signal (Faraday rotation), and for different frequencies by different amounts. The formula for the difference in rotational angle, $\theta_1 - \theta_2$, of signals at frequencies ω_1 and ω_2 is given by the equation

$$\theta_1 - \theta_2 = \left(\frac{1}{\omega_1{}^2} - \frac{1}{\omega_2{}^2}\right)\frac{2\pi e^3}{m^2c^2}\int H_{\|} n_e\, dl \tag{13}$$

where $H_{\|}$ is the component of the magnetic field along the line of sight. Combining eqs. 12 and 13, one can find the average value of $H_{\|}$, which is of the order of a few microgauss.

Fluctuations in the electron density will produce both a broadening and a steepening of the pulsar profile, as well as fluctuations in the intensity. The physical reason for the steepening and broadening is simple: it is the superposition of the parts of the signal which arrive directly and those which arrive after following a longer path because of scattering. The scattering will also lead to intensity fluctuations. A recent discussion of the whole class of problems of this kind can be formed in a paper by Lang (16), where references are given to the literature. Here we discuss only the question of scintillations. We mention in passing that a study of the scintillations caused by the density fluctuations produced by the solar wind was the object of the new radiotelescope at Cambridge with which the pulsars were discovered.

Let us consider a plane wave. Its time and space dependences are given by a factor

$$\exp\left[\frac{2\pi i}{\lambda}(r - v_{\text{gr}}t)\right] \tag{14}$$

where λ is the wavelength. If the wave passes through an irregularity with extra density Δn_e and of linear size a, there will be a phase shift, $\delta\varphi$, given by the expression

$$\delta\varphi \sim \frac{2\pi}{\lambda}\cdot\frac{1}{2}\,\frac{4\pi e^2}{m}\,\Delta n_e\cdot\frac{\lambda^2}{4\pi^2c^2}\cdot c\cdot\frac{a}{c} = \frac{e^2}{mc^2}\lambda a\,\Delta n_e \tag{15}$$

where we have used eq. 11.

If the wave passes through a path length L, the root-mean-square phase shift $\Delta\varphi$ will be

$$\Delta\varphi \sim \sqrt{L/a}\,\delta\varphi \sim \frac{e^2}{mc^2}\sqrt{La}\,\lambda\,\Delta n_e \tag{16}$$

and because of the phase shifts there will be scattering of the wave; the average scattering angle, θ_{sc}, will be of the order

$$\theta_{\text{sc}} \sim \frac{\lambda}{2\pi}\,\frac{\Delta\varphi}{a} \sim \frac{e^2}{mc^2}\sqrt{L/a}\,\Delta n_e\,\lambda^2 \tag{17}$$

In order for these fluctuations to show a diffraction pattern, which will be revealed as scintillations due to the motion of the observer through the pattern, it is necessary (a) that $\Delta\varphi > \pi$, or

$$\Delta n_e\sqrt{a} \gtrsim \frac{mc^2}{e^2}\,\frac{1}{\lambda\sqrt{L}} \tag{18}$$

and (b) that several beams interact, which means that $L\theta_{\text{sc}} > a$, or

$$\frac{\Delta n_e}{a^{3/2}} \gtrsim \frac{mc^2}{e^2}\,\frac{\sqrt{L}}{\lambda^2} \tag{17}$$

For pulsars for which $\lambda \sim 0.1$ cm, and $L \gtrsim 10^{20}$ cm, we find that conditions (18) and (19) can be satisfied for relatively small values of Δn_e and a of the order of

10^{12} cm. The time scale τ of the scintillations will be of the order of

$$\tau \sim \frac{\lambda}{\theta_{sc} v} \tag{20}$$

where v is the relative velocity of the source and the observer. From this equation one finds that this effect can produce scintillations with a time scale of hours or days, but that pulse-to-pulse variations cannot be explained by this mechanism.

Path differences from different paths will be of the order of $L\theta_{sc}{}^2$, corresponding to a phase difference of the order of $L\theta_{sc}{}^2/\lambda$. Correlations between scintillations over a bandwidth $\Delta\lambda$ will thus occur, provided (C is a constant)

$$\Delta\lambda \sim \frac{\lambda^2}{L\theta_{sc}{}^2} \sim CL^{-2} \tag{21}$$

As L is known from the dispersion measure, one can check this result for different pulsars, and relation 21 is found to hold.

That the broadening of the pulse profile (and not an intrinsic property of the radiation) is responsible for the fall in flux below about 100 MHz is made even more likely by the recent work by Higgins, Komesaroff, and Slee (12), who found at 80 MHz a radio point source at the exact position of PSR 0833 with a flux which lies exactly on a curve, extrapolated through the data at 635 and 1420 MHz.

We now must, in conclusion, consider the influence of the pulsar magnetosphere. There are two problems. First of all, assuming that electromagnetic radiation originates somewhere in the magnetosphere, we must investigate what happens subsequently to this radiation. Secondly, we want to study possible mechanisms through which the radiation can be produced.

The behavior of electromagnetic radiation in the magnetosphere depends on the dielectric tensor, ϵ_{ij}, which determines the refractive index n. If, for a wave of wave vector k and frequency ω, we write for n:

$$n = \mu + \frac{ic\kappa}{2\omega} \tag{22}$$

μ is the real refractive index, and κ the linear absorption coefficient which is related to the optical depth τ_{opt}:

$$\tau_{opt} = \int \kappa \, ds \tag{23}$$

where the integral is along the path of the ray. If η is the volume emissivity, the intensity I of the radiation which reaches us along a certain path will be given (under certain simplified conditions) by the equation

$$I = \int \frac{\eta \exp(-\tau_{opt}) \, ds}{\mu^2 \kappa} \tag{24}$$

We note in passing that one can derive similar equations for the other Stokes parameters which determine the polarization of the radiation.

The determination of the dielectric tensor is a complicated problem in plasma physics, especially when a magnetic field is present, and we shall not touch upon it here. Once it is known, one can use eq. 24 to study, for instance, the angular dependence of radiation from, say, an isotropic source emerging from an atmosphere in which there are a magnetic field and a density gradient. Even very simple-minded calculations show (8) that beaming will result and that the resultant emission outside the magnetosphere will be polarized, independently of the polarization of the original radiation.

Let us finally turn to the difficult problem of the generation of the pulsar radiation, especially of the radio emission. We must now first of all decide where the emission takes place. On this point there are essentially two schools of thought. One of them, represented by Gold, Smith, and Zheleznyakov, believes that the emission takes place well away from the pulsar at the so-called velocity-of-light cylinder, where a point corotating with the pulsar would attain a velocity equal to c (the distance of this cylinder clearly depends on the period of the pulsar). In these theories the beaming is a relativistic effect. The second school of thought, represented by Chiu and Manchester, believes that the emission takes place essentially close to the surface of the neutron star. It seems to us that, while emission far from the neutron star surface may be able to produce the pattern shown by S-type pulsars, the persistence of the complex pulse window of C-type pulsars indicates a complex magnetic field pattern which is retained over long periods and which would be difficult to envisage as occurring anywhere except near the surface.

We shall not discuss the relativistic beaming effect (which anyhow falls outside the topic of our review) or the curvature emission invoked by Pacini and Rees (20), Sturrock (22), Tademaru (23), and Goldreich and Keeley (6), although it may well be true that the criticism by Ginzburg and Zheleznyakov (5) is too severe [see Goldreich, Pacini, and Rees (7); see, however, also Bonch-Osmolovskii et al. (1)]. We shall instead concentrate on possible plasma mechanisms which may lead to the radio emission of pulsars with a power of the order of $10^{28\pm 2}$ ergs/sec. Several authors have looked into this question [Ginzburg, Zheleznyakov, and Zaitsev (5a), Kaplan and Tsytovich (14, 15), Coppi and Ferrari (3)], but I shall mention here only some recent work in Oxford by Buckee (2) and Miranda (18), who have made a careful quantitative study of various possibilities. Buckee has used Tsytovich's (25, 26) approach to a strongly turbulent plasma, while Miranda has based his work on the methods discussed by Harris (11), in which transport equations are written down, using quantum-mechanical transition probabilities. It is interesting to note that, even though doubt has been expressed about the validity of some of Tsytovich's equations, which use a kind of random phase approximation, the results obtained by the two different methods are essentially the same. The plasma considered by Buckee has a density of 10^{10} cm^{-3}, a magnetic field of about 10^{11} G, and a temperature of about 10^6 °K (this is a widely used value, but it would be helpful to determine this temperature in some kind of self-consistent manner). Miranda uses slightly different values, as he also considers plasma processes at a distance of about one stellar radius (in this case ~10^6 cm) from the neutron star surface, where the density and the field may both have dropped by one or two factors of 10.

We shall only mention some of the basic results of those calculations. First of all, it is found that, in the strongly magnetized plasma surrounding the pulsar emission at radio frequencies with a sufficiently high power will occur only if a beam of relativistic particles is present, as is also suggested by Ginzburg, Zheleznyakov, and Zaitsev and by Coppi and Ferrari. Miranda suggests a beam density of about 10^8 cm^{-3} and a beam velocity of about 10^9 cm/sec. Second, it is found that emission takes place at right angles to the magnetic field, that is, predominantly in the (magnetic) equatorial plane and with linear polarization. This is in excellent agreement with the observational results of the Jodrell Bank group [Lyne, Smith, and Graham (17)]. Miranda's analysis shows that the only mechanism which produces enough is Compton scattering by plasma waves; the plasmon + plasmon → photon interaction is too weak. It may be, however, that some of the conclusions will have to be modified when we consider a relativistic plasma.

It is interesting to note that Norman (19) has shown from an analysis of the resonance conditions of the dielectric tensor that the fact that the pulse spectrum is unchanged across the pulse window indicates that the magnetic dipole axis must be at right angles to the rotational axis.

References

1. A. G. Bonch-Osmolovskii, V. G. Makhan'kov, V. N. Tsytovich, and V. G. Shchinov.
2. J. W. Buckee, *Plasma Physics* (in press). Oxford D. Phil thesis (to be published in *Advances in Plasma Physics*).
3. B. Coppi and A. Ferrari, *Astrophys. J.*, **161**, L65 (1971).
4. V. L. Ginzburg, *The Propagation of Electromagnetic Waves in Plasmas*, Pergamon Press, Oxford, 1970.
5. V. L. Ginzburg and V. V. Zheleznyakov, Comments, *Astrophys. Space Sci.*, **4**, 464 (1970).
5a. V. L. Ginzburg, V. V. Zheleznyakov, and V. V. Zaitsev, *Usp. Fiz. Nauk*, **98**, 201 (1969); (trans.) *Sov. Phys. Usp.*, **12**, 378 (1969).
6. P. Goldreich and D. A. Keeley, *Astrophys. J.* (in press, 1971).
7. P. Goldreich, F. Pacini, and M. J. Rees, Comments, *Astrophys. Space Sci.* (in press, 1971).
8. S. Grounds, Unpublished Oxford D. Phil. thesis.

9. D. ter Haar, *Phys. Rept.*, **3** (in press, 1971).
10. D. ter Haar, *Advan. Solid State Phys.*, **11** 273 (1971).
11. E. G. Harris, *Advan. Plasma Phys.*, **3**, 157 (1969).
12. C. S. Higgins, M. M. Komesaroff, and O. B. Slee, *Astrophys. Lett.*, **9**, 75 (1971).
13. G. R. Huguenin, R. N. Manchester, and J. H. Taylor, *Astrophys. J.* (in press, 1971).
14. S. A. Kaplan and V. N. Tsytovich, *Plasma Mechanisms of Emission in Pulsars*, Lebedev Institute Preprint, 1970.
15. S. A. Kaplan and V. N. Tsytovich, *Plasma Astrophysics*, Nauka, Moscow, 1971.
16. K. R. Lang, *Astrophys. J.*, **164**, 249 (1971).
17. A. G. Lyne, F. G. Smith, and D. A. Graham, *Mon. Notic. Roy. Astron. Soc.*, **153**, 337 (1971).
18. L. C. M. Miranda, Oxford College thesis (to be published in *Advances in Plasma Physics*).
19. C. A. Norman, article to be published.
20. F. Pacini and M. J. Rees, *Nature*, **226**, 622 (1970).
21. A. J. R. Prentice and D. ter Haar, *Mon. Notic. Roy. Astron. Soc.*, **146**, 423 (1969).
22. P. A. Sturrock, *Astrophys. J.*, **164**, 529 (1971).
23. E. Tademaru, *Astrophys. Space Sci.*, **12**, 193 (1971).
24. J. H. Taylor and G. R. Huguenin, *Astrophys. J.*, **167**, 273 (1971).
25. V. N. Tsytovich, *Theory of a Turbulent Plasma*, Atomizdat, Moscow, 1971.
26. V. N. Tsytovich, *Introduction to the Theory of Plasma Turbulence*, Pergamon Press, Oxford, 1971.

Instabilities in Semiconductor Plasmas

MAURICE GLICKSMAN

Brown University, Providence, Rhode Island, USA

Abstract

Instabilities observed in semiconductors are reviewed, with emphasis on the helical instability and the pinch effect. Observations of microwave emission from indium antimonide are briefly summarized, and a number of theoretical explanations presented, corresponding to a number of different experimental configurations. Recent calculations of the helical instability under conditions of large plasma current (large self-magnetic field) are discussed and compared with experimental studies in indium antimonide.

Introduction

As in the case of gas discharges and gaseous plasmas, the experimentalist in solids has found that oscillations are relatively difficult to remove from the detecting equipment when he wishes to make other observations. Fortunately, however, in most cases such oscillations are weak in solids. Therefore the study of semiconductors and their application to practical devices has been little influenced by such problems, in contrast with the difficulties in the application of gaseous plasmas to the harnessing of fusion energy.

The plasma we deal with in the semiconductor can be either of an *immobile* type (free carriers of one sign of charge, compensated for by an equal density of ions fixed in the lattice) or of a *mobile* type, with equal densities of electrons and holes. The instabilities discussed in this paper are those that arise in a mobile solid-state plasma, with approximately equal densities of electrons and holes. There are instabilities of immobile plasmas [one (1) very useful, named the Gunn effect after its discoverer], but these will not be discussed in this review.

Although this conference is devoted to plasma theory, this paper will be restricted to a discussion of the theories of instabilities in mobile semiconductor plasmas which bear a close relationship to observations. A great many interesting calculations will not be included, since the space available for this paper is limited. There are a number of reviews of instabilities in semiconductor plasmas which should be consulted for more detail and a broader treatment (2–8).

A major part of this paper is devoted to one much-studied instability, the helical density instability. Even though this has been investigated in detail, recent results of our research indicate an incompleteness about the previous work, and some real questions about its application to many of the experiments.

I. Helical Density Instability

In 1958 Ivanov and Ryvkin reported (9) the observation of oscillations in the current passing through germanium slabs in the presence of a magnetic field. Further experimentation (10–12) showed that the same phenomenon was observed in other semiconductors (indium antimonide and silicon) and could be made very strong. In some experiments (12) the oscillating component of the current was as much as 70% of the total steady current; this was so impressive that the device was called an "oscillistor."

A theory successful in explaining these observations *and* predicting characteristics not known at the time was developed (13) as a modification of the basic treatment of the helical density instability of Kadomtsev and Nedospasov (14). This theory explained the major features of the observations and predicted that the oscillations were of a rotating, helical nature. This was quickly verified in further experiments (15–16).

Basically, the instability is the result of the interaction of carrier flow in the semiconductor with the external magnetic field. A perturbation of the density leads to radial and azimuthal electric fields (because of the carrier flow in the applied electric field), and these interact with the magnetic field to build up the perturbation in competition with the diffusion flow, which attenuates it. At sufficiently strong magnetic fields or strong current flows, the perturbation grows.

Until recently, all theoretical treatments relied heavily on the initial work of Kadomtsev and Nedospasov (14). Refinements of the theory have been published by a number of authors (17–35), and we sketch a simple approach to the calculation in what follows.

For simplicity in this discussion, we shall assume a semiconductor of cylindrical shape, containing an electron-hole plasma, with electron density $n(r)$ and hole density $p(r)$. The axial direction is assumed infinite for our initial discussion. The current densities $\mathbf{J}_e$ and $\mathbf{J}_h$ are then functions of the applied electric field $\mathbf{E}$ and magnetic field $\mathbf{B}$, as governed by the equations†

$$\mathbf{J}_e = ne\mu_e\mathbf{E} + eD_e\nabla n - \mu_e\mathbf{J}_e \times \mathbf{B} \tag{1}$$

$$\mathbf{J}_h = pe\mu_h\mathbf{E} - eD_h\nabla p + \mu_h\mathbf{J}_h \times \mathbf{B} \tag{2}$$

$$\frac{\partial n}{\partial t} = \frac{1}{e}\nabla \cdot \mathbf{J}_e + \gamma n \tag{3}$$

$$\frac{\partial p}{\partial t} = -\frac{1}{e}\nabla \cdot \mathbf{J}_h + \gamma p \tag{4}$$

$$\nabla \cdot \mathbf{E} = \frac{e}{\epsilon}(p - n) \tag{5}$$

The carrier mobilities μ and diffusion coefficient D have subscripts c and h, representing electrons and holes, respectively. The parameter γ represents the bulk generation and recombination rate; it may be a function of the density (making the equations nonlinear) when bimolecular processes are involved in the bulk recombination. When bulk generation is important, as in the case of impact ionization as the plasma source, γ will also be a function of the electric field.

Equations 1 and 2 may be solved for the average drift velocities, $\mathbf{v}_e$ and $\mathbf{v}_h$, of the electrons and holes, and these values substituted into the continuity eqs. 3 and 4. We then have three equations for the three variables, n, p, and the electric potential, $V(\mathbf{E} = \nabla V)$. In general the magnetic flux density $\mathbf{B}$ will include an external magnetic field and the magnetic field due to the current distribution in the plasma. The latter involves an integral over the density and thus makes the equations nonlinear and complicated to solve.

The boundary conditions to be applied include the condition of flow to the surface of the semiconductor: the current density in the radial direction is zero, but the particle flow to the surface is governed by the rate of recombination of particles at the surface, given by the surface recombination velocity s:

$$n_t v_{er}\,|_{r=a} = s(n - n_{\text{eq}})\,|_{r=a} \tag{6}$$

Here a is the radius of the semiconducting cylinder, and n_{eq} is the carrier density in the sample in equilibrium.

† Temperature gradients are neglected.

We first solve these equations for the steady-state condition, that is, for $\partial/\partial t = 0$. Most published studies deal with the case in which the magnetic field of the current can be neglected, and we treat this case first. We have recently investigated the case in which the self-magnetic field is considered, and this will be discussed later.

Quasineutrality is assumed, that is, the difference $n_0 - p_0$ is assumed approximately equal to the net density of fixed ions in the crystal, with small deviations supporting the plasma potential which must be present when there is a density gradient. If the steady state is assumed to have azimuthal symmetry, and the equation for the density linearized (18) by neglecting higher-order terms, the solution is a Bessel function chosen to satisfy eq. 6.

We now turn to the full equations which come from eqs. 3 and 4. To solve them, previous treatments have assumed that the equations are separable in the time and space variables and that there are eigensolutions of the form

$$\begin{aligned} n &= n_0 + n_1(r, z) \exp(-i\omega t + ikz + im\phi) \\ p &= p_0 + p_1(r, z) \exp(-i\omega t + ikz + im\phi) \\ V &= V_0 + V_1(r, z) \exp(-i\omega t + ikz + im\phi) \end{aligned} \tag{7}$$

In the earlier work (13, 14) the functions n_1, p_1, and V_1 were assumed to be Bessel functions of the form $J_1(\beta r)$. Holter (18) used finite Hankel transforms in the manner of Johnson and Jerde (36) but then retained only the first term in the expansion, which is also J_1 for the case $m = 1$. The spatial terms in eqs. 3 and 4 may then be space-averaged, finally yielding dispersion relations for the frequency ω, the wave vector k, and the phase factor m, with E_{0z} (the applied electric field) and B as parameters. It should be noted that the densities and potential have been assumed to have the same form of spatial and time dependence.

The plasma is found to be stable for $m = 0$. Otherwise, the behavior depends on the type of plasma and the values of the applied electric and magnetic fields. When the plasma is mobile ($n_0 = p_0$), an instability sets in for sufficiently large electric and/or magnetic fields, and it is of the absolute (37) kind, as shown by Hurwitz and McWhorter (21). If compensating ions are present in the semiconductor, the instability that sets in is initially of the convective type but becomes absolute at high applied fields. For parallel electric and magnetic fields, the sign of m also depends on the fixed charge present in the semiconductor. For $n_0 = p_0$ and for $p_0 > n_0$ the helix rotates in the $m = +1$ direction, while for $n_0 > p_0$ the rotation is in the $m = -1$ direction for small magnetic fields ($\mu B \ll 1$). For strong magnetic fields, the rotation is always in the $m = +1$ direction. The wave propagates in the direction of the minority carrier drift, and the direction of rotation reverses from the above when the magnetic field and current are antiparallel. All of these characteristics are in agreement with observation (15, 16).

If there are no restrictions on the wavelength caused by finite sample length, we expect that the instability sets in at the lowest values of the applied fields which allow the driving forces to overcome the diffusion damping. Results of such calculations are presented in Figs. 1, 2, and 3 for the case of an intrinsic mobile plasma (13), and in Figs. 4 and 5 for the semiconductor plasma with a neutralizing ion density present (18, 19). Figure 1 presents the values of threshold electric and magnetic fields, plotted with dimensionless variables $y_e = \mu_e B$ for the magnetic field and $E = \mu_e E_0 a/D_e$ for the electric field for

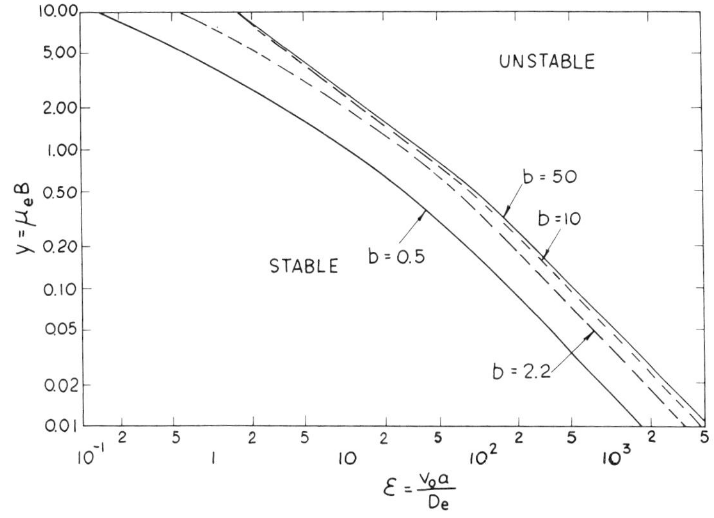

Fig. 1. Plots of the threshold curves for the helical density wave instability for a mobile plasma in a cylinder of radius *a*.

several plasmas with different ratios of electron-to-hole mobilities, $b = \mu_e/\mu_h$. The plasma is stable for values below the curves, and unstable to an $m = 1$ instability for values above the curves.

Figure 2 shows the effect of surface conditions on the instability, for the case $b = 2.2$, which is appropriate for germanium at room temperature. The plot is of the minimum wave vector for which the instability will occur, as a function of the applied magnetic field. The parameter δ is the ratio of surface plasma density to the density at the center of the cylinder, and is thus zero for the case of infinite surface recombination. We see that the instability sets in at smaller wave vectors when the radial density gradient is decreased (δ increased).

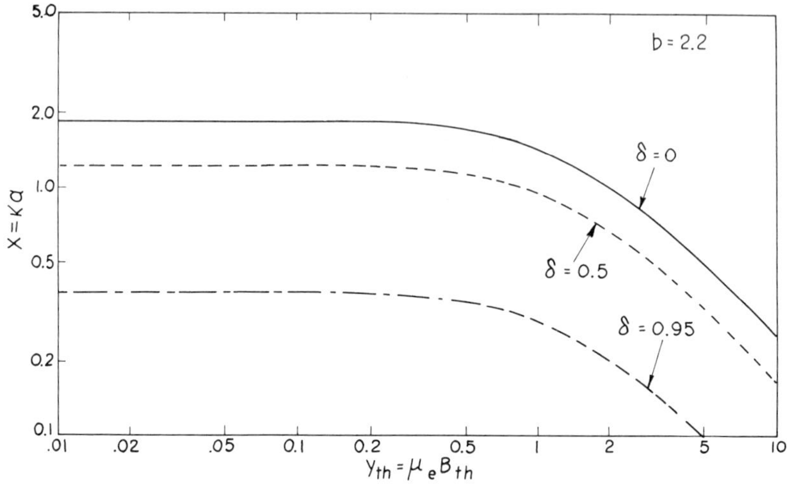

Fig. 2. Effect of surface recombination on the minimum values of dimensionless wave vector *ka* for the helical density wave instability plotted as a function of threshold magnetic field. The parameter δ is the ratio of plasma density at the surface to that at the center.

Figures 3 to 5 are plots of the dimensionless frequency, $\Omega = a^2\omega/D_e$, as a function of the square of the magnetic field ($y' = y_e y_h$), for three different types of plasma. Figure 3 is for the mobile plasma, with $n_0 = p_0$. We see that the threshold value of frequency first increases and then decreases with applied magnetic field. It should be noted that the electric field for this plot is the

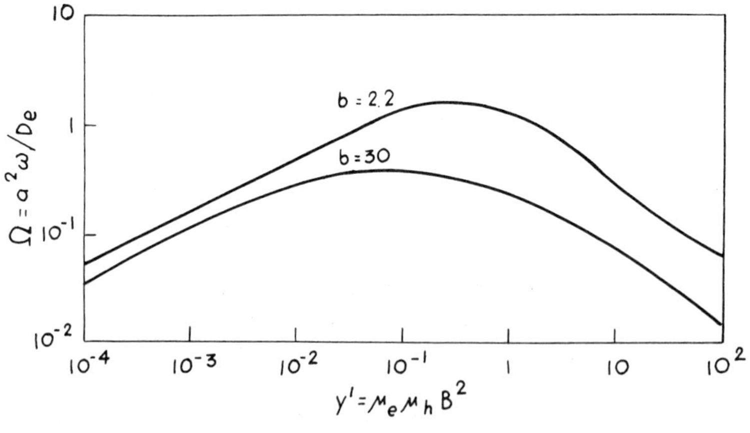

Fig. 3. The frequency of the helical density wave at threshold, as a function of magnetic field, for a mobile plasma, $n = p$.

"threshold" electric field, and is thus changing as the magnetic field is being changed.

Figure 4 is a similar plot, but for the case in which the electron density is n_0 and the hole density $p_0 = n_0 + N$, where N is the net ion density in the

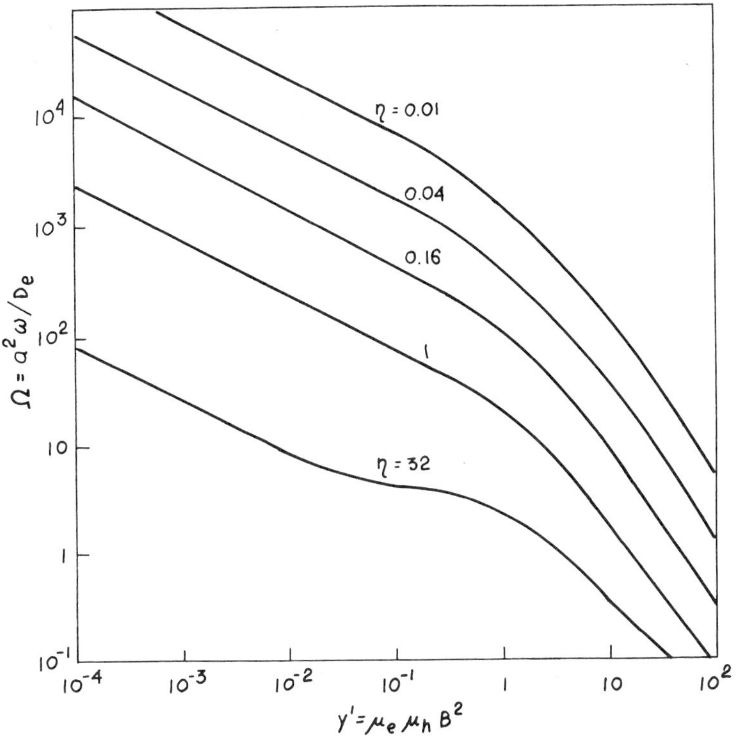

Fig. 4. The frequency of the helical density wave at threshold, as a function of magnetic field, for a mixed mobile and immobile plasma. Here η is the ratio of total electron density to the density of negative fixed ions (p-type material).

semiconductor. The parameter of the curves is the ratio of injected (or added) density to equilibrium density, $\eta = n_0/N$. We see that the frequencies are very much larger than in the mobile plasma case, when the mobile plasma density is a small fraction of the fixed ion density. The frequency is a monotonically decreasing function of the threshold magnetic field in that situation. The calculations were made for $b = 2.2$, corresponding to a germanium crystal at room temperature.

Figure 5 shows the same calculations for a plasma where the electron density ($n_0 = p_0 + N$) is larger than the hole density. In this case $\eta = p_0/N$. We see that the rotation is opposite in phase ($\omega < 0$) at low magnetic fields, but changes sign at higher ones.

Tests of the predictions of the theory have been, on the whole, qualitative or semiquantitative. The general behavior in Fig. 1 for the threshold electric and magnetic fields is seen in experiments (12, 16, 38, 39), but the magnitude of the fields will depend on the injection levels and the surface conditions, and these are not well known. Generally, agreement with experiments in germanium

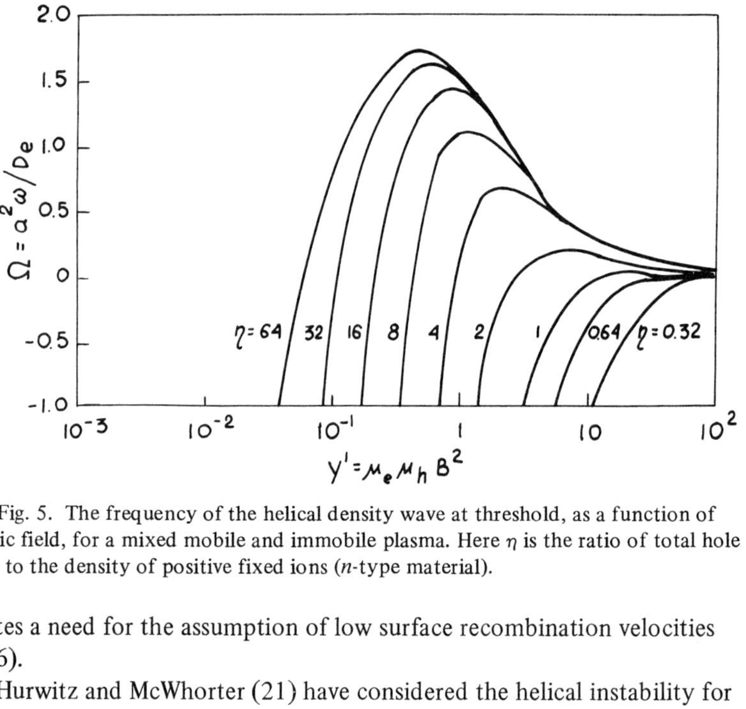

Fig. 5. The frequency of the helical density wave at threshold, as a function of magnetic field, for a mixed mobile and immobile plasma. Here η is the ratio of total hole density to the density of positive fixed ions (n-type material).

indicates a need for the assumption of low surface recombination velocities (13, 16).

Hurwitz and McWhorter (21) have considered the helical instability for injected plasmas, with the surface playing a key role in its control of the presence of a density gradient transverse to the electric and magnetic field directions. In the limit of small magnetic fields, the theory can be presented in a simple fashion; the threshold electric and magnetic fields are inversely proportional to each other, with the proportionality factor dependent on the carrier densities. In careful experiments in which the injected plasma densities were known and controlled, good agreement with the theory (20, 21) was obtained. Further studies of the same surface-controlled plasma, but with an explicit rapid variation of the density in the direction of propagation, have been made by Schulz (24) and applied to some experiments (40) with success.

The most sensitive character of the helical density waves is the frequency. Experiments in germanium done early were in rough agreement as to the magnitude of the observed frequencies (13, 16), although the difficulty with the "unknown" actual densities and surface conditions made a valid comparison difficult. Figure 6 shows a comparison of theory with experiment, from the work of Hurwitz and McWhorter (21), for the threshold frequency as a function

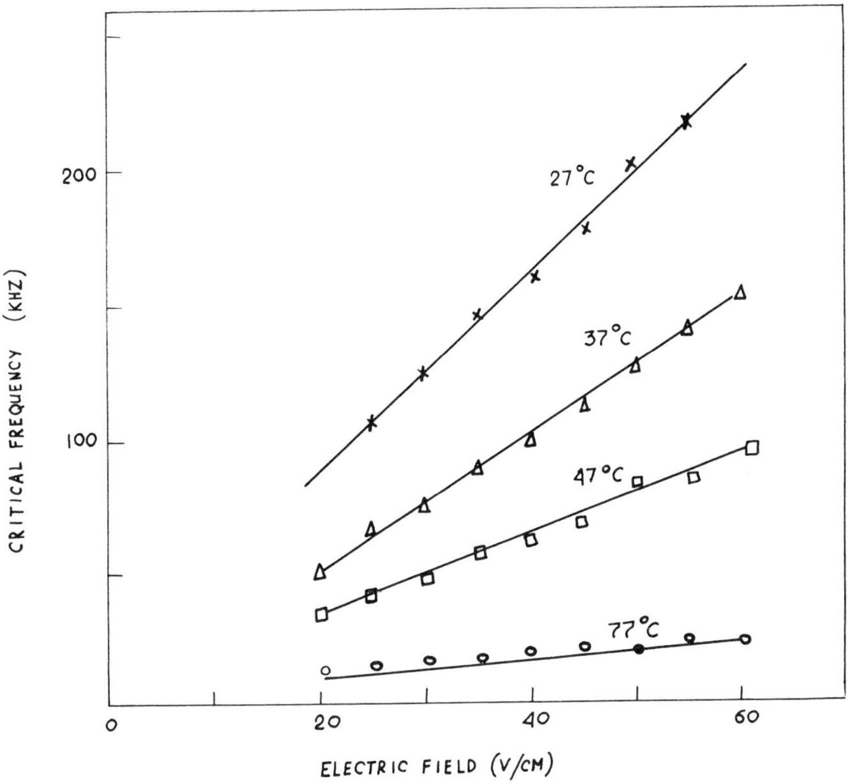

Fig. 6. Critical frequency as a function of applied electric field, for several temperatures in germanium.

of the applied electric field in germanium; the agreement between experiment and theory is excellent. In this case the material is extrinsic, and the variation with temperature shown in the figure is due to changes in the mobility with temperature. Figure 7 shows the measured (21) growth rate of a signal added to a germanium bar for conditions under which the signal should be attenuated slightly or amplified, the frequency applied being that appropriate for the growth of the helical instability. Good agreement with theory is seen. Vikulin and his coworkers (41) have investigated the case of nearly intrinsic germanium, where the frequency (in the limit of small magnetic fields) is expected to be proportional to the magnetic field (21); they verify this behavior. Dubovoi and Shanskii (41a) also check theory well in germanium, noting that in short samples wavelength restrictions modify the threshold conditions.

In indium antimonide, with a completely mobile plasma produced by impact ionization, measurements of the growth rate in time of the internally

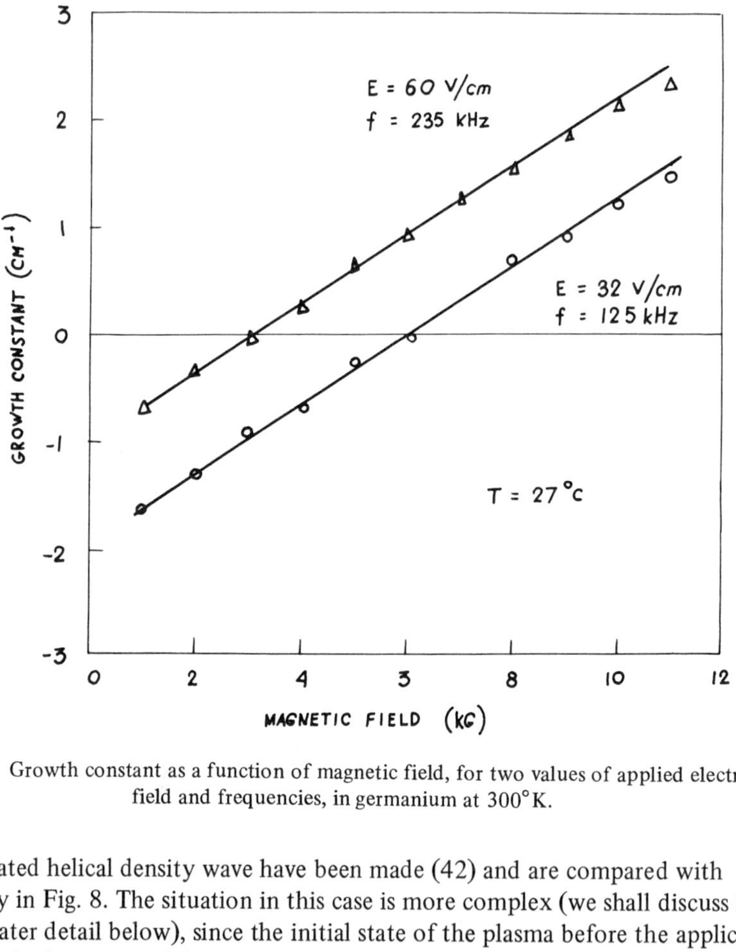

Fig. 7. Growth constant as a function of magnetic field, for two values of applied electric field and frequencies, in germanium at 300°K.

generated helical density wave have been made (42) and are compared with theory in Fig. 8. The situation in this case is more complex (we shall discuss it in greater detail below), since the initial state of the plasma before the application of a pulsed magnetic field is believed to be a pinched one. Nonetheless there is some agreement with a simple theory as sketched above (13). There is experimental evidence for higher-order modes, $m = 2$, being present. The measured threshold fields are also in general agreement, with some allowance for the finite surface recombination velocity.

Thus rather good evidence is available for the character of such an instability and its presence in a number of semiconductor plasmas. One feature which has been observed is distinctly nonlinear, that is, the hysteresis effect. Since the helical density wave involves a rotating current, when it has finite amplitude it may add a finite component to the magnetic field applied to the plasma. As a result, the external magnetic field necessary to maintain the helical

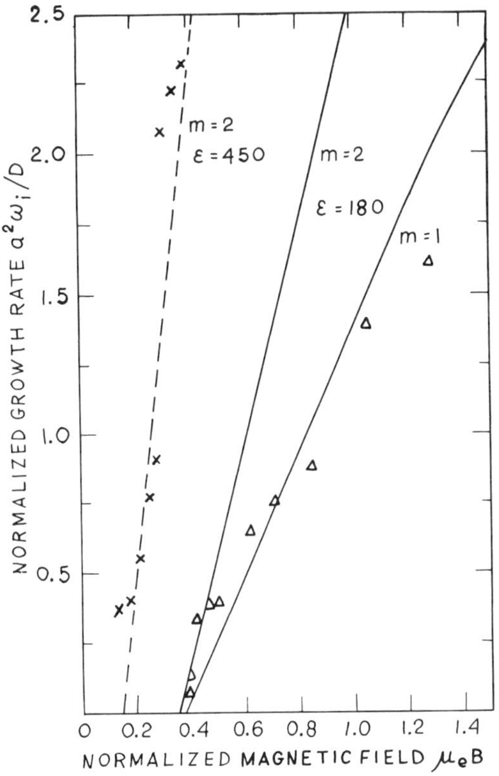

Fig. 8. Growth rate of the helical density wave in indium antimonide, for an impact-ionized plasma, as a function of magnetic field.

wave is reduced from the level necessary to initiate it. The reduction of field can be large; this has been seen in germanium (40) and also in indium antimonide (43–45), where a reduction of 50% in the external sustaining field was observed. Such a paramagnetic character to the plasma was also calculated using a theory for a finite helical density wave (23, 25, 26), but the comparison of theory and experiment has not been completely successful, since the calculated effect is about an order of magnitude less than that observed. However, the theoretical treatments have been performed with several approximations; one which may be an important limitation is the neglect of the self-magnetic field of the plasma current. Since the effect investigated is due to the "self-magnetic field" of the finite instability current, it seems clear to the author that the self-magnetic field of the plasma density distribution must be included. Holter and Johnson (25) make some estimate of the limitation on their calculation due to this neglect (which leads to a restriction to currents much less than 1 A in indium antimonide), and the limit they derive is indeed a serious restriction on the theory.

Refinements of the simple theory discussed here have been investigated for a number of special situations, and references to other studies employing much the same basic assumptions are given in one of the reviews (8). Here we turn now to some serious deficiencies in the theory that have not yet been treated in detail.

One which is most troublesome is neglect of the self-magnetic field, that is, neglect of the zero-order density and current distribution inhomogeneities due to the current flowing in the plasma. Another that is of concern is the practice, throughout the many theoretical treatments (13, 14, 25) which have been published, of ending up assuming solutions to the equations for the perturbed density and potential (eqs. 7), instead of presenting solutions to these equations.

In an attempt to resolve some of these problems Barrowclough and the present author (46) have attempted a solution of the equations for the plasma to investigate the helical density instability, using numerical methods. Although we do not yet have satisfactory solutions, some results worth noting have been obtained.

We start with the same equations (eqs. 1-5), but include the self-magnetic field in B. The equation for the electron density then has the form, for the steady state ($\partial/\partial t = 0$) and a mobile plasma ($n \cong p$),

$$\left(\mu_e n \frac{\partial V}{\partial r} - D_e \frac{\partial n}{\partial r} + \mu_e n \frac{\partial V}{\partial z} y_{e\theta}\right) \frac{\partial}{\partial r}\left(\frac{1}{1 + y_e{}^2}\right)$$
$$+ \frac{1}{r(1 + y_e{}^2)}\left(\mu_e n \frac{\partial V}{\partial r} + \mu_e n r \frac{\partial^2 V}{\partial r^2} + \mu_e r \frac{\partial n}{\partial r}\frac{\partial V}{\partial r} - D_e \frac{\partial n}{\partial r}\right.$$
$$\left. - D_e r \frac{\partial^2 n}{\partial r^2} + \mu_e n \frac{\partial V}{\partial z} y_{e\theta} + \mu_e r \frac{\partial n}{\partial r}\frac{\partial V}{\partial z} y_{e\theta} + \mu_e n r \frac{\partial V}{\partial r}\frac{\partial y_{e\theta}}{\partial r}\right) = \gamma n \tag{8}$$

where

$$y_e{}^2 = \mu_e{}^2 B_z{}^2 + y_{e\theta}{}^2, \qquad y_{e\theta} = \frac{\mu_e \mu_0}{r}\int_0^r J_z r'\, dr' \tag{9}$$

Here $y_{e\theta}$ is the azimuthal magnetic field due to the current, and we have assumed azimuthal symmetry and infinite length (i.e., homogeneity in the axial direction). The boundary condition that there is no radial electric current gives the following equation:

$$n\frac{\partial V}{\partial r}=\frac{D_e(1+y_h^2)-(1+y_e^2)D_h}{(1+y_h^2)\mu_e+(1+y_e^2)\mu_h}\frac{\partial n}{\partial r}-\frac{(1+y_h^2)\mu_e y_{e\theta}-(1+y_e^2)\mu_h y_{h\theta}}{(1+y_h^2)\mu_e+(1+y_e^2)\mu_h}\,n\frac{\partial V}{\partial z} \quad (10)$$

Substitution of this expression in eq. 8 will give us an equation for n alone, which can then be solved to obtain the steady-state plasma distribution. This differs from the Bennett (47) solution for the pinch in the inclusion of generation terms, and the requirement of a boundary condition of finite determined density at the surface (eq. 6). A calculation for the density distribution in an impact-ionized indium antimonide pinch has recently appeared (48) and should be appropriate for higher currents; in Holter's calculation (48) the Fermi degeneracy of the electrons in the plasma was found to balance the pinching force at very small radii. However, the available experimental evidence (49) (an indirect measurement of the pinched plasma radius, calculated from the plasma inductance) does not show radii as small as Holter estimates from his numerical calculations.

The numerical results for the radial distribution of plasma at small currents (0.5 to 1.5 A for indium antimonide) are shown in Fig. 9. These were calculated for a surface recombination velocity (50) of 10^3 m/sec. We see appreciable concentration of the plasma at currents of 1 A and above, indicating that the self-magnetic field may not be ignored at such currents. As the current is increased, in our isothermal theory the plasma collapses in the classical Bennett pinch. We will discuss this situation briefly a little later, in our discussion of the pinch effect instabilities in semiconductor plasmas.

We then used the solutions of Fig. 9 (obtained by an iterative numerical technique) as the steady-state distributions and looked for solutions to eqs. 3 and 4 having the form of eqs. 7. The frequency was assumed to be complex, and the electron and hole transport-continuity equations were investigated for solutions of the helical wave form, with the applied longitudinal magnetic field as parameter. As boundary conditions we required that the perturbed density (helical wave) be zero on the axis, since there are no perturbing forces there; we found that arbitrary conditions on the gradient of the perturbed density gave no differences in the resulting solutions for the dispersion of the wave, but affected only the magnitude of the perturbed densities and potential. Arbitrary values of the perturbed density at the surface could then be matched with no effect on the frequency or wave vector.

Our technique of solution was not exact, however, since we solved a truncated set of equations—and found that, under some conditions, we could not find single eigenvalue solutions. This may be the result of our truncation

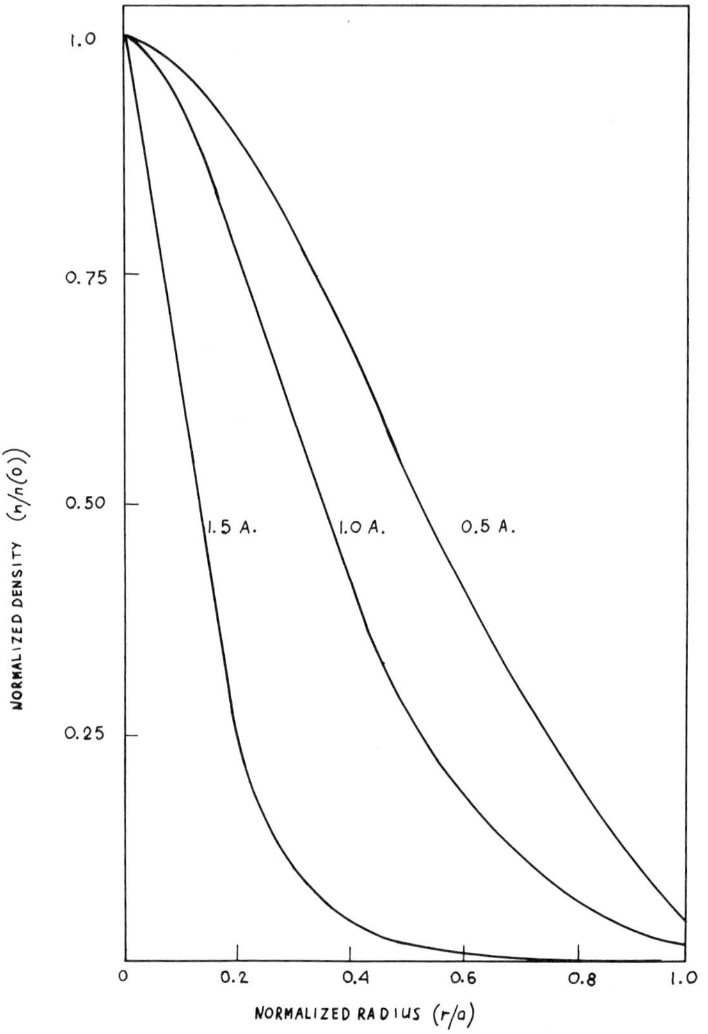

Fig. 9. Spatial distribution of plasma in indium antimonide, for currents of 0.5, 1.0, and 1.5 A. An applied longitudinal magnetic field had a negligible effect on this distribution, for fields up to 1000 G.

(including the assumption that n_1, p_1, and V_1 are real). However, there *are* solutions for the case in which the initial functions are Bessel functions; the strong variation with radius in our functions, due to the self-magnetic field, and the self-magnetic field terms in the equations for the instabilities may lead to the difficulty in finding single eigenvalue solutions that are unstable. What occurred in our approximate numerical solution was the appearance of a radial dependence to the (assumed) eigenvalue, ω. This occurred with the imaginary part of ω only, when the pinched plasma was used as the steady-state n_0, but the self-magnetic field terms were omitted from eqs. 3 and 4; the real part of ω showed such behavior also when the self-magnetic field terms were included.

Since the assumed form is incorrect, the actual solution is at most of limited validity, in the limit of vanishing times, since some spatial derivatives of the terms exp $(-i\omega t)$ were ignored. With this restriction understood, we looked at what an average of the plasma wave parameters over radius would look like, as a function of the applied magnetic fields, currents, and wave vector. The result of such averaging gave the following behavior:

1. The presence of an instability (at large *ka*) in the plasma in the limit of vanishing external magnetic field is observed, provides that the pinch is not too strong. This has to be investigated in more detail. There is evidence that a weak longitudinal magnetic field stabilized this instability. In the limit of a vanishing self-magnetic field, the plasma is stable.
2. For given external magnetic and electric fields, the increase of the current (the self-magnetic field) stabilized the plasma against the helical density instability. This agrees with observation (42). It then requires an external magnetic field larger than the self-magnetic field to cause the helical density instability to grow.
3. An increase in the growth rate of the instability with increasing applied longitudinal field is predicted, also in agreement with observation (42).

A proper analysis of the equations, with self-magnetic fields included, is in progress—with the approximations of truncated variables and equations removed. Since the observations indicate a very well defined wave instability (42), a proper solution should show that one of the frequency eigenvalues should be unstable alone, at threshold.

II. The Pinch Effect Instabilities

The pinch effect is well known in gaseous plasmas and has been studied in much detail. Its occurrence in semiconductor plasmas was noted by Steele and the present author (51), and a number of treatments of the theory† have been published (48, 52–59). It is an instability in two respects. First, generally the time of contraction of the electron-hole plasma (there can be very little contraction of the immobile plasma) is long, being determined by the carrier ambipolar mobility, so that the plasma initially has a rather more homogeneous

† See Ref. 8, pp. 388–398, for earlier references to theoretical work.

form and subsequently changes to the concentrated shape distinctive of the final pinched state. Since, for moderate current in the limit of infinite time, this process no longer involves a time-varying property, it is a "temporary" instability—a change from an unstable state to the steady state. This process was studied by the author in experiments (60) in indium antimonide, which seem to be satisfactorily explained by a simple theory (61).

The second respect is most interesting. It involves the instabilities which occur once the pinching process has been set in motion and which persist (theoretically) indefinitely with time. Although there have been experimental reports of oscillation (2) in indium antimonide, only in the most recent work (62, 63) has an attempt been made to compare theory with observation. In these studies, oscillations were observed to occur after "pinching" of the electron-hole plasma. Such oscillations are seen on long time scales, that is, microseconds after the production of the plasma, as compared to the time for the pinching process itself, which is less than 0.1 μsec for the currents employed in the indium antimonide experiments. Thus appreciable heating of the lattice occurs. As a result, a magnetothermal theory for the plasma is used (62, 64), in which the plasma is assumed to have different temperatures at different parts of the crystal (a hot inner core, and a cooler outer core), but to be in a steady-state balance with the lattice at the same temperature. Such a plasma can be shown to be unstable to $m = 0$ mode oscillations, and these are observed to be present with a wavelength equal to twice the sample length. The theory has not been extended to higher-order modes, $m = 1$ and larger, although these have been observed in the high-density plasma in indium antimonide. There are several consistency checks in the theoretical interpretation which provide some confidence in regard to its application to the experiments. It is also clear that the actual situation is considerable more complex than the theoretical model at the present time. Many experimental results are available, awaiting the development of a more complete theory. Recent measurements of the growth rate of these oscillations and evidence for their feedback stabilization have been presented (63).

III. Other Instabilities in Solid-State Plasmas

A great deal of attention has been given to many observations over the past seven years of high-frequency radiation from the semiconductor indium antimonide, upon the application of electric and (sometimes) magnetic fields. The studies were reviewed several years ago (7). At that time it was noted that a number of sources for some of the observations had been definitely identified. One of these (7) is the acoustoelectric interaction: in this case, the electric current interacts with sound waves in the crystal, generating acoustic domains containing high-frequency components, and these in turn generate high-frequency electromagnetic radiation, which can be observed when an external circuit is available to couple out such radiation. However, the interaction is not very strong, so that samples with appreciable length seem necessary for this mechanism to be used to explain the broad emission seen.

A second mechanism that has been effectively tied to experiments involves the two-stream, collision-induced instability of Swartz and Robinson (65, 66). They are able to show good correlation between theory and experiment, under appropriate geometries and field arrangements for the observation of their instability, which yields a narrow frequency range of oscillations when some geometrical parameter that defines the wavelength is present.

More recently, there have been several other suggestions for the broad-band emission seen (megahertz up to 100 GHz). These involve basically the generation of noise due to current fluctuations (67, 68), located either in the vicinity of the contacts or at some inhomogeneity in the bulk of the indium antimonide, and connected with the impact ionization process locally. Although the evidence is suggestive, it is not as yet conclusive.

Observations of narrow-band emission have also been related to the helical density wave instability (69–71). In these cases, some estimate is made of the small size of filaments (10^{-5} m) necessary, and plausibility arguments are then presented for relating the necessity of these to their possible existence. The dependence of the observations on threshold parameters is similar to that of the helical density wave. However, definite correlation of the observations with such waves requires observation of the postulated small filaments. This has not yet been available, and would indeed be very difficult, since the filament size is of the order of the wavelength of the recombination radiation which should be emitted if an electron-hole plasma is present there.

A suggestion made recently (72) that some (or all) of the observations may be related to the generation of galvanomagnetic waves in the semiconductor plasma merits further study. Since such a wave includes a well-defined mechanism for coupling to the external field, a study of the effect of the coupling arrangement on the observed radiation may be helpful in determining the presence of such waves in indium antimonide.

IV. Conclusions

From the discussion above, it is clear that the study of oscillations and instabilities in solids, particularly in semiconductors, offers a very fruitful technique for understanding the physical nature of plasma instabilities and testing the validity of theory. As an experimentalist, I continue to be astonished by the applicability of theory which is so hedged with assumptions that its validity is generally questionable. The increasing sophistication of experimental studies continues to stimulate the theorist to seek valid solutions to models which more closely approximate the real world. A continuation of this process will be important in the future, as a means of contributing to the development of plasma theory in general.

References

1. J. B. Gunn, in *Plasma Effects in Solids, Paris, 1964*, Dunod, Paris, 1965, p. 199.
2. M. Glicksman, in *Plasma Effects in Solids, Paris, 1964*, Dunod, Paris, 1965, p. 149.
3. B. Ancker-Johnson, in *Semiconductors and Semimetals*, ed. by R. C. Willardson and A. C. Beer, Vol. 1, Academic Press, New York, 1966, p. 379.
4. J. Bok, *J. Phys. Soc. Japan*, Suppl., **21**, 685 (1966).
5. M. C. Steele and B. Vural, *Wave Interactions in Solid State Plasmas*, McGraw-Hill, New York, 1969.
6. H. Hartnagel, *Semiconductor Plasma Instabilities*, American Elsevier, New York, 1969.
7. M. Glicksman, *IBM J. Res. Dev.*, **13**, 626 (1969).
8. M. Glicksman, *Solid State Phys.*, **26**, 275 (1971).
9. Yu. L. Ivanov and S. M. Ryvkin, *Zh. Tekh. Fiz.*, **28**, 774 (1958).
10. J. Bok and R. Veilex, *C. R. Acad. Sci. Paris*, **248**, 2300 (1958).
11. R. D. Larrabee and M. C. Steele, in *Proceedings of the International Conference on Semiconductor Physics, Prague, 1960*, Czechoslavian Academy of Sciences, Prague, 1960, p. 227.
12. R. D. Larrabee and M. C. Steele, *J. Appl. Phys.*, **31**, 1519 (1960).
13. M. Glicksman, *Phys. Rev.*, **124**, 1655 (1961).
14. B. B. Kadomtsev and A. V. Nedospasov, *J. Nucl. Energy*, Part C, **1**, 230 (1960).
15. F. Okamoto, T. Koike, and S. Tosima, *J. Phys. Soc. Japan*, **17**, 804 (1962).
16. T. Misawa and T. Yamada, *Jap. J. Appl. Phys.*, **2**, 19 (1963).
17. T. Misawa, *Jap. J. Appl. Phys.*, **1**, 67, 130 (1962).
18. Ø. Holter, *Phys. Rev.*, **129**, 2548 (1963).
19. Ø. Holter, *Arbok for Universitet I Bergen*, Mat-Naturv. Serie No. 8, 1963.
20. L. E. Gurevich and I. V. Ioffe, *Fiz. Tverd. Tela*, **6**, 445 (1964).
21. C. E. Hurwitz and A. L. McWhorter, *Phys. Rev.*, **134**, A1033 (1964).
22. V. V. Vladimirov, *Zh. Eksp. Teor. Fiz.*, **49**, 1562 (1965).
23. Ø. Holter and R. R. Johnson, *Phys. Lett.*, **27A**, 642 (1968).
24. M. Schulz, *Phys. Status Solidi*, **25**, 521 (1968).
25. Ø. Holter and R. R. Johnson, *Phys. Rev.*, **183**, 503 (1969).
26. Ø. Holter, *Phys. Norvegica*, **4**, 125 (1969).
27. E. Z. Meilikhov, *Fiz. Tekh. Poluprovod.*, **4**, 237 (1970).
28. I. A. Gilinskii, *Fiz. Tekh. Poluprovod.*, **4**, 377 (1970).
29. Yu. I. Tsipivka, G. F. Karavaev, and I. M. Vikulin, *Fiz. Tekh. Poluprovod.*, **4**, 508 (1970).
30. M. K. Balakirev and S. V. Bogdonov, *Fiz. Tverd. Tela*, **12**, 1414 (1970).
31. L. E. Gurevich, I. V. Ioffe, and A. A. Tursunov, *Fiz. Tverd. Tela*, **12**, 1566 (1970).
32. B. V. Paranjape and K. C. Ng, *Phys. Rev.*, **B2**, 413 (1970).
33. L. E. Gurevich, I. V. Ioffe, and A. A. Tursunov, *Fiz. Tekh. Poluprovod.*, **4**, 1232 (1970).

34. V. V. Vladimirov, L. V. Dubovoi, and V. F. Shanskii, *Zh. Eksp. Teor. Fiz.*, **58**, 1580 (1970).
35. V. V. Vladimirov, *Zh. Eksp. Teor. Fiz.*, **59**, 162 (1970).
36. R. R. Johnson and D. A. Jerde, *Phys. Fluids*, **5**, 988 (1962).
37. A. Bers and R. J. Briggs, MIT Research Laboratory of Electronics, Quart. Prog. Rept. No. 71, 122–131, 1963; R. J. Briggs, *Electron Stream Interaction with Plasmas*, MIT Press, Cambridge, Mass., 1964.
38. B. Ancker-Johnson, in *Proceedings of the International Conference on Semiconductor Physics, Exeter, 1962*, Institute of Physics and the Physical Society, London, 1962, p. 141.
39. S. Nakashima and Y. Noguchi, *Jap. J. Appl. Phys.*, **2**, 307 (1963).
40. M. Schulz and E. Voges, *Z. Angew. Phys.*, **25**, 141 (1968).
41. I. M. Vikulin, L. L. Lyuze, and V. A. Presnov, *Fiz. Tekh. Poluprovod.*, **2**, 1138 (1968).
41a. L. V. Dubovoi and V. G. Shanskii, *Zh. Eksp. Teor. Fiz.*, **56**, 766 (1969).
42. K. Ando and M. Glicksman, *Phys. Rev.*, **154**, 316 (1967).
43. B. Ancker-Johnson, *Appl. Phys. Lett.*, **3**, 104 (1963).
44. B. Ancker-Johnson, *Phys. Rev.*, **135**, A1465 (1964).
45. B. Ancker-Johnson, *Phys. Rev.*, **135**, A1423 (1964).
46. G. F. Barrowclough and M. Glicksman, article to be published.
47. W. H. Bennett, *Phys. Rev.*, **45**, 890 (1934).
48. Ø. Holter, *Phys. status solidi*, (*b*) **45**, 433 (1971).
49. H. Morisaki, *Phys. Lett.*, **31A**, 211 (1970).
50. A. R. Beattie and R. W. Cunningham, *Phys. Rev.*, **128**, 533 (1962).
51. M. Glicksman and M. C. Steele, *Phys. Rev. Lett.*, **2**, 461 (1959).
52. I. I. Boiko, V. V. Vladimirov, and A. P. Shotov, *Zh. Eksp. Teor. Fiz.*, **57**, 567 (1969).
53. V. V. Vladimirov and V. M. Chernousenko, *Zh. Eksp. Teor. Fiz.*, **58**, 1703 (1970).
54. V. V. Vladimirov, *Fiz. Tverd. Tela*, **12**, 1296 (1970).
55. Y. L. Igitkhanov and B. B. Kadomtsev, *Zh. Eksp. Teor. Fiz.*, **59**, 155 (1970).
56. V. N. Dobrovolskii, O. S. Zinets, and Fan van An, *Fiz. Tverd. Tela*, **12**, 2945 (1970).
57. Yu. N. Yavlinskii, *Fiz. Tverd. Tela*, **12**, 3385 (1970).
58. I. I. Boiko, *Phys. status solidi*, **43**, 483 (1971).
59. I. I. Boiko and V. I. Pipa, *Fiz. Tverd. Tela*, **13**, 579 (1971).
60. M. Glicksman and R. A. Powlus, *Phys. Rev.*, **121**, 1659 (1961).
61. M. Glicksman, *Jap. J. Appl. Phys.*, **3**, 354 (1964).
62. W. S. Chen and B. Ancker-Johnson, *Phys. Rev.*, **B2**, 4468, 4477 (1970).
63. B. Ancker-Johnson, H. J. Fossum, and A. Y. Wong, *Phys. Rev. Lett.*, **26**, 560 (1971).
64. J. E. Drummond and B. Ancker-Johnson, in *Plasma Effects in Solids, Paris, 1964*, Dunod, Paris, 1965, p. 173.
65. G. A. Swartz and B. B. Robinson, *J. Appl. Phys.*, **40**, 4598 (1969).
66. G. A. Swartz, *J. Appl. Phys.*, **40**, 5343 (1969).
67. J. E. King, *J. Appl. Phys.*, **40**, 5350 (1969).
68. A. Bers and R. N. Wallace, *Phys. Rev. Lett.*, **25**, 665 (1970).
69. T. Musha, S. Ohnishi, and M. Hirakawa, *Phys. Rev. Lett.*, **22**, 1254 (1969).
70. M. Tacano and S. Kataoka, *J. Appl. Phys.*, **42**, 494 (1971).
71. M. Tacano and S. Kataoka, *J. Appl. Phys.* (to be published).
72. L. E. Gurevich and I. V. Ioffe, *Zh. Eksp. Teor. Fiz.*, **58**, 2047 (1970); **59**, 1409 (1970).

Author Index

Numbers in parentheses are reference numbers and show that an author's work is referred to although his name is not mentioned in the text. Numbers in *italics* indicate the pages on which the full references appear.

Subject Index

Advances in Plasma Physics

Cumulative Index, Volumes 1-5

Author Index

Subject Index